Between Understanding and Trust

Between Understanding and Trust

The Public, Science and Technology

Edited by

Meinolf Dierkes and Claudia von Grote

Wissenschaftszentrum Berlin für Sozialforschung
Germany

ho
ap
harwood academic publishers
Australia • Canada • France • Germany • India • Japan
Luxembourg • Malaysia • The Netherlands • Russia
Singapore • Switzerland

Amsteldijk 166
1st Floor
1079 LH Amsterdam
The Netherlands

British Library Cataloguing in Publication Data

Between understanding and trust : the public, science and technology
1. Technology – Social aspects 2. Science – Public opinion
3. Mass media and technology 4. Technological innovations –
Government policy – Citizen participation
I. Dierkes, Mienolf II. von Grote, Claudia
303.4′83

ISBN 90-5823-007-4

Contents

PART ONE
The Historical and Political Context of the Discussion Surrounding the Public Understanding of Science and Technology

PART TWO
Comparative Analysis: Results of and Reflections on Surveys as Methodological Instruments

PART THREE
"Science" and "The Public"—Revised Concepts

PART FOUR
Informing the Public and Debating Science and Technology

CONCLUSION

List of Figures

List of Tables

FOREWORD

Elisabeth Noelle-Neumann

The collection of articles in this book is devoted to an issue that is truly of universal significance today. Many people are now aware of the tremendous pace of scientific and technological progress as well as of the accompanying dangers posed by environmental damage and overpopulation. Given the high pace of change, the increasing complexity of the world, and the visible threats to humanity, it is crucial, according to the social philosopher Hermann Lübbe of Zürich, for people to develop trust–trust in competent persons and institutions—so that the difficulties can be surmounted and the best solutions found.

Yet exactly the opposite has happened. According to the findings of surveys conducted by the Allensbach Institute since 1947, the population at large, at least in the western world and particularly in Germany, has increasingly lost trust in institutions and scientific authority. Moreover, this view is shared by that group charged with informing the population: the journalists, who have lost faith in authority in general as well. When asked whether scientists are united or divided on the whole, either in general or in connection with one of the important issues of the day, a great majority of the German population has regularly responded that "scientists are divided." Another key question, which can be posed for a variety of issues, reads: "If you think about all the people who talk about the issue of the energy supply, whom do you trust will give you good and thorough information?" In July 1986 first place went to television, which was selected by 47% of the respondents. An alternative, "I trust scientists," was chosen considerably less often, by 35%.

About 15 years ago, the American communication researcher Stanley Rothman of Smith College in Massachusetts developed a question model that has been used continuously ever since in the United States and Germany. A controversial issue of the day, such as the hole in the ozone layer, population growth, or the safety of nuclear energy, is presented in the same wording to scientists and experts in that field and to journalists specializing in scientific topics, star journalists in general, politicians, and the population at large. The findings in both countries are almost as consistent as clockwork. The responses that scientists and experts give to questions about controversial issues are located at one end of the spectrum, the journalists' responses are at the other extreme, and the responses of the general population lie in close proximity to those of the journalists. Once, when I presented findings of this kind to a gathering of journalists, a member of the audience called out: "How do you know that the journalists aren't right?"

That question is precisely what I mean by the loss of trust. Decades of study, teaching, and research at universities and major research centers are of no significance. A journalist who deals with an issue now and then is completely confident that his or her opinion is better. And the population has practically no other choice but to follow the journalists' lead when forming its opinions on a particular issue. Under these circumstances, Allensbach surveys have repeatedly found that matters of great significance are not decided on the merits of the concurring opinions of the scientific community but instead on the climate of opinion. As the prime minister of Lower Saxony, Ernst Albrecht, once said at a public hearing on using the salt mines of Gorleben as a disposal site for nuclear waste: "The solution proposed by the experts may in fact be the best one, but it is not politically feasible."

The pressure created by the loss of trust in experts represents a serious menace to science, as evidenced by the findings of another question posed by the Allensbach Institute. Respondents were asked in 1996 to decide which of 16 listed fields of research should be supported and which should be halted. A majority of the population favored stopping research in 9 of the 16 fields. Scientists are faced with the nightmarish prospect of research being forbidden by plebiscite.

What can be done about this possibility? It is here that social research can make a decisive contribution. One highly promising model is the *Media Monitor*, which is published by the Center for Science, Technology, and Media in Washington, D.C. In regular news bulletins, the findings of systematic media content analyses on controversial scientific issues are presented and evaluated. *Media Monitor* also includes the findings of surveys conducted among scientists in the fields in question. The respondents are selected according to standard sampling principles, and no more than 50 interviews are necessary. Frequently, it turns out that scientists are not divided at all. On the contrary, they often exhibit a high level of consensus on a particular issue. This majority opinion also often runs counter to the media tenor. In such instances, it ought to be possible to persuade journalists and politicians to believe the experts and to inform the population accordingly. In Germany, there already exists one organization that can be taken as a model for this function: *Stiftung Warentest*, which tests consumer products and enjoys a great degree of public trust.

There must be some way to restore people's faith in scientific authority. The problems identified in this book and the discussion of key issues in analysis of the public understanding of science and technology could make an important contribution to the achievement of that goal.

PREFACE

The dynamics of development in new technologies and scientific knowledge as well as the growing perception of the risks they entail have brought about a considerable shift in the relation between technological development and the general population in a significant number of industrialized societies. Since the mid-1970s the strong unexpected resistance that large groups have mounted against the spreading and unquestioned use of nuclear energy has undoubtedly had something to do with the rise of academic and political interest in the public's attitude toward scientific and technological innovations and the underlying perception of societal consequences of technological development.

The interest of political actors in the public's attitude is obvious. In democratic societies political decision-makers cannot afford—at least in the long run—to ignore the attitudes and reactions of large segments of the population. This fact is particularly relevant because the economic prosperity and competitiveness of a country seems to be closely tied to its technological and scientific development. The political elites therefore tend to presume that skepticism toward new technologies, or what is often called hostility toward technology, reduces a country's economic competitiveness.

The wide scope of research and theoretical work on the changed relation between society, technology, and science have been accompanied, and to no small degree also shaped, by this political interest. Within the European Union (EU), comparative research on the perception and public discussion of technological and scientific developments has been accorded great political weight. Apart from funding research and collecting data about the public understanding of science and technology, the Commission of the European Communities sponsored an international conference at the Wissenschaftszentrum Berlin für Sozialforschung (WZB) in 1979 to discuss a key topic in the relationship between science, technology, and society: the risks posed by modern technologies (*Technological Risk: Its Perception and Handling in the European Community,* ed. by Meinolf Dierkes, Samuel Edwards, and Rob Coppock).[1] Approximately 15 years later, the European Commission again supported and co-organized a conference at the WZB on the European population's perception and assessment of technological and scientific developments. This conference, 'The Public Understanding of Science and Technology—A Critical Examination of Current Results and Methods', took place from 30 November through 2 December 1995 and was organized by Meinolf Dierkes and Claudia von Grote in cooperation with the

---

[1]Cambridge, M.A.: Oelgeschläger, Gunn & Hain, 1980.

Directorate General for Science, Research, and Technology Development (Directorate General XII) of the European Commission. The purpose of this second gathering was to take stock of the then-current state of the research as represented by the large-scale surveys conducted under the auspices of this Directorate in the framework of Eurobarometer monitoring of the European public's attitudes toward science and technology. For the first time, such survey research was to be critically examined for its conceptual, substantive, and methodological dimensions and evaluated in relation to other research instruments and concepts being used to enhance the academic understanding of the formation of the public's attitudes, knowledge, and behavior in the fields of science and technology.

This volume represents the participants' desire to increase attention to this topic and to preserve the level of discussion achieved at the conference. While preparing this event, we found the subject matter to be anything but a well-defined and coherent field of research. Instead, what we found was a broad spectrum of research traditions and theoretical issues with little cross-fertilization thus far despite the common ground and overlapping fields of endeavor involved. By taking the public understanding of science and technology as the focal point of the various approaches, methods, and results, scholars representing theory of science, history of science, history of technology, sociology of technology, risk research, technology assessment, risk perception, risk communication, and media analysis were able to meet, bring their differing theoretical and methodological perspectives to bear, and discuss them. In these discussions the Eurobarometer-type of large-scale survey research—the largest concerted effort in the field-served as a reference point and a baseline for benchmarking.

This volume is intended to enhance the highly fruitful linkages between the different approaches and to foster the establishment of the study of the public understanding of science and technology as a distinct field of research. It reflects the fact that the conference prompted the authors to explore selected topics further and to pursue special questions in this field. The following chapters thus constitute a selection of independent essays based mostly on the conference presentations but elaborated, strengthened, and revised for this publication through a referee-type editorial process.

The contributions to this volume are grouped into four parts. Each part is introduced with an explanation of the particular shared aspect from which the authors take up the current agenda for studying the public understanding of science and technology. In the first part of the book, policy-centered and historical stances are adopted for the question about the manner in which the relation between society, science, and technology has shaped the investigation into the public understanding of science and technology and about how this self-analysis can be used to change the research agenda. In the second part of the book, comparative data taken from the Eurobarometer's European survey study serve as the foundation for discussing ways of creating and using indicators to interpret data and for dealing with theoretical and methodological shortcomings that affect surveys of public understanding. The third part is devoted to detailed treatment of conceptual issues arising from the manner in which researchers have applied the concepts of public understanding, science, and technology. Consideration of the latter two concepts shows that it is necessary for the object of research, whatever it might be, to be seen in a context broader than has hitherto been the case. The authors in the fourth part of this book address one of the most important assumptions underlying the traditional approach to studying the relation between knowledge and attitudes in the public understanding of science and technology. They also discuss how public understanding for new technological developments is formed and what significance institutional and informational processes have therein. The conceptual, substantive, and methodological state of the art of the public understanding of science and technology is summarized in the concluding chapter and integrated into a formulation of prospects for a future research agenda.

Because this book ultimately grew out of the conference initiated by the Directorate General for Science, Research, and Technology Development of the European Commission, we take this opportunity to thank Dr Gabolde, Director of this Directorate, and Dr Barbara Rhode, who was responsible for the conference. They not only made the event financially possible but also lent their ideas and organizational support. We also express our appreciation to David Antal for his invaluable work in translating and editing the chapters of this book and to Friederike Theilen-Kosch for her technical assistance in managing the volume's layout and design.

PART ONE

THE HISTORICAL AND POLITICAL CONTEXT
OF THE DISCUSSION SURROUNDING
THE PUBLIC UNDERSTANDING OF
SCIENCE AND TECHNOLOGY

INTRODUCTION

Meinolf Dierkes and Claudia von Grote

The question that has led to this reader, namely, the current status of the discussion, measurement, and analysis of the public understanding of science and technology, is put into a broad and basic context by the essays of this first part. As their lines of argumentation indicate, a proper answer to this question requires the general structure and history of the relation between science, society, and political systems themselves to be analyzed for their effect on the public understanding of science and technology.

By setting the past investigation and use of the public understanding of science and technology into their specific political and historical context, these essays make the "shaping" of the research topic itself the subject of inquiry. Beginning with the historical development of discourses on science and technology and related topics (U. Felt), this section then ranges from the analysis of key patterns of orientation behind political action as a basis for shaping the understanding of science (S. Jasanoff) to the interests and science policies of political actors in the member states of the European Union (EU) (M. E. Gonçalves). This metaperspective provides a basis on which new research perspectives and concrete proposals for conceptualizations and research questions are derived.

Taking a historical approach, Felt addresses the question of where the discussion of the public understanding of science and technology currently stands by asking the complementary question: Why should the population understand science? This strategic and provocative reformulation allows her to analyze the patterns and lines of reasoning encountered in the astonishingly protracted discourse about the state of scientific and technological knowledge in the population at large. Drawing on historical material about the representation of science in the Austrian daily press, educational programs of the Austrian labor movement in the early 20th century, and the arguments provided in the politically influential 1985 report on the public understanding of science and technology as issued by the Royal Society of London, Felt reconstructs the strategies and concepts of the academic and political actors behind the efforts to popularize science and technology. She examines, for instance, what the public and what experts are defined to be, what "understanding" has been taken to mean, and what political function these rationales had for popularization.

From this reconstruction of conceptual terminology and its historical functions, Felt comes to a conclusion relevant to research. She finds that the emergence or disappearance of certain topics in the political and academic elites on public understanding cannot simply be explained in terms of the general development of science and technology. The way in which science and technology issues are dealt with in the public domain is governed to a large extent by the specific political context. From the outset, therefore, such contexts have to be included in the analysis of the public understanding of science and technology.

More radically, Jasanoff asks about the way in which the research field relating to the public understanding of science and technology is shaped: Why is it that this relatively new field of research has emerged in political programs as a subject of inquiries, and what use is being made of this fact? Her main focus is not on political interests or motives of the actors involved. Instead, Jasanoff devotes her attention to the structural principle underlying such political relations. She frames the question about the public understanding of science and technology as a question about the characteristics of western democratic political culture that affect the citizen's access to science.

In her analysis of the United States after 1945, Jasanoff identifies the culture of individualism as the political order's fundamental pattern of orientation shaping the way in which citizens understand and develop scientific knowledge. Jasanoff concludes that one does not ascertain the public understanding of science and technology by asking about the population's degree of knowledge but by answering the interpretive question of what understanding means and why it is important. These observations are echoed in the contributions by other authors dealing specifically with the key concepts of the public understanding of science and technology (see part 3).

Another way of looking at the political aspects that shape the public understanding of science and technology is found in the third essay in this section (Gonçalves). It focuses on the interests and motives of the political actors relevant to this field of research. Gonçalves begins by asking to what extent the interest that the state shows in the understanding its citizens have of science and technology stems from its responsibility to educate the population and to what extent that interest serves primarily to legitimate the state's own policies on science and technology. To answer this question, she tracks the different courses taken by the political actors in the industrialized and nonindustrialized countries of Europe. Taking Portugal as an example, Gonçalves discusses the issues marking the various political points of departure for dealing with science. This analysis leads her to hypothesize a connection between the politician's attitudes toward science and the population's attitude toward and trust in the everyday relevance of science. The survey that she and others conducted of members of parliament reinforces her view that policy-makers' varied understanding of science must itself be studied in order to analyze the aforementioned relation more precisely than has been the case. The relations that Gonçalves finds to exist between political institutions and their actors on the one hand and the population's confidence in political and scientific institutions on the other are specially treated in part 4 (E. Jelsøe).

CHAPTER 1

WHY SHOULD THE PUBLIC "UNDERSTAND" SCIENCE? A HISTORICAL PERSPECTIVE ON ASPECTS OF THE PUBLIC UNDERSTANDING OF SCIENCE

Ulrike Felt

It has become routine in the 20th century to encounter science alongside politics, sports, and other focal points of news in newspapers, magazines, radio programs, and television broadcasts. Scientists have come to play important roles developing and justifying public policies and legislation in the political and economic domain (see Jasanoff, 1987, 1990; Nelkin, 1975, 1987). The interwoven relation between science and technology has made them important actors in the industrial domain. At many different levels of everyday life, people now need to have a basic understanding of science and technology when making choices, be they individual or global. From the general public's point of view, science may be less "popular" than other subjects, but both science and the public discourse about science have become an integral part of modern industrial society.

The growing importance and influence of science in the political, economic, and social sphere has been accompanied, however, by every greater skepticism and ambivalence. Particularly in the second half of this century, it has become obvious that science and technology can no longer be equated naively with economic and social progress. Indeed, science and technology have often been the source of many, increasingly complex problems. People became aware that the science system could not keep expanding at the speed witnessed since the 1940s. The soaring cost of scientific research seemed to outpace national budgets more and more. It was no longer possible to regard the pursuit and acquisition of knowledge as valuable enough in itself to justify pouring money into basic research. Evaluation and accountability for the funds spent became an indispensable part of science policies. Even in basic research, pressure mounted to anticipate possible benefits, potential negative impacts, and the reaction of the public (see de Solla Price, 1963/1986; for an analysis of the present changes in the science system, see Holton, 1986; Ziman, 1994).

Yet the same science that people have identified as a source of problems has also been called upon to solve them. The public attitude toward science has thus not been exclusively negative; it has been marked by ambivalence and instability. It has been subject to sudden change, and the public has seemed more easily alerted than it used to be. The legitimacy of science as an independent project has been questioned. Consequently, the role of the public in science has had to be reconsidered. The public cannot be regarded anymore as an anonymous crowd of people eager for any kind of scientific or technology progress. Scientists and science policy-makers alike have had to convince members of the public, win them as allies, or manage to hold out against them.

In the history of science these views and worries are not new. To varying degrees and with changing focuses, they played an important role in the institutionalization and professionalization of science from the 18th century on. Communication of science to the public has always been closely inter-twined with the development of science. Researchers from different disci-plines and with different methodological approaches and research perspec-tives have addressed the topic of the relations between science and the pub-lic. But no dominant and clearly perceivable paradigm has emerged so far. Moreover, academic interest in this question has oscillated. The topic attracts intense attention for a time, fades, and is then rediscovered from a variety of angles upon some other occasion when boundary conditions change (Felt & Nowotny, 1993). This long history of discourse on the relation between science and society, particularly the popularization of science, and the public understanding of science, shows the issue's remarkable persistence. And the fact that research has offered valuable insights into the mechanisms at work has only added to its vitality.

This chapter begins with the assumption that the popularization of science and the public uptake of science are merely two aspects of the same interac-tion process. In my view it is an undirected process with many layers of communication and experience that overlap and criss-cross, ultimately forming a variety of attitudes and images of science in the public space. The relation between science and the public thus cannot be described according to a logic of interaction. One cannot think about effectiveness of interaction or reflect more generally on the sense of science-related communication without making a number of presuppositions about the functions of the rela-tion between science and the public and without taking context into account. These different functions and the rhetoric pertaining to them are the central subject of this chapter. Hence my leading question: Why should the public *understand* science?

Elaborating the main lines or argumentation, I take the report by the Royal Society of London (1985), *The Public Understanding of Science*, as the point of reference and as a fine example of a particular approach to the question of interaction between science and the public. It is true that the report's underlying assumptions and some of the research being conducted under its auspices have been strongly criticized and that it represents only one view among many, but the document has had a powerful political impact and has paved the way to numerous comparative cross-national studies of what has been labeled the public understanding of and attitudes toward science (e.g., Institut National de Recherche Agricole [INRA] & Report International, 1993; see also the journal *Public Understanding of Science*). The results of quantitative survey research have been widely published, with reputable journals such as *Nature* printing summaries (Durant, Evans, & Thomas, 1989) and daily newspapers carrying the most newsworthy outcomes.

The feature common to all these reports and articles is a complaint about an apparent lack of understanding with regard to scientific issues—meaning that the public has been unable to reply "correctly" to questions about "scientific facts." Increased efforts at education have seemed to be the preferred remedy for the suboptimal functioning of the interface between science and the public. However, criticism of methodology has been aimed, for example, at the practice of taking survey respondents out of their social environment when being questioned, of completely decontextualizing knowledge and understanding, and of allowing the questions that are posed to reproduce certain assumptions (e.g., about scientific methods). As Wynne (1995) has said: "Evidence of internal coherence among survey data is not itself evidence of wider validity—only of consistence. Too often the latter is mistaken for the former" (p. 370).

Without wishing to pursue the methodological debate further here, I ask instead what consequences are to be drawn from statements that large segments of the population do not know "basic" scientific facts? How should one handle results purporting to show that "men tend to give correct answers more than women do" (INRA & Report International, 1993, p. 60). What does it mean for democratic participation if people are told that an important percentage of the public believes that acid rain is linked to nuclear power plants? These questions point out the power of the discourse that has formed around surveys results, a discussion that could sway science policy.

However, many of the current issues discussed and the reasoning that has developed can be traced back through the history of the popularization of science, particularly that of the 20th century (see Béguet, 1990; Felt, 1996, 1997; LaFollette, 1990; Raichvarg & Jacques, 1991; Tiemann, 1991; in this chapter I draw particularly on research results from *Wissenschaft und Öffentlichkeit in Wien*, 1994, a project funded by the Austrian Science Foundation). Even though this historical contextualization must remain rudimentary, it reveals astonishing continuity in the discourse as well as in the motivations and issues raised and enables one to identify major changes. It shows that the reason why at given moments some considerations move to the fore while others disappear depends very much on the local context and not so much on the overall development of science and technology. This contextualization will help one understand the dynamic of the multifaceted relation between the popularization of science and the public uptake of science on the one hand and the political, social, and economic support system of science on the other.

The notion of the public and its relation to a science system that is becoming increasingly specialized and fragmented will be revisited in the first step of my analysis. When drawing a line between scientists and non-scientists, I look at the mechanisms at work and identify the major roles attributed to the public. The second step centers on the ambiguous and multifaceted notion of understanding. Discussion of that topic will shed light on the diversity and vagueness of expectations attached to efforts to popularize science. Lastly, I investigate the popularization of science as a "political" act with regard to the different functions it has been expected to fulfill. That examination will afford insights into the reasons why the public has been expected to know about science.

"Defining" the Publics

Discussion and investigation of relations between science and the public was long based on a so-called linear model consisting of a sender, a receiver, and a mediator. Scientists were regarded as producers of genuine knowledge, which was then simplified. The public was perceived as a body of rather undifferentiated, passive consumers of knowledge. The mediators were forced into the role of translators. Strong hierarchies were inherent in this model, which predefined a body of scientific knowledge strictly separated from popular knowledge and regarded as superior to it. Information flowed in a single direction, namely, from the producers to the receivers. Science was to set the basic standards to which the public was to aspire (but never completely attain), and the scientists had a monopoly on the status as experts in the public domain. the relation between science and the public was thus inevitably imbalanced and unequal. The communication of science was reduced to a process of translation, so many of the theoretical considerations focused on transfer media, their structural limitations, and the possibilities and restrictions that language imposes on the popularization of scientific knowledge.

Not until the 1970s did people begin systematically questioning these prevailing views and start deconstructing the processes of knowledge dissemination. The rigid demarcation between genuine and popular knowledge turned out to be problematic, for nonexperts appeared to have their own models and representations of the world surrounding them, and these conceptions could not simply be ignored or declared excessively simplistic. Consequently, the dichotomy between scientific texts and popularized accounts gave way to the idea of a continuum of different kinds of texts. Popularization started to be understood increasingly as the negotiation of meaning, and it was underlined that both the very act of popularization and popular knowledge would be fed back into the process of knowledge production and thus have an impact upon the cognitive dimension of science itself (see, for example, Hilgartner, 1990; Shinn & Whitley, 1985).

However, this turn in the understanding of popularization cannot be regarded as complete or definite. Indeed, some of the contemporary discussions surrounding the public understanding of science still implicitly suggest that it is possible to demarcate clearly between science and the lay public. With the increasing involvement of science in the political, economic, and social sphere, this question of drawing a distinction obviously becomes an important political issue and a sensitive point in the debate about the public understanding of science: The legitimation to claim expertise is at stake. Only by drawing frontiers, by barricading the science system and regulating access to it, can the power of expertise be kept in the hands of scientists (and some policy-makers).

What role does popularization and the public uptake of science play in drawing this line between science and the public domain? Answers to this question require a closer look at the consumers of knowledge (the public). How is scientific knowledge handed over to them, interpreted and rearranged by them in their existing context of knowledge and experience? What do different publics want to know or not want to know about science? What are their expectations when they encounter science? These questions are only a few of those that need to be tackled. Qualitative field research (e.g., Irwin & Wynne, 1996; Michaels, 1992; Wynne, 1992) has complex and multilayered processes of interaction between science and diverse publics. In particular, quantitative survey research, known from the U.S. context since the 1950s, has started to play an important role in Europe (see Lewenstein, 1995b).

This shift in research focus has meant that the public can no longer be conceptualized as a hypothetical group of knowledge consumers. There is a need to define the public more precisely and understand its historical development more clearly than in the past. Habermas (1962/1990) provided a detailed account of the public's complex transformation from a small, critically discussing public in the 18th century to a public sphere dominated by mass media and mass culture and imbued with power in mass democracies. According to him, these gradual and successive changes were closely linked to the formation of classes, growing urbanization, cultural mobilization, and new communication structures. In a certain sense, "public is in itself nothing more than a socially empty field with free entry. Free entry for everyone is its constituting feature, and depends on a multitude of conditions that are, historically, rarely given"[1] (Neidhardt, 1993, p. 340). Using the term *public* in relation to science in the 20th century opens a large range of possible meanings. Comparatively abstract constructions like *public opinion*, which as mainly strategic and legitimating purposes, coexist with rather differentiated and specialized publics. At the same time institutions also represent the members of the public in their capacity as users of knowledge, spectators, referees, and many other functions. I would therefore agree with Neidhardt (1993), who stated that "in modern democracies the public plays an important role—but nobody seems to know exactly what the public is" (p. 339).

In most cases the public for which a popularization of science is intended can only be inferred. Occasionally, however, discussion of a scientific issue does briefly reveal the intended target audience more directly, as happens in editorial statements in popular science journals. The speeches and addresses at the openings of popular universities or museums also suggest the envisioned addresses of such educational measures, as do parts of popular science books. The formulations chosen in these instances are rather general, however. They have far more rhetorical and strategic character than concrete meaning. Expressions such as "science for everyone," "really popular science," and "science for a broad public" have often contradicted the subsequent message, which is clearly directed to a much more restricted and specialized public. The "everybody" has actually stood for specialized segments of bourgeois society, members of the working class, women excluded from higher education, or other groups thought to be of interest as either consumers or allies. With increasing commercialization in the popularization of sci-

---

[1] Unless otherwise specified, all translations of originally non-English passages are by U. Felt.

ence, the power of the clientele's wishes and expectations grew, and the relation between producers and consumers of popular science became ever more closely intertwined (see Béguet, 1990; Luhmann, 1996).

All constructions of "the public" have two basic assumptions in common: (a) the public is ignorant of scientific knowledge, and (b) the public simultaneously has the wish to know. Indeed, a specific condition distinguishing the popularization of science from many other enterprises of knowledge transfer is that the supposed audience is always perceived as a mixture of ignorance and something that has often been labeled "natural curiosity," a *libido sciendi*. Such descriptions, which are found in numerous accounts, particularly with regard to workers taking courses offered by the popular universities, seem to form an important part of the myth. It is unsure, though, how valid this general presupposition is or why and to what degree the different publics "consumed" popular science accounts. Two examples of research that have focused on those questions are notable for the Viennese context of the early 20th century. Studying the motives that workers in Vienna and Berlin had for taking the courses offered at popular universities, Siemering (1911) identified three. First, many members of the working class had left the church and were looking for an alternative value system, with science being expected to replace religion in some way. Second, the living conditions of the working class were miserable, and working class people hoped to improve their lot by acquiring knowledge about hygiene, health, the human body, and related topics. Third, they were confronted by a rapidly changing work environment and hoped to acquire the additional knowledge they needed in order to cope with it. The second example is a study by Schreder (1936), who examined newspaper-reading habits among the Viennese population and concluded that the less education readers had, the more they read popular science articles and regarded them as a means of education. Although neither case study is representative, each shows that a variety of driving forces are at work and that the public uptake of science has to be understood in the framework of people's living conditions and aspirations.

To clarify what is meant by the term the public, one can also investigate the process of drawing the line between scientists and nonscientists. Four aspects seem important. The first concerns the notion of lay public. Even a very simplistic definition of *scientist* as one who belongs to the institutional setting of science and of *lay public* as people excluded from that setting quickly runs into conceptual difficulties. With increasing specialization and differentiation within scientific disciplines, the ideal of the generalist becomes a fiction and the line between science and the lay public becomes increasingly blurred. Lévy-Leblond (1992) brought the issue aptly to the point:

> When discussing the public understanding of science, a serious, but current
> fallacy is to equate the "public" with "laypeople," that is, "non-scientists."
> However, it must be recognized that we all, scientists and non-scientists
> alike, share a common "public misunderstanding of science." Indeed given
> the present state of scientific specialization, ignorance about a particular
> domain of science is almost as great among scientists working in another
> domain as it is among laypeople. (p. 17)

Awareness of this blurring was already present in the early 20th century, as in
the Viennese context. The issue was discussed several times, particularly in
the *Arbeiter-Zeitung*, a social democratic newspaper whose editors were
deeply concerned with the functioning of the interface between science and
the public. One of the first examples, dated 1900, stressed the enormous
growth in many different areas of knowledge, an expansion that made it
"impossible to follow up progress even in related areas. The physicist can
hardly keep up with developments in chemistry and technology, not to mention zoology and botany, and the same holds for engineers with regard to
science" ("Deutscher Naturforscher und Aerztetag," 1900, p. 26). The matter
was expressed even more clearly a quarter of a century later in the same
newspaper: "[B]ecause of the vast and multilayered expansion of the sciences, it has become impossible to know everything nowadays. Each scientist today is a specialist in some domain, and in all the other fields of knowledge is a lay person" (Koenig, 1925, p. 26).

An important further consequence of this specialization in science was
stressed by the Spanish philosopher Ortega y Gasset (1930/1993) in *La
rebelión de las mases* (The revolt of the masses):

> In former times it was easily possible to divide people into those who knew,
> those who did not know, and those who knew more or less. But the specialist
> does not fit into any of these categories. ... We will have to call him a
> knowledgeable ignorant, and this is a very serious matter because it implies
> that he will behave with regard to all questions he does not know anything
> about with the arrogation of a man who is an authority in his specialty.
> (pp. 117–118; translated here from the German version)

In other words, the question will increasingly be who has the right to speak
for what kind of science and who can claim expertise in the public space. It
will be difficult to handle these ever more subtle differentiations.

A second important aspect of the process that has separated scientists from
nonscientists is people's images of science. Though there is little direct historical evidence on how people have perceived and defined science, more
recent empirical studies on the public understanding of science have demonstrated convincingly that there is a certain tendency to define science as what
it was *not* rather than what it was (e.g., Michaels, 1992). Not being able to
understand, being excluded from the institutions of science, missing familiar

things, and not being able to follow lines of argumentation seem to be important facets of this image. However, this negative definition does not mean that people feel disenchanted or disinherited in the face of science but that they try to maneuver discursively around science in a variety of trajectories. Not knowing about science can thus not be regarded as a lack of knowledge pure and simple; it can also be interpreted as deliberately not wanting to know. Moreover, it has been shown convincingly that people hold completely different positions on science, depending on whether they are thinking of "science-in-general" or "science-in-particular," that is, of a specific scientific or technologic problem confronting them. The line between science and the public is thus drawn by means of mutual negotiations and exclusion and has to be seen as a steady process of change (e.g., Michaels, 1992; Wynne, 1992).

A third aspect of the separation between science and the public is the manner in which scientists are portrayed in popularized accounts in magazines, newspapers, and other media. "Many of our strongest beliefs about science come from what we think scientists are like, principally because those beliefs include assumptions about how the appearance, personality, and intellect of scientists relate to the importance and consequences of this work" (LaFollette, 1990, p. 66). There are two strategies being pursued that, taken together, introduce a great deal of ambiguity into the relation between science and the public. One strategy is to construct a strong myth that scientists are different in terms of attributes like intelligence, patience, and endurance. The myth also incorporates descriptions of the scientist's cold scientific mind and remoteness. It is suggested that all these attributes are especially prominent in scientists and are largely lacking in the lay public. On the other hand scientists are also depicted as absent-minded and incapable of facing everyday problems. Features of their physical appearance are used to evoke the image of a "normal" human being and thus to instill the feeling of familiarity. This double strategy has been pursued throughout the 19th and 20th centuries, with certain variations in the basic stereotypes of scientists and national differences when it comes to using the description of scientists when reporting scientific findings.

The fourth means of constructing a barrier around science is, paradoxically, the very act of communicating with the public. At the moment the public is brought "closer" to science, it is also confronted with the difficulties and, partly, with the incapacity to grasp how science works. In particular, when conflicts related to scientific and technological issues escalate beyond the institutional boundaries of science, an important strategy seems to be to construct "intellectual distance between the forms of discourse prevailing on two sides in [the] dispute" (Dolby, 1982, p. 271). This strategy proves to be effective "when linked to acceptable hierarchical notions of relevant expertise" (p. 271). Dolby used the controversy over Darwin's theory of evolution as an example of how the biologist could successfully claim expertise and thus win the conflict with the religious establishment.

Some of the active functions that the public has with regard to science have existed intermittently in history in various forms and contexts. At a theoretical level rather specific tasks the public is supposed to perform can be identified, but one must keep in mind that a mixture of roles is the rule. The multiplicity makes it difficult to gain a clear impression of the public's role vis-à-vis science. In many cases members of the public slip into the role of "naive" spectators meant to be fascinated, amused, and impressed by science rather than enlightened. Science is presented as unique, magical, powerful, and promising, the important part being the message, the image, that is conveyed rather than the scientific or technological information that is imparted. Spectacular scientific demonstrations, partly exhibitions, popular science books of fictions or nonfiction, radio performances, films, and theater plays have all been vehicles of popular science that are aimed at attracting this kind of attention. The public is clearly perceived as a consumer, and the popularization of science became a good of mass consumption.

The second role attributed to the public has been that of supporters of science. Either science has been "sold" as a general cultural good—a packaging that has proved to be increasingly suspect—or its practical applicability has been underlined. Once convinced of the importance of scientific knowledge, the public was supposed to be an ally in arguing for more funding or even exerting direct pressure for investment in particular scientific domains. An example in the medical sector has been the formation of self-help movements, which, in turn, try to influence research in their particular areas (see Gizycki, 1987). Though the phenomenon of "selling science" has been linked in many analyses to the rising cost of science since World War II, it is observable much earlier. In the Viennese context, for example, much of the debate surrounding additional funding for university research in the public arena was argued by advocates who pointed out the positive effects such research would have on the development of local industry and hence for the welfare of all citizens.

Third, the members of the public had the function of being witnesses, a role that has existed in various forms and for various reasons since the advent of modern science. The public (often selected according to suitable criteria) was used to testify to experimental results and thus to assure the credibility and authority of the author. It was the administration of the scientific proof in which the public played a decisive part. With increasing institutionalization and the differentiation of the science system, the reward system, too, became more formalized and standardized. However, members of the public have continued to serve as witnesses at moments when the conventional procedures of the science system threaten to break down. At those junctures scientists tend to use the press or other mass media to announce their scientific findings well before their research is published or submitted to the critical eye of colleagues. Recent instances of such behavior have been the furor over cold fusion and high-temperature superconductivity (see Felt, 1993; Lewenstein, 1995a; Nowotny & Felt, 1997).

Lastly, the public has also been a participant in the science system, a role performed in some disciplines and with great national variations until the end of the 19th century (Sheets-Pyenson, 1985). With increasing professionalization and specialization, major effort was spent on drawing frontiers between scientific and amateur opinion, between the amateur and the real scientist. This supersession of amateur science can thus be seen as a gradual process that inevitably became linked to the development of science. In the end it left the power of expertise entirely with the establishment.

The Multiple Meanings of Understanding

Investigating the discourses on popularization of science and the public understanding of science, one is struck by the inherent vagueness of the term *understanding*. Throughout the 20th century many scientists involved in the educational sector, politicians, and public forums have dealt more or less explicitly with the meaning of this word. It has represented for them, as I shall show, an important rhetorical means of creating a context for their efforts to popularize science. I also show that the notion of understanding acquires meaning only in an applied sense and that one is often confronted with partly contradictory meanings. A detailed analysis of the different standpoints would go far beyond the scope of this chapter, but I shall try to convey an impression of the large spectrum of concerns that this term subsumes. Is the word *understanding* intended to mean "the content of scientific knowledge or the nature of science as a cultural enterprise" (Collins, 1987, p. 690)? Does it mean that the public should become familiar with "how, with what confidence, and on what basis, scientists come to know what they do" (Shapin, 1992, p. 28)? Is the word intended simply to ensure the public's appreciation (and support) of science? Or is it used to mean the transmission of knowledge that is then applicable in the context of work or everyday life?

In general one could say that in the 19th and early 20th century there existed a strong belief that any science could in principle be made understandable to a larger public. As stated, for example, in the French popular science journal *Causeries scientifiques* in 1862: "Never believe those who pretend that science is a riddle impossible to decipher. . . . Everything presented here is simple, accessible to everybody, to everybody without exception" (as quoted in Béguet, 1990, p. 16). In the following decades and well into the 20th century there was regular reference to the fact that it was a characteristic of real science that even complex facts could be described in a popular manner. Though most authors hid behind expression such as "science for everybody" and "understandable science," they often had more distinct and restricted ideas about the public. Increasingly popularizers stressed that it was not easy to acquire knowledge and that it would need courage and audacity:

> [I]f it seems too easy to the reader, he has been cheated. Only those who live through the deep difficulties of modern physical thought can achieve understanding. This is why I ask the reader no to give up because of difficulties, but to become suspicious if something seems too easy. It should be possible for science to become "understandable" for everybody, but not easily understandable. (Hopf, 1936, p. vi)

The idea was thus to give the reader an impression of both the complexity of knowledge production and of the effort made to acquire this knowledge. Essentially, then, the different degrees of the public's understanding of science were mainly seen in relation to the quality of popularization and not to the complexity of the subject matter. In line with this logic, rules were formulated for "good popularization," which included guidelines on speaking, choice of subject, use of language, and much more (Hartmann, 1897, p. 159; as cited in Altenhuber, 1995, p. 143).

"Understanding" could also mean the comprehension of science as a cultural enterprise, as an act of trust in an elite and as admiration for their work and not as a complex body of knowledge. Science began to be regarded as an essential part of national culture and was used for identity-building. To read popular science books, to listen to talks of scientists, and to be carried away by the fascination of scientists' ideas seemed to be at center stage. Mach (1896/1910) expressed this view succinctly in the introduction to his *Popular-science lectures*: "Given the knowledge required and the time available, popular-science lectures can only be modestly instructive. Nevertheless, a suitable choice of subject can convey the romantic side and poetry of research" (p. viii). As clearly stated by Walter Benjamin (1935), a prominent German philosopher who was also engaged in public education:

> To bring science closer to larger segments of the public, . . . more than knowledge is needed. . . . [The popularizers of modern physics] involve the reader in the game and give him the certainty that he is progressing. This certainty need not be related to the content—no reader will have practical use

> for relativity theory. But he will profit from something else: With knowledge, he acquires a thought that is new, but not only to him. (pp. 450–451)

Members of the public were meant to participate in the fascination and complexity of scientific thought even though they could not grasp the details. This perspective is consistent with the conclusion arrived at by Cloître and Shinn (1986) after studying contemporary articles in French popular magazines:

> As far as the popularization of science is concerned, the language, the reasoning, and the images do not manage to elucidate the phenomenon. Quite the contrary, there is a tendency to create conceptual incomprehension. . . . Popularization therefore does not constitute an efficient instrument for transmitting better knowledge about the physical world. Its force and its pertinence lay in the links that it establishes between a scientific subject and the social sphere. (p. 163)

The transmission of scientific knowledge via popular articles was also expected to lead to the understanding necessary for survival in a rapidly changing professional and private environment. There was a need for a kind of instrumental knowledge that could greatly relax the approaches to technology. Knowledge was supposed to enable people to increase the efficiency of their control over and manipulation of their environment. This utilitarian attitude was expressed regularly upon such occasions as the announcement of popular lecture series for members of the working class in the early 20th century. These courses were explicitly conceived to offer "the basis we absolutely need in order to understand the effects of natural forces, particularly their application in technologies" ("volksthümliche Universitätskurse," 1901, p. 4). It was hoped that more knowledge would indirectly lead to a more positive way of dealing with fundamental changes due to science and technology. In addition, orientation was needed in an increasingly complex world, and scientific knowledge was supposed to be able to offer it.

During the 20th century, however, a change occurred. The necessity of public trust in scientists and their findings became the focus of increasing attention, whereas the claim that people should understand the details of popularized science was gradually pushed into the background. In a report of Einstein's visit to Vienna, for example, the author wrote:

> His achievement could not be understood by mankind in all its details, could not be fathomed in all its consequences and meanings. But the audience understood that its aspirations and thought were borne away from the low, the ordinary, to the cosmic and eternal. (Galten, 1921, p. 1)

Doubts were expressed about the belief that massive educational efforts would solve the problems accompanying scientific and ethnological progress. In the 1930s it was already clearly perceived as questionable to think that "progress in science and the spreading of education would guarantee and promise perfect coexistence" (Huizinga, 1935, p. 54). The overabundance of knowledge was judged to carry the danger of hindering judgment and free development (Huizinga, 1935). Throughout the 20th century and especially from the 1970s on, this vision came to be shaken fundamentally. Increasing scientific knowledge by no means always ensures support for the sciences; it could also breed skepticism and uncertainty.

This fact makes it astonishing that the argument has not disappeared but has found its way into the report of the Royal Society of London (1985):

> Ignorance of elementary science cuts off the individual from understanding many of the tools and services used every day. Some basic understanding of how they function should make the world a more interesting and less threatening place. . . . Scientific literacy is becoming an essential requirement for everyday life. (p. 10)

Lastly, the omnipresent ambiguity in the use of the word *understanding* partly reflects scientists' ambivalence toward the act of popularizing their research. On the one hand, simplification and the use of images borrowed from everyday contexts has always been scientific practice (Fleck, 1935/1979). Speaking to colleagues from other disciplines or to students is part of that simplification process. On the other hand, scientists have always expressed worry about the image thus created of their discipline. The very act of simplification seems to threaten their role as experts and jeopardize the mystic flair of science. Some scientists even went so far as to link increasing popularization to the decline of science. As Spengler (1918–1922/1939) claimed: "our 'popular' science is without value, *detraque*, and falsified" (p. 328). The desire of science for "wide effect" (p. 328) was thus judged as a first sign of "the commencing and already perceptible decline of Western science. That the severe esoteric of the Baroque Age is felt now as a burden is a symptom of sinking strength and of the dulling of that distance-sense which *confessed* the limitation with humility" (p. 328).

Upon closer look at the interaction of science and the public, one does in fact discern a complex process of negotiating the meaning and value of scientific knowledge. The social context and the relational networks people live in impinge upon the ways they perceive scientific knowledge handed down from institutions as though it had already been validated. Science has to be seen as attracting social interest and as having an impact upon existing relations, identities, and value systems. However, when it is realized how little actual discourse reflects past experience in the field of science popularization, I cannot but agree with Wynne (1995): "science appears to be unable to recognize these social dimensions of its own public forms or the fact that public readiness to 'understand' science is fundamentally affected by whether the public feels able to identify with science's unstated prior framing" (p. 377). We observers thus assist the encounter between two cultures: the scientific culture, which tends to reduce issues to those of control and prediction, and social worlds, which are much more open than scientific culture.

It is necessary to improve the understanding of the presuppositions in the concept called the public understanding of science and to spell out what is expected of the research carried out in its name. The more one clarifies the presuppositions and expectations of the concept, the more relevant the results and conclusions will be. Thus it seems important to try and take a closer look at the "political paradigm . . . that shapes a particular framing of 'the PUS problem'" (Wynne, 1995, p. 361). What is this political paradigm behind much of the concern publicly expressed about the low degree of scientific literacy and of the research funded to achieve insights in this domain?

Popularization of Science: What for?

The popularization of science is a practice clearly marked by a multiplicity of motivations, aims, and driving forces, by a set of actors that is not strictly defined, and by a relatively open institutional framework. A closer sociological or historical look at the various processes by which scientific knowledge is disseminated reveals complex and often even contradictory rationales that rarely relate to the openly stated aims behind the popularization of science. What are the different functions that such popularization is supposed to fulfill, and what are the underlying motivations for it?

Popularization as Boundary Work

If one believes that the meaning of *science* is subject to negotiation, and if there is no obvious set of criteria to distinguish science from nonscience, what are the essential components with which to construct and maintain a borderline? And who is involved in the process of negotiation? Science being embedded in society and exerting manifold influence on both the private and public domain while being shaped by its environment, one surely cannot reduce science to a body of factual knowledge and methodological rules accepted by the core of the scientific community or to the mere fact of belonging to the institutional setting of science. It is necessary instead to broaden the perspective and investigate how science is perceived in the public space, how scientists themselves construct the delimitations of their territory, and what role popularization plays in it.

Within the scientific community considerable effort is devoted to formalization and standardization precisely in order to delimit science from non-science and even to distinguish between disciplinary territories within science. This "boundary work," as Gieryn (1995) has called it, "occurs as people contend for, legitimate, or challenge the cognitive authority of science" (p. 405). If there is a social interest in "claiming, expanding, protecting, monopolizing, usurping, denying, or restricting the cognitive authority of science" (p. 405), then pragmatic demarcations of science from non-science seem important. Seen from this perspective, science is

> nothing but a space, one that acquires its authority precisely from and through episodic negotiations of its flexible and contextually contingent borders and territories. Science is a kind of spatial "marker" for cognitive authority, empty *until* its insides get filled and its borders drawn amidst context-bound negotiations over who and what is "scientific." (p. 405)

A central way to give meaning to the notion of science is through the act of public communication of science.

The preoccupation with demarcating between science and other forms of cultural knowledge production—whether it is called folk knowledge, pseudo-knowledge, or charlatanism—and the idea of fighting superstition was prevalent in the discourse about the popularization of science throughout the 19th and 20th centuries. However, it did not become an openly expressed issue until the existence of the one was effectively threatened by the other. Werner von Siemens voiced his concern, for example in an 1886 lecture he delivered at the meeting of German natural scientists and physicians. At that prestigious and highly visible gathering, he strongly pleaded for keeping the conviction "that the light of science that is penetrating deeper and deeper into human society will effectively fight humiliating superstition and destructive fanaticism, which are the greatest enemies of mankind" (Siemens, 1891, p. 495). Mach (1896/1910) sounded a similar note in his popular science lectures held at the turn of the century. He took advantage of these occasions to urge better access to education for women, for he perceived that access to be the only solution that would sustain scientific and technological progress.

> The uncivilized woman carefully cultivates and preserves all kinds of usual superstitious beliefs, down to the fear of the number 13 and spilled salt, transfers them conscientiously to future generations, and is thus always an object of attack for all movements of regression. How can humans advance in security if not even half of them are walking on enlightened paths! (p. 355)

Not only superstition but folk knowledge, too, was seen as the enemy of science and a threat to the scientific establishment. What makes folk knowledge powerful in the public domain is its style, which is

> more suited to the popular forum than to that of scientific orthodoxy. It is wide ranging in coverage, offering insight into and explanations of many popular issues. It is visionary and programmatic rather than rigorous and testable. It gains its support without the mediation of the scientific expert. (Dolby, 1982, p. 272)

For these reasons the scientific establishment found it desirable to construct barriers, and "effective" popularization of science was seen as an important way to counter folk knowledge in spheres where it was still powerful.

The negotiations about what was to be considered science and what not had thus clearly moved to hybrid spaces where the public became a relevant actor that had to be convinced. Meteorology was one of them. As a rather young field at that time, it had to "fight" the lively domain of folk knowledge when it came to weather forecasts. This struggle proved extremely hard because expectations of what science could achieve were high and meteorology was obviously not capable of providing strictly reliable local weather forecasts. Astronomers, too, found themselves confronted by difficulties— on two fronts. First, they had to differentiate their field clearly from the superstitions of astrology, which was still regarded as scientific and which was strongly present in the public sphere. It is interesting to note that even in the 1990s interviewed people still rated astrology as being scientific (INRA & Report International, 1993). Second, a line had to be drawn between the scientific opinion of astronomers, whose field was becoming increasingly professionalized, and the lay opinion of amateur researchers. Having long been central to astronomy, the latter group henceforth tended to find itself excluded from core knowledge production.

The difficulty of negotiating the boundary between professional and lay knowledge is even more evident in the medical sector. Elaborate information campaigns were waged to inform as large an audience as possible to establish what was then the scientifically authorized ideas about health and illness. But even though the amount of information on health and illness grew rapidly throughout the 20th century, people tended to retain private notions about their state of health and the causes of their illnesses. Later in the century the spreading of education, improved distribution of knowledge throughout society, and popular demand for a voice in matters that concern people as patients even led to forms of organization opposing the scientific establishment. In the medical domain, for instance, one finds a high degree of what used to be called protoprofessionalization, and patients confronting the professionals with their own versions of health and illness (Nowotny, 1993).

The 1985 report by the Royal Society of London tackled these problems of delimiting science and nonscience in the strong conviction that increasing the level of information about science will make the task easier.

> Greater familiarity with the nature and the findings of science will also help the individual to resist pseudo-scientific information. An uninformed public is very vulnerable to misleading ideas on, for example, diet or alternative medicine. An enhanced ability to sift the plausible from the implausible should be one of the benefits from better understanding of science. (p. 10, 1985)

But does not the view expressed in this quotation set my argument right back where it started? Do not the authors of the 1985 report pretend that there is general agreement about what science is, what can be classified as pseudo-science, and whom to believe and trust and whom not (see also Collins, 1987)?

Work to establish boundaries between science and nonscience is also linked to the important question of who has the legitimacy to speak for science. Should these intermediaries be scientists, or should the professionalization of science journalism in the first half of the 20th century be greeted as a way of gaining a clearer and perhaps more critical view on scientific issues from the outside? To whom should it be let to define "patterns of cognition, interpretation, and presentation, of selection, emphasis, and exclusion" and thus to choose one version of reality? (LaFollette, 1990, p. 47).

Bridging the "Gap"

The communication of science to a wider public has long been based on and legitimated by an assumption that large segments of a given population lack scientific knowledge. The image of a gap or a gulf between professional and public understanding of science, or between those who know and those who do not, is generally used to describe this separation. In this image, ignorance (attributed to the public) and knowledge (attributed to scientists) are concentrated on the opposing sides of the gap. According to this logic, every advance in science increases the number of those who don't know, so ignorance proliferates. To bridge the gulf or fill the gap, other images are used as well to describe the work of science popularizers.

Throughout the 20th century science and technology have by and large been seen as a motor of social and economic progress (although clear articulation of doubts about the direction and speed of development began in the early 1970s). Consequently, at least the broad lines of science and technology were thought to be understood by a wider public. But how is scientific and technological understanding to be appropriated? Accelerated and socially more balanced access to educational programs has often been regarded as *the* main approach to bridging the apparent gap between science from a wider public. This idea has been applied in different ways from one national context to the next. To illustrate one of those ways, I propose to examine the developments linked to the working-class movement. As has been shown, the workers did not fight their battles *against* science, as one could naively imagine, but rather with and around science, which was considered objective and value-free (Bayertz, 1983, 1985). They judged science to be an ideal basis for a well-functioning society, and scientific views were assimilated in the world view of members of the working class.

The workers' intense efforts to create their own educational programs show the ambiguity of the educational solution for bridging the gap. On the one hand, it was, as Ferdinand Lassalle put it, "the destiny of our time is to do what dark decades behind us have not been able to think of: Bring science to the people" ("volksthümliche Universitätskurse," 1909, p. 9). Much of this "bringing to the people" was thought of in terms of improved education for all those excluded, particularly members of the working class and women. Education was perceived as a unique chance for social promotion: "Today, popularization is an effective means of improving the understanding of different classes and bridging the gap between the educated and noneducated" (Loos, 1911, p. 315). Social justice was at stake in the issue of raising the level of scientific literacy.

Lack of understanding with regard to science was thus attributed mostly to a lack of general education. Physicists and science popularizers argued that they could achieve a better understanding of Einstein's relativity theory if "the necessary preparatory work were done by the high schools. ... Only then will educated people appreciate the achievements of this man in all their importance, liberated from all mysticism" (Thirring, 1920, p. 4).

On the other hand, the Viennese context also reveals that the spreading of scientific ideas among workers has to be seen as a collusion of many different interests. Though social democrats founded the popular universities in Vienna that were to allow the workers access to knowledge, and though associations for educating the working class were founded in Germany and were thought to contribute to the emancipation of the workers, these efforts were partly financed and partly influenced by the liberal bourgeoisie. The interest of that class was obvious. With growing industrialization, qualified workers were desperately needed, and knowledge about science and technology was judged essential to the full integration of people into a work process ever more reliant on technology. The efforts to increase scientific literacy can also be interpreted as an effort to discipline and control, however:

> The humanitarians and priests who advocated popular education were nothing but henchmen for capitalist industry, which demanded from the state that it provide them with qualified workers. . . . It was not a question of giving way to a "written culture" or of emancipating human beings. Another kind of progress was meant. It consisted of taming the illiterates, that lowest class of human beings, and of exploiting not only the power of their muscles and their manual skills but also their intellect. (Enzensberger, 1988, p. 65)

It was also hoped that improvement in the education of oppressed workers and the new perspectives it opened to them would, as a welcome "by-product," weaken their power to mobilize themselves.

Roqueplo (1974) pointed out yet another side to the ideology of science communication. If modern societies are considered technocratic, he argued, then the knowledge necessary for mastering the increasingly complex environment is transformed in the hands of those who possess it into an ideology of competence and into real power over people. According to this logic, the transfer of knowledge is not necessarily in the interest of those who possess it. Knowledge withheld from the public domain can be used to define the boundaries and keep power in the hands of those who possess that knowledge (see also Jeanneret, 1994). As scientists saw it, sharing knowledge would be advantageous only if the power of expertise remained in their hands.

The idea of emancipation through scientific education thus quickly turned out to be an all too idealistic and naive concept. Popularization has neither fully democratized knowledge nor reduced the differences in the level of education. It even seems that differences have increased with the explosion in the number of media and places where scientific knowledge is traded. The more that consumers have as intellectual starting capital, the more they profit from the information offered, an effect that has been called the increasing knowledge gap. This relation points to the utopian character of many of the projects launched at the beginning of the century and of many contemporary discussions (see Jeanneret, 1994).

Ultimately, there remains the question of whether knowledge can really be attributed only to one group and ignorance to the other. Have not studies such as Wynne's (1992) on Cumbrian sheep farmers shown that ignorance of folk knowledge was widespread in the scientific community, whereas the nonscientific group possessed rather refined knowledge of their environment? A less dichotomized, more balanced image of the distribution of knowledge and ignorance makes the popularization of science more a discursive than an educational tool.

Making Society Function According to Scientific Principles

In addition to electrification, progress in the chemical sciences, communication technologies, and transportation were but the most outstanding changes that marked the beginning of the 20th century. In many ways the turn of the century thus meant the opening of a new area in which science and technology would occupy an important place in society. But scientific knowledge and technological artifacts were not the only contributions to be greatly admired. Science as a way of thinking, as a method with which to approach problems, and as an "ideally" functioning system also attracted keen interest. Though general reflections on these developments appeared throughout popular literature in many different countries, Vienna's liberal political climate in the early 20th century makes the Viennese context at that time a suitable focus for intense investigation of the hopes that had been placed in science. Being conceptualized as value-free, objective, and "perfect" in its internal logic, science appeared to be an ideal basis for the functioning of society. Science as a source of truth and as a morally superior social enterprise was believed to been an ideal foundation for all political and ethical judgments. Marcelin Berthelot, French chemist, popularizer, and statesman, expressed this assessment well in a letter to Moritz Szeps, the initiator of the first popular science weekly in Austria, *Das Wissen für Alle* (Szeps-Zuckerkandl, 1939). Berthelot placed great weight on the essential advantage of a sustained effort to popularize science and emphasized the important moral consequences this popularization could have for all society.

Many traces of this ideology appear in the *Arbeiter-Zeitung*. In one article it was argued that "a Weltanschauung or a political party" that would "refrain from scrutinizing its basic principles and claims with respect to scientific findings and from keeping them in harmony . . . would sign its own death sentence" ("Afterwissenschaft," 1903, p. 1). Even *der neue Mensch* (the new human being) envisioned as the outcome of the social democratic policies was intended to be shaped mainly by educational efforts in which science was to play a central role. It was assumed that educational transformation was necessary to create the new consciousness within the existing state. Seen from this ideological perspective, science was obviously considered an essential part of culture. Consequently, scientific ideas and discoveries had to be shared like other elements of cultural patrimony.

The aspect of sharing knowledge is particularly important when issues of control and participation are involved in public controversies over science and technology—precisely the kind of dispute that began to erupt ever more frequently in the course of the 20th century. Blind trust in science vanished as science grew and negative impacts appeared. Simultaneously, information and public forums of exchange on science and technology have come to play a new and rather different role. With the establishment of democratic structures and with public opinion becoming a major factor in the making of national policies, a new argument was heard: Because science is a dominant shaping force in society, democratic structures can work only if people know enough about science. "[I]t is . . . important that individual citizens, as well as the decision-makers, recognize and understand the scientific aspects of public issues" (Royal Society of London, 1985, p. 10).

Though all the presented arguments seem rather straightforward and obvious, one should not disregard a more critical issue posed by Lévy-Leblond (1992). He stressed that, after all,

> the requirement . . . that people should be experts, or at least fluent, in science . . . before giving their view about it . . . is contrary to the basic tenet of our democratic societies. Democracy is a *bet*: the bet that conscience should take precedence over competence. We do *not* require an expert, nor even an 'amateur' level of knowledge in constitutional or criminal law before allowing citizens to use their voting rights or participate in a jury. Why should we be more demanding concerning technical and scientific matters? (p. 20)

Bearing in mind the democratic standpoint and the arguments about bridging the gap, one finds the analysis by Lévy-Leblond convincing: "The problem we face is not so much that of a knowledge gap which separates people from scientists but that of the power gap that puts scientific and technical developments outside democratic control" (Lévy-Leblond, 1992, p. 20).

Accountability and Legitimation

Although the perception of science popularization as a politically relevant activity can be traced back well into the 19th century, the legitimating function of science popularization has been at its most visible since the end of World War II. Linked to the enormous expansion of the system, to increasing specialization, to the growing cost-intensiveness of research, and, later, to the obvious adverse impacts of science and technology, accountability for the funds spent and ever better strategies for justifying allocation of additional money have become part of science policy. The popularization of science must therefore be discussed also in terms of power, political usefulness, and accountability.

A major reason to foster sustained popularization of science early in the 20th century was surely the existing tension between the process of democratization and the fact that access to the system of higher education was restricted to a narrow elite. Although universities were the central knowledge-producing institutions and although they were publicly financed, few people knew what happened inside them. The general population had little or no idea of science as a profession, so popularization was a way to compensate for the lack of transparency and familiarity. At the beginning of the century, for example, the University of Vienna had only about 6,000 students, whereas the city had about 2.1 million inhabitants. As overall financial and infrastructural conditions deteriorated, comparatively open-minded and active scientists realized they needed public support. Ludo Moritz Hartmann, himself a scientist and the initiator of the popular university courses in Vienna, stressed as early as 1895 that

> no critical observer can overlook the fact that the University of Vienna has become really popular only through these public courses and that the majority of the Viennese population today no longer sees it as a place for the education of the privileged classes but as a matter for all parts of the population. . . . In our democratic times it is meaningful enough when large segments of the population take part in activities of the university and regard the demands of the university and its rights as their own concern. But only through these popular courses is it possible to spread respect not only for science in general but for research and the scientific work of universities in particular. (Hartmann, 1910, p. 126)

However, much of what was initially thought of or "sold" as a general educational project often turned out to be publicity for science instead of any critical and informative discourse on it, as illustrated in Paris in the 1930s by the construction of the *Palais de la Découverte*, a science museum promoted mainly by Jean Perrin. In negotiating the project, Perrin brilliantly adapted his use of the term *education* to his respective audience. At one end of the spectrum, he argued that science should be understood in its ideological implications (i.e., that scientific progress is synonymous with material, social, and moral progress). At the other end, he stressed the genuinely didactic concern of remedying the deficiencies of the educational system. Eidelmann (1985) concluded that the Palais was "more the main-spring of an effort to legitimate the power of French science than a vehicle for the diffusion of scientific culture or for a 'sharing of knowledge'" (p. 195).

Popularization was thus definitely perceived as a measure to counter decline "when . . . scientists and their enterprise were experiencing a loss of prestige" (Forman, 1971, p. 6). These "countermeasures [were] in general attempts to alter the public image of science so as to bring that image back into consonance with the public's altered values" (p. 6). But as Forman clearly emphasized, "if this is not mere image projection, then such alterations of the image of the scientist and his activity will also involve an alteration of the values and ideology of the science" (p. 6). Popularization not only produced a different image in the public's eye but also had scientific consequences.

As the 20th century continued, it became increasingly obvious that the science system, particularly the development of technologies, needed not only ideological but financial resources as well. Popularization was meant to contribute in this regard. Lippmann (1913) made that point explicit in an editorial statement to a French popular science journal:

> What is more, science is horrifyingly expensive, even in times of peace. We also need considerable incomes in order to bear the increasing financial burden of billions. This means a strong industry. It means a large number of enlightened industrialists, of businessmen who understand their century. In short, we need an educated bourgeoisie. (pp. 104–105)

It was important to educate an entire society to be ready to support science and invest funds. In countries where the science system was in financial straits—as in Austria—it was considered important to remind the public regularly of the general impact that science had on the development of society.

> The prosperity of a nation is coupled most closely to progress in science. Governments that let scientific research become stunted are short-sighted. Science is paying back with major gains everything that has been invested. The death of science means not only stagnation but the regression of all progress, including that in the social sphere. (Abderhalden, 1920, p. 1)

Numerous other statements could be added, some of them ending with rather histrionic statements like "a nation abandoning its scientific research is abandoning itself" (Abderhalden, 1920, p. 1).

Popularization as a Creative Space for Science

Much has been said about the functions of popular science literature, ranging from education to legitimation. But what consequences does the popularization of science have for development within science itself? Why did some scientists feel it necessary from the very beginning of modern science to spread their knowledge beyond institutional boundaries? At least three aspects are relevant. First, by moving to the public arena of communication and leaving the core of the science system, which is rather formal about what should be dealt with and how, scientists gained latitude for creativity. Highly controversial or speculative topics could be discussed there far more easily than would have been possible through normal channels within the science system. Relativity theory, quantum physics, and psychoanalysis, for example, were discussed much earlier than was the case in the traditional university setting. In France, interdisciplinary work by historians or philosophers on relativity theory and quantum mechanics started much earlier in the public realm than did open discussion inside the respective scientific communities.

Second, the popularization of science was seen as a chance to rethink the relations between different complexes of knowledge and to place them in a more general context. Hartmann (1910) pointed out the relevance of this fact to public lectures on science in Vienna:

> There is no doubt that university teachers draw many ideas from their occupation with public lectures. . . . There is no better way to achieve an understanding of the relativity of our scientific modes of expression[.] . . . [N]othing sharpens the sense of clarity better than the obligation to present research results and their significance to people who are not used to thinking abstractly. (p. 126)

This fact was also elaborated upon about 25 years later by Fleck (1935/ 1979):

> No matter how a given case may be described, the description is always a simplification permeated with apodictic and graphic elements. Every communication and, indeed, all nomenclature tends to make any item of knowledge more exoteric and popular. . . . Certainty, simplicity, vividness originate in popular knowledge. That is where the expert obtains his faith in this triad as the ideal of knowledge. Therein lies the general epistemological significance of popular science. (pp. 114–115)

Lastly, at a moment in time when differentiation and specialization was steadily growing in the science system, popular accounts served as an important means for cross-fertilization between different disciplines and areas of research. Because people who are scientists in one discipline are laypeople in another, popular accounts were one way to keep in touch with developments in related areas.

Concluding Observations

I have attempted to argue that any judgment of the popularization of science or the public uptake of science has to be seen in the light of the functions this practice is supposed to fulfill. I thus posed the central question of why the public should understand science. Why have scientists wanted to share their knowledge, their concepts, and their ways of thinking and working with those they regard as laypersons and outsiders? I started by showing major conceptual difficulties with the concept of *the public* as used by the different actors, and I have argued that there have been two phenomena. On the one side there is a specialization and differentiation into various publics. The public uptake of science clearly depends on the particular group's own social context and experiences and on other available knowledge. The vagueness of what is meant by understanding was then discussed from various perspectives showing the coexistence of many implicit and sometimes even contradictory assumptions. The only clear and steady shift perceptible was that the claim of making people knowledgeable about science has been supplanted by the idea of building trust between science and wider publics. Lastly, I discussed five main functions of popularization and explained the different logics behind them in order to point out that popularization definitely cannot be thought of as translation, that there is no simple remedy against the deficiencies of knowledge, and that popularization must be understood as a discursive space where the meaning and position of science is negotiated.

In closing, I wish to make a few general observations about contemporary discussion of the public understanding of science. First, it is a domain confronting a central paradox. Throughout the 20th century media for imparting scientific information have multiplied and differentiated rapidly. More images of science are traded now than ever before, numerous formal and informal spaces where the public and science meet have been opened up, and the number of people with access to information about science has clearly increased. Popularization, which has been a predominantly written culture, has been enriched during the 20th century by new media like radio and TV. Popular science accounts have thus definitely become a good of mass consumption. The much-discussed gap between science and the public has apparently not been bridged, however. Quite the contrary. If analysts are to be believed, there seems to be great disorientation and incomprehension in the public (see, for example, Shapin, 1990). The more familiar science has become to wider publics, the more those publics have become aware of the complexity of the issues at stake and, indirectly, of the power that lies in scientists' hands. The feeling of familiarity with and understanding of science has been offset by a growing awareness of one's own inability to follow their complex arguments. Science has increasingly demanded that members of the public believe what they are told, trust the information and interpretation they are given, and accept the expertise of scientists, which has seemed omnipresent and reserved only unto them. The democratization of science has, in other words, increased the remoteness and authority of science.

A much improved qualitative understanding of the interaction between science and diverse publics; a more refined picture of how people choose among different kinds of knowledge offered to them; and an understanding of how they integrate it into their existing set of knowledge, experiences, and beliefs is needed. Finally, a more grounded discussion should be started on how to handle the previously mentioned paradox at a time when new media are invading public terrain and making more information accessible than ever before.

Second, the discussion above makes it clear that one cannot speak sensitively about national concepts of similarities and differences in the public understanding of science without reflecting upon the larger cultural context and the local environments of the people questioned. Science that has an international system of values and norms is clearly at odds with the public uptake and attitudes toward science, which are conditioned by the local context. There is, therefore, a need for in-depth examination of gender, age, and educational background, especially of personal experience, social background, economic status, work environment, and general ideological background.

Furthermore, historical studies on popularization of science in several national contexts teach that the intensity and general tone of popularization vary in ways that often seem to follow their own logic. The discussion on the breakdown of the science system, for example, can be traced in several European countries to the last years of the 19th century, but it emerged at a time when the development of science and technology had never been faster (see Béguet, 1990, for instance). In national contexts one finds no trace of such considerations at that time.

A third point about contemporary discussion of the public understanding of science is that the development of science has to be seen as a product of tensions and competition within science as well as with domains surrounding it. The popularization of science and the public uptake of science are a key part of this process, and scientists, the public, and mediators are actors with individual or collective interests. In this sense, "the very fact that this discourse exists and the framework it offers practitioners is more important than the question of whether the information that is processed is right or wrong" (Jacobi & Schiele, 1988, p. 14). What is at stake with the popularization of science is the power of presenting and interpreting. Consequently, the understanding of science cannot be reduced to a set of textbook questions and a particular ideology about methodological rules. Instead, it is necessary to create awareness about motives for diffusing and taking up information and for gaining a clearer view about "political" expectations of programs like the public understanding of science.

Lastly, the history of the 20th century teaches that increased knowledge about science has not necessarily increased support for science. Quite the contrary, the more information that has been diffused to the public, the more evident it has become that decisions on issues related to science and technology are linked to uncertainties and variation in scientific interpretation. Blind support for science can no longer be expected. As Collins (1987) put it, only distance lends enchantment.

References

Abderhalden, E. (1920, May 28). Teurung und Wissenschaft [Inflation and Science]. *Neue Freie Presse* (Vienna), p. 1.

Afterwissenschaft [Spurious science]. (1903, January 15). *Arbeiter-Zeitung* (Vienna), pp. 1–3.

Altenhuber, H. (1995). *Universitäre Volksbildung in Österreich 1895–1937* [University adult education in Austria, 1895–1937]. Vienna: Österreichischer Bundesverlag.

Bayertz, K. (1983). Naturwissenschaft und Sozialismus: Tendenzen der Naturwissenschafts-Rezeption in der deutschen Arbeiterbewegung des 19. Jahrhunderts [Natural science and socialism: Trends in the reception of natural science in the German labor movement]. *Social Studies of Science, 13*, 355–394.

Bayertz, K. (1985). Spreading the spirit of science: Social determinants of the popularization of science in nineteenth-century Germany. In R. Whitley (Managing Ed.) and T. Shinn & R. Whitley (Vol. Eds.), *Sociology of the sciences: Vol. 9. Expository science: Forms and functions of popularization* (pp. 209–227). Dordrecht: Reidel.

Béguet, B. (1990). *La science pour tous 1850–1914* [Science for everyone, 1850–1914]. Paris: Bibliotheque du Conservatoire National des Arts et Métiers (CNAM).

Benjamin, W. (1935). Volkstümlichkeit als Problem [Popular culture as a problem]. In W. Benjamin, *Gesammelte Schriften* (Vol. 3, pp. 450–452). Frankfurt on the Main: Suhrkamp.

Cloître, M., & Shinn, T. (1986). Enclavement et diffusion du savoir [Restriction and diffusion of knowledge]. *Information sur les Sciences Sociales, 25*, 161–187.

Collins, H. M. (1987). Certainty and the public understanding of science: Science on television. *Social Studies of Science, 17*, 689–713.

de Solla Price, D. J. (1986). *Little science, big science . . . and beyond.* New York: Columbia University Press. (Original work published 1963)

Deutscher Naturforscher und Aerztetag [The Conference of German Natural Scientists and Physicians]. (1900, September 16). *Arbeiter-Zeitung* (Vienna), pp. 26–27.

Dolby, R. G. A. (1982). On the autonomy of pure science: The construction and maintenance of barriers between scientific establishments and popular culture. In R. Whitley (Managing Ed.) and N. Elias, H. Martins, & R. Whitley (Vol. Eds.), *Sociology of the sciences: Vol 6. Scientific establishments and hierarchies* (pp. 267–292). Dordrecht, Reidel.

Durant, J. R., Evans, G. A., & Thomas, G. P. (1989). The public understanding of science. *Nature, 340*, 11–14.

Eidelmann, J. (1985). The cathedral of French science—The early years of the Palais de la Découverte. In R. Whitley (Managing Ed.) and T. Shinn & R. Whitley (Vol. Eds.), *Sociology of the sciences: Vol. 9. Expository science: Forms and functions of popularization* (pp. 195–208). Dordrecht: Reidel.

Enzensberger, H. M. (1988). *Mittelmaß und Wahn. Gesammelte Zerstreuungen* [Mediocrity and madness: Collected musings]. Frankfurt on the Main: Suhrkamp.

Felt, U. (1993). Fabricating scientific success stories. *Public Understanding of Science, 2*, 375–390.

Felt, U. (1996). "Öffentliche" Wissenschaft. Zur Beziehung von Naturwissenschaft und Gesellschaft in Wien von der Jahrhundertwende bis zum Ende der Ersten Republik ["Public" science: On the relation between natural science and society in Vienna from the turn of the century to the end of the First Republic]. *Österreichische Zeitschrift für Geschichtswissenschaften, 1*, 45–66.

Felt, U. (1997). *Wissenschaft auf der Bühne der Öffentlichkeit. Zur "alltäglichen" Popularisierung von Naturwissenschaften in Wien, 1900–1938* [Science on the stage of public life: On the "everyday" popularization of natural sciences in Vienna, 1900–1938]. Unpublished postdoctoral dissertation, University of Vienna.

Felt, U., & Nowotny, H. (Eds.). (1993). Science meets the public—A new look at an old problem [Introduction to a special issue]. *Public Understanding of Science, 2*, 285–290.

Fleck, L. (1979). *Genesis and development of a scientific fact* (F. Bradley & T. J. Trenn, Trans.). Chicago: University of Chicago Press. (Original work published 1935)

Forman, P. (1971). Weimar culture, causality, and quantum theory, 1918–1927: Adaptation by German physicists and mathematicians to a hostile intellectual environment. *Historical Studies in the Physical Sciences, 3*, 1–116.

Galten, F. (1921, January 11). Einsteins Gegenwart [Einstein's presence]. *Neue Freie Presse* (Vienna), pp. 1–2.

Gieryn, T. F. (1995). Boundaries of Science. In S. Jasanoff, G. E. Markle, J. C. Petersen, & T. Pinch (Eds.), *The handbook of science and technology studies* (pp. 393–443). Thousand Oaks, CA: Sage Publications.

Gizycki, R. von. (1987). Cooperation between medical researchers and a self-help movement: The case of the German Retinitis Pigmentosa Society. In R. Whitley (Managing Ed.) & S. S. Blume (Vol. Ed.), *Sociology of the sciences: Vol. 11. The social direction of the public sciences* (pp. 75–88). Dordrecht: Reidel.

Habermas, J. (1990). *Strukturwandel der Öffentlichkeit* [Structural change of the public]. Frankfurt on the Main: Suhrkamp. (Original work published 1962)

Hartmann, L. M. (1897). *Berichte über volksthümliche Universitätsvorträge im Studienjahr 1896/97* [Reports on university popular lectures about national traditions, 1896–1897]. Universitätsarchiv, Vienna.

Hartmann, L. M. (1910). *Das Volkshochschulwesen (seine Praxis und Entwicklung nach Erfahrungen im Wiener Volksbildungswesen). 66. Flugschrift zur Ausdruckskultur* [Adult education (Its practice and development in light of experience in the Viennese adult education system): 66th pamphlet on the culture of expression]. Munich: Dürer Bund.

Hilgartner, S. (1990). The dominant view of popularization: Conceptual problems, political issues. *Social Studies of Science, 20*, 519–539.

Holton, G. (1986). *The advancement of science, and its burdens*. Cambridge, England: Cambridge University Press.

Hopf, L. (1936). *Materie und Strahlung* [Matter and radiation]. Vol. 30 of the series *Verständliche Wissenschaft*. Berlin: Springer.

Huizinga, J. (1935). *Im Schatten von morgen. Eine Diagnose des kulturellen Leidens unserer Zeit* [In tomorrow's shadow: A diagnosis of the cultural malaise of our time]. Bern: Gotthelf.

Institut National de Recherche Agricole (INRA), & Report International. (1993, June). *Europeans, science and technology—Public understanding and attitudes* (EUR 15461). Brussels: Commission of the European Communities.

Irwin, A., & Wynne, B. (Eds.). (1996). *Misunderstanding science? The public reconstruction of science and technology.* Cambridge, England: Cambridge University Press.

Jacobi, D., & Schiele, B. (1988). *Vulgariser la science—Le procès de l'ignorance.* Seyssel: Champs Vallon.

Jasanoff, S. (1987). Contested boundaries in policy-relevant science. *Social Studies of Science, 17,* 195–230.

Jasanoff, S. (1990). *The fifth branch—Science advisers as policymakers.* Cambridge, MA: Harvard University Press.

Jeanneret, Y. (1994). *Ecrire la Science—Formes et enjeux de la vulgarisation* [Writing science—Forms of popularization and what is at stake]. Paris: Presses Universitaires de France.

Koenig, O. (1925, January 1). Volksbildung. *Arbeiter Zeitung* [Vienna], pp. 26–27.

LaFollette, M. (1990). *Making science our own: Public images of science, 1910–1955.* Chicago: University of Chicago Press.

Lévy-Leblond, J.-M. (1992). About misunderstandings about misunderstandings. *Public Understanding of Science, 1,* 17–21.

Lewenstein, B. (1995a). From fax to facts: Communication in the cold fusion saga. *Social Studies of Science, 25,* 403–436.

Lewenstein, B. (1995b). Science and the media. In S. Jasanoff, G. E. Markle, J. C. Petersen, & T. Pinch (Eds.), *The handbook of science and technology studies* (pp. 343–360). Thousand Oaks, CA: Sage Publications.

Lippmann, G. (1913). La science et la vie. *La Science et la Vie, 1,* 104–105.

Loos, J. (Ed.). (1911). *Enzyklopädisches Handbuch der Erziehungskunde.* Vienna: A. Pichlers Witwe & Sohn.

Luhmann, N. (1996). *Die Realität der Massenmedien* [The reality of the mass media]. Opladen: Westdeutscher Verlag.

Mach, E. (1910). *Populär-wissenschaftliche Vorlesungen* [Popular-science lectures] (4th expanded and edited ed.). Leipzig: Johann Anbrosius Barth. (Original work published 1896)

Michaels, M. (1992). Lay discourse of science: Science-in-general, science-in-particular, and the self. *Science, Technology, and Human Values, 17,* 313–333.

Neidhardt, F. (1993). The public as a communication system. *Public Understanding of Science, 2,* 339–350.

Nelkin, D. (1975). The political impact of technical expertise. *Social Studies of Science, 5,* 33–54.

Nelkin, D. (1987). *Selling science: How the press covers science.* New York: Freeman.

Nowotny, H. (1993). Socially distributed knowledge: Five spaces for science to meet the public. *Public Understanding of Science, 2,* 307–319.

Nowotny, H., & Felt, U. (1997). *After the breakthrough: The discovery of high-temperature superconductivity and its consequences.* Cambridge, England: Cambridge University Press.

Ortega y Gasset, J. (1993). *Der Aufstand der Massen* [Revolt of the masses] (H. Weyl, Trans.). Stuttgart: Deutsche Verlags-Anstalt. (Original work published 1930)

Raichvarg, D., & Jacques, J. (1991). *Savants et ignorants—Une histoire de la vulgarisation des sciences* [Those who know and those who don't: A history of the vulgarization of the sciences]. Paris: Seuil.

Roqueplo, P. (1974). *Le partage du savoir. Science, culture, vulgarisation* [Sharing knowledge: Science, culture, popularization]. Paris: Editions du Seuil.

Royal Society of London. (1985). *The public understanding of science* (Report of the Ad Hoc Group). London: Royal Society.

Schreder, S. (1936). *Der Zeitungsleser, Eine soziologische Studie mit besonderer Berücksichtigung der Zeitungsleserschaft Wiens* [The newspaper reader: A sociological study with special focus on the newspaper readership of Vienna]. Unpublished doctoral dissertation, Basel University.

Shapin, S. (1990). Science and the public. In R. C. Olby, G. N. Cantor, J. R. R. Christie, & M. J. S. Hodge (Eds.), *Companion to the history of modern science* (pp. 991–1007). London: Routledge.

Shapin, S. (1992). Why the public ought to understand science-in-the-making. *Public Understanding of Science, 1*, 27–30.

Sheets-Pyenson, S. (1985). Popular science periodicals in Paris and London: The emergence of a low scientific culture, 1820–1875. *Annals of Science, 42*, 549–572.

Shinn, T., & Whitley, R. (Vol. Eds.). (1985). *Expository science: Forms and functions of popularization.* In R. Whitley (Managing Ed.), *Sociology of the sciences* (Vol. 9). Dordrecht: Reidel.

Siemens, W. (1891). Das naturwissenschaftliche Zeitalter [The scientific age] (Lecture no. 59 at the Meeting of German Natural Scientists and Physicians, September 18, 1886). In W. Siemens, *Wissenschaftliche und technische Arbeiten* (Vol. 2, pp. 491–499). Berlin: n.p.

Siemering, H. (1911). *Arbeiterbildungswesen in Wien und Berlin. Eine Kritische Untersuchung* [Workers' education in Vienna and Berlin: A critical study]. Karlsruhe in Breisgau: Braunsche Hofbuchdruckerei und Verlag.

Spengler, O. (1939). *The decline of the West* (C. F. Atkinson, Trans., special edition). New York: Alfred A. Knopf. (Original work published 1918–1922)

Szeps-Zuckerkandl, B. (1939). *Ich erlebte fünfzig Jahre Weltgeschichte* [I experienced 50 years of world history]. Stockholm: Bernmann-Fischer.

Thirring, H. (1920, January 18). Die Gravitationstheorie Einsteins [Einstein's theory of gravitation]. *Neue Freie Presse* (Vienna), pp. 3–5.

Tiemann, K. H. (1991). Institutionen und Medien zur Popularisierung wissenschaftlicher Erkenntnisse in Deutschland zwischen 1800 und 1933 (ein skizzenhafter Überblick) [Institutions and media popularizing scientific knowledge in Germany between 1800 and 1933 (an outline)]. *Probleme der Kommunikation in den Wissenschaften, Kolloquien des Instituts für Theorie, Geschichte und Organisation der Wissenschaft, 75, 165–185.*

volksthümlichen Universitätkurse im Jänner und Februar, Die [The popular-science university courses in January and February]. (1901, January 11). *Arbeiter-Zeitung* (Vienna), p. 4.

volksthümlichen Universitätkurse im Jänner und Februar, Die [The popular-science university courses in January and February]. (1909, January 3). *Arbeiter-Zeitung* (Vienna), p. 9.

Wissenschaft und Öffentlichkeit in Wien: 1900–1938 [Science and the public in Vienna: 1900–1938] (Project No. P-10050HIS). (1994). Austrian Science Foundation, Vienna.

Wynne, B. (1992). Misunderstood misunderstanding: Social identities and public uptake of science. *Public Understanding of Science, 1,* 281–304.

Wynne, B. (1995). Public understanding of science. In S. Jasanoff, G. E. Markle, J. C. Petersen, & T. Pinch (Eds.), *The handbook of science and technology studies* (pp. 361–388). Thousand Oaks, CA: Sage Publications.

Ziman, J. (1994). *Prometheus bound: Science in a dynamic steady state.* Cambridge, England: Cambridge University Press.

CHAPTER 2

THE "SCIENCE WARS" AND AMERICAN POLITICS

Sheila Jasanoff

The public understanding of science can be seen either as an objective phenomenon to be measured, monitored, and, if possible, manipulated, or as a social construct to be interpreted for the light it sheds on science–society relations in democratic societies. Adopting the latter perspective, I examine the recent breakdown in the U.S. public's so-called postwar social contract with science, as reflected in the contemporary debates labeled "science wars." I suggest that public support for science and technology in America was driven historically by individualist, instrumental, and largely anticollectivist tendencies.

Four constitutive features of the politics of science in America are discussed in this chapter: the insulation of basic research, the technocratic discourse of politics, the rise of experts and counterexperts, and the undertheorized nature of scientific and social progress. It will be pointed out that there is a need for new deliberative forums and discourses to build visions of the future that can be accepted by scientists as well as their fellow citizens.

Science in a Season of Change

Like other powerful institutions of modernization in the late 20th century—universities, corporations, the state—science, with its dense penumbra of new technologies, seems uneasily poised between the Scylla of modernity and the Charybdis of postmodernity. Science stands accused of having failed society by selling its insights into, and its transforming power over, nature at too high a price. Manifestations of disorder that were never part of society's bargain with science, such as pollution, environmental degradation, threats from nuclear power, genetic manipulation, new epidemics, and weapons proliferation, offer to critics of modernity humbling evidence of humanity's loss of control over its technological projects (Beck, 1992a, 1992b). Equally, the march of science and technology is held responsible for deterioration of civic life in contemporary societies, where anomie too often prevails over engagement (Putnam, 1995) and where fragmentation, loss of the self, and a relentless instrumental rationality impede collective meditation on the aims of citizenship (Bauman, 1991; Ezrahi, 1984, 1990; Giddens, 1990; Habermas, 1973/1988). Scientific experts appear increasingly removed from the concerns of ordinary citizens, and the reassurances they offer ring hollow to publics grown mistrustful of any authority that cannot be validated against people's everyday experiences of living in the world.[1]

Others see the problem of contemporary science in a starkly different light. To believers in the program of modernity, the greatest worry flows not from science's occasional detrimental impacts on society but from the more subversive impact of social disintegration on science. Postmodern denials of western values and commitments are blamed for a retreat on many fronts from the clear light of reason and the liberal conception of progress:[2] through threats to free inquiry from confused, self-indulgent, political or social extremists fighting to restore a romanticized and unreal past (Gross & Levitt, 1994; Weinberg, 1995); through Luddite attacks on industry by greedy, technically untutored plaintiffs and their equally irresponsible legal agents (Huber, 1991); and through costly regulatory barriers to innovation that are erected in response to ignorant and fearful citizen demands (Breyer, 1993). Mass education and tough legal and political measures are needed (on the pressure for legal reform, see Jasanoff, 1995b), so say the rationalists, to keep at bay the forces of darkness and superstition that threaten to engulf three centuries of scientific and technological achievement.

In one guise or another, these fundamental disagreements about science and modernity resurface in many of today's most salient political debates in democratic societies, signaling sharp divisions in the thinking of scientists,

---

[1] At the time this paper was first written in the spring of 1996, a crisis of extraordinary proportions was unfolding in Britain over the contamination of beef cattle by Bovine Spongiform Encephalopathy (BSE) and the associated risks to human health. Governmental experts who less than a decade earlier had dismissed the likelihood of BSE migrating across species, from cattle to human beings, could not rule out the possibility that a new strain of Creutzfeldt-Jakob Disease in human patients provided evidence of just such transmission. The social costs of the earlier mistaken prediction have proved to be staggering, not only in economic terms but also in public loss of confidence in state-controlled expertise. By contrast, people's faith in common-sense responses, such as the banning of beef in many British schools and nationwide trends toward lower beef consumption and vegetarianism, have only grown stronger.

[2] The contours of scientists' discontent about their declining cultural and political authority were perhaps most strikingly manifested at a three-day conference hosted by the New York Academy of Sciences in May–June, 1995. A proceedings volume bearing the same title as the conference was published by the Academy. See Gross, Levitt, and Lewis (1996).

policy-makers, and academic social scientists. Did the modern welfare state fatally misconceive its goals by wrongly ascribing to nurture (that is, to the social environment) defects that are more basically determined by nature's laws (Duster, 1990)? Did modernization theories fail in developing countries because those theories underestimated the role of institutions and social capital (Putnam, 1992) in sustaining technological progress? Are human activities placing unsustainable strains on the global environment and must states renounce their sovereignty to meet the challenge? And on what normative or theoretical basis, if any, might governments today adopt proactive stances in steering the courses of technology (Sclove, 1995; Willke, Kruck, & Thorn, 1995)?

A relatively recent issue on the policy agenda—or, perhaps better, a complex of issues—bearing the label *public understanding of science* has begun to crystallize the contradictions inherent in current debates about science, technology, and the ambiguous gifts of modernity. As if encoding Max Weber's logic of bureaucratization, the public understanding of science has emerged in recent years as an object of social and political scrutiny, a real-world phenomenon to be measured, monitored, and, if possible, manipulated, and as a professional discourse around which to organize meetings, journals, and productive scholarly agendas. The topic has sprouted its own funding opportunities within governmental agencies and its own experts who control literatures, survey instruments, and data sets not generally accessible to observers from other areas of social and political inquiry. Resisting this bureaucratizing trend, however, are less instrumental, more reflective lines of scholarship that question the objective status of the public understanding of science (Wynne, 1995). For this research school, the public understanding of science is rather a trope to be interpreted for its social meanings, a construct to be taken apart for insights into its origins and uses, and a political domain whose power structure must be understood because it at once reflects and reinforces broader sociopolitical currents in contemporary liberal democracies.

It is in the latter stream of analysis that I wish to locate this article, approaching the public understanding of science from a standpoint that particularly emphasizes the issue's political and cultural dimensions. Sociological perspectives on the public understanding of science have laid important groundwork for my analysis by stressing the problematic nature of all three concepts that comprise this issue: *public* (is there one or many, and how are the boundaries between experts, individuals, and publics constructed?); *understanding* (what forms of knowing count as understanding, and is understanding more appropriately seen as a product of intersubjective negotiation or individual cognition?); and, not least, *science* (what kinds of knowledge constitute science, how is such knowledge organized and legitimated, and how is scientific activity embedded in society?) (Wynne, 1995). Although I, too, accept the need to question the meaning of each component of the public understanding of science, my more central concern is with the political relationships that shape or constrain the capacity of citizens in liberal democracies to develop a critical awareness of science. This manner of looking at the public understanding of science throws into relief the features of political culture, such as constitutional divisions of state power or norms of citizen participation, that influence people's access to science, their readiness to mobilize scientific resources for political objectives, and, in turn, the material and symbolic currencies with which science purchases its authority in the public domain.

My focus here is on the public understanding of science in the specific context of science–society relations in the United States, but this case, I suggest, holds wider implications for all who are concerned with the contemporary public alienation from science and with associated questions about the vitality of industrial democracies. I begin with a peculiarly American political development of the 1990s: the much discussed and widely lamented breakdown in the stable consensus governing state support for science since World War II—the so-called social contract with science—and the consequent eruption of a set of debates about the public representation of science, popularly tagged as the "science wars." I then trace some of the roots of the present malaise to distinctive characteristics of democratic self-expression in America that relate to the use of and support for science. These observations provide the basis for a politically centered reanalysis of the "problem" posed by the apparent deterioration in the public understanding of science.

Rents in the Social Contract

In 1945, presidential adviser Vannevar Bush issued an influential report, *Science—The Endless Frontier* (Bush, 1945), that supplied the cornerstone for what two generations of scientists and science policy analysts have come to regard as America's "social contract" with science. The contract, simply put, promised increased federal support for university-based research (Bush used the term "basic science")[3] with considerable autonomy for scientists to

---

[3] I use this term rather than *basic science* in order to call attention to the primary institutional locus of autonomous scientific activity, namely, universities. The term also avoids the problematic boundary issues raised by the concept of basic science at a time when work in many scientific fields is closely tied to goals of industrial pro-

decide how and where to spend the money, in return for discoveries that translate into new technologies and general enhancement of the nation's economic, social, and physical well-being. Bush, for all his perspicacity, displayed remarkably little interest in the actual mechanics of science's translation from academic research laboratories into tangible social goods. His brief report sketched little more than a simple input-output model— federal funding in, trained personnel and beneficial technologies out—with no detailed elaboration of what went on in the black box in between. Yet for decades after the end of World War II, neither U.S. scientists nor U.S. citizens had cause to doubt the adequacy of this pared-down model. On the surface, the social contract worked for twenty years much as Bush had presumed it would. Federal support for R&D grew each year from 1953 to 1965, drawing added legitimacy in 1957 from the Soviet Union's launch of Sputnik and the ensuing space race. Even the latter half of the 1960s were golden years for university-based science. Although many other troubles beset the U.S. polity, government commitment to funding science did not waver until the end of the decade, and scientists had little reason to question their relations with the state or their fellow citizens (Smith, 1990, pp. 39– 40).

Signs of trouble began accruing in the early 1970s and accumulated rapidly after the fateful events that realigned world politics in 1989. Economic analysts have noted the decline of U.S. technological leadership as increasing mobility of knowledge, capital, and training eroded the foundations of American preeminence in science and technology (Nelson, 1990). But despite growing competitive threats from Japan and, to a lesser degree, Germany, cold-war tensions and perceived military needs precluded fundamental questioning of the postwar social contract through most of the Reagan era. New issues requiring urgent scientific attention (most significantly, environmental protection) appeared on the public agenda, while old issues, most notably defense against the "evil empire" of the Soviet Union, retained high priority.[4] There was as yet little warning that particle physics

---

duction. Of course, the term university-based science raises its own boundary problems, such as the question of how the boundary between universities and industry is constructed, but these issues are not central to the argument in this chapter.

[4] Since the establishment of the National Institutes of Health, American science policy has drawn a sharp institutional distinction between health-related research and other forms of basic science. Political support for biomedical research remained strong even after the rightward swing in the 1994 U.S. national elections and subsequent governmental downsizing. Subsequent economic prosperity helped restore support for other areas of university-based research in the later 1990s.

might one day have to be robbed to pay for the space shuttle or the budget for climate modeling be reduced to pay for mapping the human genome.

A distinctly bleaker picture confronted university-based science in the 1990s. Physicists, suddenly in the vanguard of people deploring their fellow citizens' alleged flight from reason, saw omens of disaster in the cancellation of federal funding for the Superconducting Supercollider and the conversion of its scientifically incapacitated tunnel into a space for growing mushrooms (Park & Goodenough, 1996). The landslide victory of Republicans in the 1994 U.S. congressional elections confirmed the physicists' worst fears, only with the threat now extended to university-based science as a whole: massive and possibly deepening cuts in funding for basic research; a targeted attack on funding for the social sciences at the National Science Foundation; and selective elimination of research programs unpopular with conservative politicians, from advanced technologies to climate change. Few actions so eloquently testified to the changed political fortunes of science as Congress's almost whimsical decision to abolish its own Office of Technology Assessment. This advisory body of modest size and solid accomplishment had been admired and emulated in many other countries for its capacity to integrate technical analysis with a sober evaluation of legislative policy options (Bimber, 1996).

In this newly adversarial environment, scientists expanded their quarrel with Congress to take on the wider public, whose lack of understanding for science and technology they held to be the root cause of the legislature's short-sighted budgetary axe-wielding. Some of the battle lines were drawn in what may well be remembered as the science wars of the 1990s. Controversy centered, to begin with, on museum displays. Veterans' organizations succeeded in getting the National Air and Space Museum to adopt their version of history in exhibiting the *Enola Gay,* the infamous aircraft that had transported the atomic bomb to Hiroshima. Protests by activist women, historians, and psychologists led to the postponement of a retrospective on Freud at the Library of Congress (Grossman, 1996; Weeks, 1996). At the same time, powerful coalitions of chemists and physicists clamored for science and technology to be represented in a fittingly celebratory light in the Museum of American History's exhibition entitled "Science in American Life" (Forman, 1997; Gieryn, 1996; Gross, 1996; Park & Goodenough 1996). A normally quiescent scientific community aroused itself to denounce the national museum's inappropriately negative portrayal of science, a consequence they attributed to poor judgment and even poorer advice from nonscientists involved in the planning process.

Suspecting conspiracy and malign ideological change, some scientists went on the offensive against an oddly assorted array of academic commentators on science—social scientists, feminists, and postmodern literary theorists—whose dishonest politics and shabby scholarship, they argued, had provided intellectual cover for the nation's "antiscience" revolt (Gross & Levitt, 1994; Gross et al., 1996; Holton, 1993; Weinberg, 1995). In the most celebrated of these attacks, a little-known physicist named Alan Sokal published in *Social Text*, a nonrefereed journal of cultural studies, an article purporting to establish quantum gravity as a "postmodern science" through links to fashionable ideas in critical theory (Sokal, 1996). The entire journal issue had been devoted, ironically enough, to exploring the science wars. Sokal's subsequent declaration that his article had been planted as a conscious hoax made him an instant culture hero. That a physicist had successfully appropriated some of murkiest of contemporary scholarly discourse to make fun of its fashionable progenitors only made his spoof the more appealing to the reading public. Sokal's escapade triggered a flurry of admiring reports in the mass media (for instance, in the *New York Times, Newsweek,* and Britain's *Daily Telegraph*) as well as specialist journals, and it was widely received as a devastating exposé of the bankruptcy of humanistic and social critiques of science. By mid-decade, rancor had replaced self-satisfaction as the dominant chord in American scientists' conversations with the rest of society, including, not least, significant segments of their own intellectual community.

The Politics of Understanding Science

How should one explain this unraveling of the postwar consensus on U.S. science policy at century's end? Public misunderstanding of science, the explanation increasingly favored not by scientists alone (Gross et al., 1996) but also by lawyers and politicians defending "rational" public policy (Breyer, 1993; Huber, 1991) scarcely seems equal to the task. There is little evidence, for example, that the public's knowledge of basic scientific facts or its appreciation for technologically assisted improvements in the quality of life has diminished noticeably in recent years. U.S. citizens remain quite willing to master the intricacies of technical discourse when required to do so for meaningful purposes, such as in choosing among medical treatment options or building the scientific basis for compensation claims (Jasanoff, 1995b). Public opinion polls reveal that the fear of technological disaster, though widely held, is counterbalanced by continued faith in science's ability to conquer disease and hunger and provide tangible material improvements. When asked to weigh the benefits and risks of scientific research in

1995, 72% of Americans said the former outweighed the latter (National Science Board, 1996, pp. 7–17).

More important, the thesis of individual scientific incapacity (or *scientific illiteracy*) provides no explanation for the particular expressions of public alienation from science or the associated conflicts between science and society that are observable in different political cultures. Why, for instance, are American scientists most vocally concerned about the rise of antiscience, only distantly followed by colleagues in Britain,[5] when academic scientists everywhere in the industrial world are confronting similar pressures from shrinking economies and the post-cold-war retrenchment of state support? Why have these quarrels been framed in the United States, but not in most other countries, as an attack on academic studies of science and technology? As tempting as it may be to blame the members of the U.S. public individually and collectively for scientific illiteracy, it seems more plausible that the causes of the alleged flight from science lie elsewhere, embedded more plausibly in the institutional and cultural foundations of American democracy.

Any attempt to provide a genuinely political interpretation of the public understanding of science in the United States has to begin with the distinctive texture of the country's political dynamics and to show how they, in turn, affect understandings between science and society. Thus, acute observers of the U.S. scene since Tocqueville have noted the culture of individualism that permeates the nation's social and political life. In a study of science and religion in America, Price (1983) associated the individualist and perfectionist leanings of American religious movements with a concomitant distrust not only of government but even of family and community. Discursively reaffirmed and reproduced from one generation to the next, the culture of individualism has made Americans peculiarly resistant to the welfare state and its institutional trappings. Although individual autonomy, arguably, has been a universal desideratum of modernity, the tension between autonomy

---

[5] A debate between scientists and academic analysts of science also took place in Britain, where it was less overtly coupled with issues of science funding, museum displays, or the public understanding of science. Britain's emergence as an important site for the science wars is worth speculating on from the standpoint of the political sociology of knowledge. Note that Britain, like the United States, made heavy commitments to defense spending during the cold War, with consequent strong support for science. Under Thatcher's prime ministership, however, relations between British science and the state cooled markedly following severe funding cutbacks that scientists saw as imperiling the health of university-based research. Note also that Britain for decades has been home to a lively critical discourse on the sociology of scientific knowledge, a discourse that some British scientists came to see as unfriendly to their interests. To pursue these suggestive leads and provide a fuller political and cultural account of the science wars in Britain would go well beyond the scope of this essay.

and collectivization, or the "discourse of disciplinization" as Wagner (1994, p. 6–8) terms it, has been manifested differently in Europe and the United States. The idea of a strong public sphere did not take hold in America until the New Deal, despite some earlier efforts to construct agencies responsive to the concerns of the military and women (Skocpol, 1992). The extreme disenchantment with the federal government in the 1990s was accompanied by a resurgence of individualist ideology with force enough to destabilize the pillars of the New Deal settlement. How, one may ask, has the public understanding of science and technology been caught up in the construction and reconstruction of American individualism, and what implications, if any, does this cultural propensity have for the triadic relationship of science, citizens, and the state?

If one looks at the multitude of forums in which science engages with social needs (e.g., technological warfare, biomedical research, regulation of risks, litigation), one finds the most unproblematic patterns of commitment in two areas of U.S. science policy: (a) where science and technology have served to ratify the military superiority of the American state and (b) where scientific discovery has promised to enhance the capacity and competence, whether real or perceived, of the consuming public. Thus, U.S. citizens historically have given their assent to grand, instrumental uses of technology by the state, such as the atomic bomb, the moon landing, and the space station. Ezrahi (1990) argues that such compelling demonstrations are essential for the survival of a liberal democracy that maintains its hold on its members by the promise of transparency. The state's legitimacy depends crucially, in Ezrahi's view, on enabling each individual American to assess the benefits of membership in the larger polity. Successful demonstrations of technological might, whether in Hiroshima, Libya, or Baghdad, reassure U.S. citizens of the continued effectiveness of their national government. In an evocative passage, Ezrahi (p. 42) describes how "the body of Saturn 5, a gigantic space leviathan whose carcass lies open in a didactic gesture toward curious taxpayers," is transformed at the Kennedy Space Center into an object lesson in the benefits of democracy. A guide conducts the visitors' gaze over the rocket's anatomy, detailing how each small part was made in a different state, creating in sum a potent image of collective action to which all can pay homage and from which all can derive a reinforced sense of national identity.

At the other extreme, state investment in science commands public assent where the objective is to place valuable commodities on the market. Even before the adoption of the Bush contract and the rise of publicly supported science, U.S. science policy was guided by a desire for practical benefits (Price, 1983). Researchers in the biomedical sciences, for example, long ago learned to live with the hypocrisy of doing "pure" research while promising useful applications to society.

Between the instrumentalist state, with its sporadic public displays of strength, and the ungoverned free-for-all of the market, scientific research proceeds independent of carefully articulated societal goals. The U.S. Constitution protects scientists' freedom of inquiry but provides few guidelines on how to envision collective purposes toward which inquiry should be directed (Goldberg, 1994; Sclove, 1995). Governmental institutions, including the courts, are more hospitable to policies that promote individual scientific and technological initiative (e.g., through patent laws) than to claims for the redistribution of wealth and benefits. When courts have employed science and technology in support of redistribution (e.g., in mass tort or affirmative action cases), their work has frequently been criticized on grounds of scientific incompetence. In a pluralistic society with exceptionally decentralized sources of knowledge (Jasanoff, 1990), faith in subjective understandings of the world have taken precedence over trust in superior authorities; science is appropriated not to cement a broad social consensus but to advance selective political interests and microutopian visions (Ezrahi, 1984). The daily mechanics of an adversarial political culture work better at producing citizen experts and counterexperts knowledgeable about specific issues (Balogh, 1991; Brickman, Jasanoff, & Ilgen, 1985) than at defining common purposes for the nation as a whole.

It would be strange indeed if such fundamental features of a nation's political order as these failed to affect the ways in which its citizens understand and deploy scientific knowledge, much as if a contract were to be drawn up without respecting the syntax and lexicon of the language in which it is composed. The rights, duties, constraints, and opportunities that define citizenship in a complex society necessarily shape the pathways by which its members grope for greater understanding, through science, of their circumstances in the world. Contextualized within a political framework, the public understanding of science emerges not as an objectively measurable index of scientific literacy but, more important, as the aggregate of what people wish to know from science in order to inform their daily choices and the ways in which they wish to deploy this knowledge for policy.

For U.S. citizens, in particular, the right to be skeptical of authority, to associate freely, to define new social identities, to express political, religious, and sexual preferences without fear, and to sell the products of their wit and ingenuity on the open market are all inextricably bound up not only with an abiding faith in American exceptionalism but also with the understanding of science. Relatively few institutional barriers stand in the way of individuals or groups who wish to empower themselves (or, as a corollary, disempower others) through science and technology (Jasanoff, 1996).[6] In the process, the contract that U.S. society negotiates with science becomes contingent, particularistic, and directed toward localized needs. Except in the context of military deployment and related "big science" projects, science becomes an instrument that even the state exploits principally for purposes deemed worthwhile by the market. Citizens and interest groups meanwhile claim science as their own to advance personal causes, from AIDS research (Epstein, 1996) to exoneration in the courtroom, but disavow support when the aims of science disappoint hope or elude understanding.

Postmodern Alienation and the (A)social Contract

What are the politically salient features of America's contract with science?[7] Four of its elements deserve to be briefly sketched because of their relevance to the public understanding of science: (a) the insulation of university-based, or "basic," scientific inquiry from politics, (b) the tendency for technical discourse to usurp political debate, (c) the reduction of complex political differences to binary disputes between experts and counterexperts, and (d) the politically undertheorized character of scientific and technological progress.

The Insulation of Basic Research

To the extent that American science policy has a central dogma, it is that the pure, curiosity-driven, disinterested forms of inquiry thought to be conducted in academic science departments should proceed with as little inter-

---

[6] Claims of privileged knowledge by members of problem-focused religious, political, or social groups strike professional scientists as antiscientific because they are overtly interest-laden. The frequency of such claims in the United States, however, reflects a widely shared acceptance of the social importance of science. As Bennett Berger, the reviewer of Gross and Levitt's (1994) book in *Science,* shrewdly observed, much more of the supposedly antiscientific activity that these authors decry "reveals awe at the power of science (with an attendant desire to control it) than hostility to it" (Berger, 1994, p. 986).

[7] My reading here diverges significantly from the rather literal interpretation of the so-called social contract that dominates much recent writing on U.S. science policy. The conventional view stresses the material exchange between science and society: public funding in return for trained personnel and new technologies. By contrast, I am primarily interested in exploring how the science-society relationship (the contract) reflects American politics writ large and, more particularly, how it bears on the public understanding of science.

vention as possible from the state, even when some 50% of this work is publicly funded. Anyone presuming to question this principle is instantly confronted with the specters of state-regulated science in Nazi Germany and Stalinist Russia, neither of which, quite deservedly, has lost the power to offend liberal sensibilities after more than 50 years. Less frequently discussed, however, are the not inconsiderable political costs of academic science's long independence from public scrutiny.

Paradoxically, the supposed autonomy of university-based science may have rendered it more vulnerable to the escalating demands of postmodern politics than it otherwise might have been, as scientists unused to trafficking in the political realm find themselves increasingly less able to explain the benefits of their independence to fellow citizens. Traces of this political weakness have surfaced in recent U.S. controversies over science policy. Pressed to cut budgets, legislators feel a new need to justify expenditures on science but can agree on no unifying public vision to guide them. Incoherence and uncertainty are the predictable results. Liberals, questioning the very concept of state-supported research with no well-articulated public aims, want scientists to be more explicitly "mission-oriented." Conservatives shy away from scientific activity too closely tied to missions that they abhor, ranging from environmental regulation to R&D initiatives perceived as interfering with the market (Norman, 1992; Park & Goodenough, 1996).

Research on health, one of the few "goods" that American society unequivocally endorses, is the primary beneficiary, as evidenced by continued high appropriations for the National Institutes of Health. By contrast, support for the social sciences, which demand almost by definition a prior consensus about the nature and causes of social problems, remains wavering and insecure.

Cutting across party lines is a rising skepticism about the integrity of research, with consequent tightening of monitoring and reporting requirements in the conduct of science. Scientists themselves appear curiously defenseless against these contradictory and confusing challenges. The language of public accountability has grown threadbare through disuse. The moral authority that atomic physicists once commanded in their crusade against nuclear weapons has eroded, and global environmental phenomena, such as overpopulation or climate change, have proved to be too contentious to endow scientists with comparable authority. In reproaching the public for misunderstanding science and espousing antiscience sentiments (Gross & Levitt, 1994; Holton, 1993), the scientific community discloses its own discursive poverty and the two-dimensionality (one is either for science or against science) of its political self-image. Dennis (1997), a historian of 20th-century science, has argued that the history of science emerged as a separate discipline in postwar America to provide a rhetoric of justification for the science of Vannevar Bush's era. Today, however, episodes like the Sokal hoax and the refusal in May 1997 of the famed Institute for Advanced Study at Princeton to offer a post to the historian of science Norton Wise[8] show scientists at war with the scholars who are trying to write their histories. These events, each a skirmish in the ongoing science wars, are not of merely academic interest. They underscore a disquieting inability on the part of scientists to engage fruitfully with the very discourses that could form the basis for their renewed accommodation with society.

The Predominance of Technical Discourse

One of the ironies of American political life is that strenuous efforts to preserve the purity of scientific inquiry are coupled with a sweeping reliance on science to legitimate the slightest political demands (Ezrahi, 1984; Jasanoff, 1995b, 1996). In courts and before regulatory agencies, in state governments and in Washington, nationally and internationally, political positions are continually articulated in terms that incorporate science. New frontiers of inquiry, often generously supported by public funds, open up where existing natural and social sciences seem not to hold adequate answers for people's perceived needs. Examples include the war on cancer, the violence initiative, and the institutionalization of alternative medicine at the National Institutes of Health; the interagency program on human dimensions of global change; the ethical, legal, and social implications (ELSI) program of the Human Genome Project; and the emerging methodologies of environmental equity analysis. Hardly any issue rises to political prominence in the United States without drawing science into its ultimate resolution.

What are the wider political consequences of these developments? How does the ever increasing permeation of technical discourse into politics affect the public understanding of science? Briefly put, the effect is both to engage and alienate lay citizens. Alienation prevails when technical methods and practices become overly specialized, fragmented, rule-bound, and closed to almost everyone except bureaucratic decision-makers, as has arguably happened with quantitative risk assessment for environmental protection (Breyer, 1993; Irwin & Wynne, 1996; Winner, 1986). The appeal of science

---

[8] Trained as both physicist and historian, Norton Wise, a professor at Princeton University, appeared to have impeccable credentials as a scholar in the history of science. His rejection by members of the institute faculty was attributed, in part, to a letter that Wise had written criticizing the physicist Steven Weinberg for his public stance on the Sokal affair.

diminishes under these circumstances, except for the few powerful actors who can afford the high entry costs associated with arcane, tightly black-boxed domains of scientific expertise. By contrast, the prospect of using science to further new, micropolitical agendas (Beck, 1992b; Ezrahi, 1984) continually invites public reengagement. U.S. citizens become insatiable, and often expert, consumers of scientific knowledge when the goal is instrumental and personal: to cure a life-threatening disease, to ward off perceived environmental risks, to win a lawsuit, or, more generally, to create an objective foundation for rights and benefits claimed from the state.

Experts and Counterexperts

The themes of skepticism and distrust are often linked in the work of con-temporary social theorists (Bauman, 1991; Beck, 1992a; also see Irwin & Wynne, 1996) to citizens' alienation from the rationalistic culture of moder-nity. But civic engagement, too, can breed its own forms of distrust when scientific knowledge is brought to bear on political ends. In an accessible and adversarial policy culture such as that of the United States (Brickman et al., 1985; Jasanoff, 1990, 1995b), it is relatively easy for interested players to muster sufficient expertise to participate in highly technical debates. By the same token, however, no expert viewpoint enjoys complete immunity from challenge, and professional credentials alone are seldom sufficient to ensure public deference. U.S. policy controversies commonly unfold as battles between contending experts, with each side possessing enough knowledge, skill, and credibility to expose weaknesses in the other's argu-ments. One need never look far behind the scenes to find the constructed character of policy-relevant science openly on display (Bimber & Guston, 1995).

Distrust of scientific authority, then, is a paradoxical consequence of the U.S. public's heavy reliance on science. Scientific knowledge is valued for its legitimating power, but individual claims and even established bodies of knowledge are always open to question (for example, attacks by religious fundamentalists on evolutionary theory and by political conservatives on climate-change science). Scientists are respected as a body, but well-regarded scientists may find their credibility demolished on the witness stand or in bruising congressional hearings.

Expert knowledge is seen as the key to winning important legal and politi-cal battles, but expertise is also widely recognized as being available for hire. The idea that science can be molded by interests is part and parcel of every U.S. citizen's intuitive understanding of science, counterbalancing any abstract faith in science's impartial authority. In this way, the interest-driven appropriation of science by competent citizens contributes to the dissolution of objective, expert authority, a phenomenon that Beck (1992a) and others have noted as one of the markers of "reflexive modernity."

The Limits of Progress

From Vannevar Bush's famed report to the present day, the promise of progress has been held forth as the primary justification for public investments in science and technology. The rhetoric of progress is embraced as much by scientists as by their political sponsors. The central notion of "progress," however, is so deeply taken for granted in American life that it merits little or no public discussion. Unexamined and undertheorized in political discourse, progress typically is equated, as in the following quotation, with new developments in science and technology, particularly when they promise practical applications that will enhance individual well-being and liberty:

> In this century, science has doubled the life span of Americans, freed them from the mind-numbing drudgery that had been the lot of ordinary people for all of history, given them the means to indulge in whatever activity they find rewarding, and put all the knowledge of the world at their fingertips. (Park & Goodenough, 1996, p. 12)

Bill Gates, the technological wizard and head of Microsoft Corporation, imagines the ideal home of the future as a place where people will move from room to room accompanied by a sensor that instantly calls up their favorite television show or photographic image. This is an atomizing vision. It ties technology to individual needs and preferences. A question rarely asked about the brave new world to be created by further developments in science and technology is what kinds of *societies* should people it. The unstated assumption is that it will be a society composed of equally informed and equally affluent consumers with highly developed appetites for technology. Not on the political agenda is the disciplining effect of a high-tech market that will prosper only if the tastes and preferences of consumers themselves are as standardized as the products offered by technology.

U.S. public policies for biotechnology offer a telling example of this phenomenon. In two decades of public debate, genetic engineering has been cast as the archetype of a progressive technology: "as the century draws to a close, molecular biology is unraveling the secrets of life itself" (Park & Goodenough, 1996, p. 12). But the progress envisioned by most U.S. policymakers barely glances beyond the immediate consumer satisfaction of their fellow citizens. Biotechnology is hailed as a virtually inexhaustible font of new products and new capabilities for individual human beings, such as gene therapy, genetic enhancement, and, most recently, human cloning; distributive concerns receive relatively short shrift, except for an occasional nod at the possibility of relieving world hunger by means of new, genetically engineered crop varieties. U.S. regulatory policy toward biotechnology has been framed primarily as a question of assuring the safety of new products, thereby virtually excluding any meaningful public steering of a technology that disrupts the very ordering of nature and society.

This narrowing of the technopolitical imagination reflects, as I have argued elsewhere (Jasanoff, 1995a, 1995b), deeply entrenched features of the political culture of liberal democracies. The formal discourses of law, risk analysis, and even bioethics empower public deliberation, but they do so at the cost of fragmenting social concerns into tightly delimited policy issues that can be addressed by existing administrative and bureaucratic structures in accordance with established practices or modes of reasoning. Concerns that do not easily lend themselves to such packaging are dismissed as too vague, remote, unfocused, hypothetical, or otherwise inappropriate for the decision-making forum. Thus, the U.S. Supreme Court refused to hear generic arguments about risk when approving the patentability of new, genetically modified life forms. Similarly, the U.S. Food and Drug Administration considered only risks to human health, not wider implications for agriculture or animal husbandry, in allowing the marketing of bovine growth hormone. The possible effects of bioengineering on global biodiversity, a matter of concern to ecologists and environmental activists, have received no systematic attention in U.S. policy forums.

Toward a New Politics of Public Understanding

A politically grounded analysis of the public understanding of science inexorably shifts attention from the bureaucratic challenge of how to monitor and regulate that understanding to interpretive questions of what understanding means and why it matters to the conduct of democracy. For most people, understanding science no longer is, if indeed it ever was, a product of free choice or formal training. It is a necessary condition of modernity. Almost every facet of life in advanced industrial societies is mediated by scientific knowledge and its far-flung applications. Science interpenetrates the very categories of thought by which people organize their impressions of the world. Indeed, the public understanding of science, by virtue of its capacity to channel political debate, has itself become such a category. Scientific inquiry continues to provide a powerful model for critical, self-reflexive democracy, and the knowledge generated by science is an indispensable resource in the political realm. Technological artifacts replace, enhance, modify, and open the way to new human capabilities. People live in a world of hybrids and cyborgs, where human and nonhuman entities are joined in functionally inseparable networks of action and mutual support (Callon, 1987; Haraway, 1991; Latour, 1993). Can there be a person competent to carry out the work of citizenship in such a world who does not in a profound sense "understand science"?

To be sure, this understanding is not achieved by acquiring and retaining miscellaneous scientific facts (although capable citizens of industrial societies inevitably do have a good number of such facts at their disposal). Nor is it manifested by openly reverencing the scientific method, accurately calculating mathematical probabilities, knowing the internal workings of a computer program or jet engine, or recalling the order of the major breakthroughs in knowledge in a given scientific field. Rather, it is a keen appreciation of the places where science and technology articulate smoothly with one's experience of life, of moments at which to turn to science for help and those at which to look elsewhere, and of the trustworthiness of expert claims and institutions (Irwin & Wynne, 1996). For most of modern humanity, understanding science, like the end-product of most successful processes of learning, is less a question of knowing things than of knowing when, where, and for what purposes to look them up.

It is not, however, a level or uncontoured playing field on which scientific understanding is constructed, neither for scientists nor for laypeople. Political and legal traditions, culture and national identity, and the discourses and practices of power all constrain the ways in which people seek to question the world around them. In any modern polity these fundamental commitments influence not only the favored directions of scientific research but the institutional and political terms of science's daily interactions with society. One cannot speak, then, of the public understanding of science in the abstract, as if it were a physical constant like the speed of light. Rather, it mirrors the particularities of time and place, of history and culture. Quarrels about the public's understanding of science are most satisfactorily interpreted as surface indicators of disharmony in the underlying social and political relationships that shape the very possibility of such understanding. As such, debates about the public understanding of science offer a unique vantage point for understanding science's place in modern society.

The threats to science and reason that American scientists find so troubling (Gross & Levitt, 1994; Gross et al., 1996; Holton, 1993; Park & Goodenough, 1996; Weinberg, 1995) can be reexamined in the light of this analysis. A politically informed inquiry into the framing of the public understanding of science as an issue of concern reveals certain systematic tendencies that I have outlined above. We have seen that U.S. citizens energetically pursue scientific knowledge and willingly support science for instrumental purposes sanctioned by the market and the national security state. Although there is little interest in the acquisition of scientific facts for their own sake, laypeople quickly metamorphose into experts when they have to fight localized battles for health, safety, autonomy, and well-being. There is a broad consensus on a vision of technological progress that harnesses scientific discovery to the enhancement of individual capacity and identity. Yet missing from this picture is a truly social understanding of science. The American polity contracts with science and technology on an individual basis, not as a collective entity with a clear sense of where its historical commitment to science derives from and where future commitments should ideally lead it.

The deep stresses and realignments of the past decade have brought to light the dangers of America's disconcertingly asocial contract with science. Neither the public nor its elected representatives are any longer satisfied with a program of unrestricted support conceived in an expanding economy and geared to the production of technological miracles on a grand scale. The receding threat of Armageddon has liberated people to think again about issues that are much more local and immediate and for which the universalist project of modernity does not hold obvious answers. The contradictions of modernity have in any case become too obvious for informed publics to subscribe to monolithic dreams of progress through science. The institutions and processes of American politics are all too hospitable to the inward-turning, micropolitical tendencies of the present era. New collective problems, such as reduced economic competitiveness, interdependence, and global change, have not yet acquired sufficient hold on people's minds to motivate a robust new politics of science.

Scientists, for their part, have been locked away for so long in the safe (if only imaginary) havens of basic research that they have lost the power to project a persuasive vision of the future to a newly skeptical and fragmented society. Policy vacillation and the rise of so-called antiscience rhetorics flag the uncertainties of the moment. The challenge for scientists and lay citizens alike in this changing social order is to find new languages of communication and new forums where they can deliberate together on the shape of things to come. Specialists in the public understanding of science can play an invaluable role in diagnosing the causes of past miscommunication and suggesting ways of moving beyond them. Scientists, too, will have to sit at the same discursive table, but their participation will prove most effective if they can turn a more sympathetic ear to stories that tell not only of science's heroism but also of its essential humanity.

References

Balogh, B. (1991). *Chain reaction: Expert debate and public participation in American commercial nuclear power, 1945–1975.* New York: Cambridge University Press.

Bauman, Z. (1991). *Modernity and ambivalence.* Ithaca, NY: Cornell University Press.

Beck, U. (1992a). *The risk society: Towards a new modernity* (M. Ritter, Trans.). London: Sage. (Original work published 1986)

Beck, U. (1992b). From industrial society to the risk society: Questions of survival, social structure and ecological enlightenment. *Theory, Culture and Society, 9,* 97–123.

Berger, B. M. (1994). Taking arms. *Science, 264,* 985–986, 989.

Bimber, B. (1996). *The politics of expertise in Congress: The rise and fall of the Office of Technology Assessment.* Albany, NY: State University of New York Press.

Bimber, B., & Guston, D. (1995). Politics by the same means: Government and science in the United States. In S. Jasanoff, G. E. Markle, J. C. Petersen, & T. Pinch (Eds.), *The handbook of science and technology studies* (pp. 554–571). Thousand Oaks, CA: Sage Publications.

Breyer, S. (1993). *Breaking the vicious circle: Toward effective risk regulation.* Cambridge, MA: Harvard University Press.

Brickman, R., Jasanoff, S., & Ilgen, T. (1985). *Controlling chemicals: The politics of regulation in Europe and the United States.* Ithaca, NY: Cornell University Press.

Bush, V. (1945). *Science, the endless frontier.* Washington, DC: U.S. Government Printing Office.

Callon, M. (1987). Society in the making: The study of technology as a tool for sociological analysis. In W. Bijker, T. Hughes, & T. Pinch (Eds.), *The social construction of technological systems: New directions in the sociology and history of technology* (pp. 83–103). Cambridge, MA: Massachusetts Institute of Technology (MIT) Press.

Dennis, M. A. (1997). Historiography of science: An American perspective. In J. Krige & D. Pestre (Eds.), *The sciences in the 20th century* (pp. 1–26). Amsterdam: Harwood Academic Publishers.

Duster, T. (1990). *Backdoor to eugenics.* New York: Routledge.

Epstein, S. (1996). *Impure science: AIDS, activism, and the politics of knowledge.* Berkeley: University of California Press.

Ezrahi, Y. (1984). Science and utopia in late 20th century pluralist democracy—With a special reference to the U.S.A. In R. Whitley (Managing Ed.) and E. Mendelsohn & H. Nowotny (Eds.), *Sociology of the sciences, Vol. 8. Nineteen eighty-four: Science between utopia and dystopia* (pp. 273–290). Dordrecht: Reidel.

Ezrahi, Y. (1990). *The descent of Icarus: Science and the transformation of contemporary democracy.* Cambridge, MA: Harvard University Press.

Forman, P. (1997). Assailing the seasons [Review of the book *The flight from science and reason*]. *Science, 276,* 750–752.

Giddens, A. (1990). *The consequences of modernity.* Stanford, CA: Stanford University Press.

Gieryn, T. C. (1996). Policing STS: A boundary-work souvenir from the Smithsonian exhibition "Science in American life." *Science, Technology, and Human Values, 21,* 100–115.

Goldberg, S. (1994). *Culture clash.* New York: New York University Press.

Gross, P. R. (1996). Reply to Tom Gieryn. *Science, Technology, and Human Values, 21,* 116–120.

Gross, P. R., & Levitt, N. (1994). *Higher superstition: The academic left and its quarrels with science.* Baltimore: Johns Hopkins Press.

Gross, P. R., Levitt, N., & Lewis, M. W. (Eds.). (1996). *The flight from science and reason.* New York: New York Academy of Sciences.

Grossman, R. (1996, January, 7). The past ain't perfect. *Chicago Tribune,* Section 2, pp. 1, 7.

Habermas, J. (1988). *Legitimation crisis* (T. McCarthy, Trans.). Cambridge, England: Polity. (Original work published 1973.)

Haraway, D. (1991). *Simians, cyborgs, and women: The reinvention of nature.* New York: Routledge.

Holton, G. (1993). *Science and anti-science.* Cambridge, MA: Harvard University Press.

Huber, P. W. (1991). *Galileo's revenge: Junk science in the courtroom.* New York: Basic Books.

Irwin, A., & Wynne, B. (1996). *Misunderstanding science? The public reconstruction of science and technology.* Cambridge, England: Cambridge University Press.

Jasanoff, S. (1990). American exceptionalism and the political acknowledgment of risk. *Daedalus, 119*(4), 61–81.

Jasanoff, S. (1995a). Product, process, or program: Three cultures and the regulation of biotechnology. In M. Bauer (Ed.), *Resistance to new technology—Nuclear power, information technology, biotechnology* (pp. 311–331). Cambridge, England: Cambridge University Press.

Jasanoff, S. (1995b). *Science at the bar: Law, science and technology in America.* Cambridge, MA: Harvard University Press.

Jasanoff, S. (1996). Beyond epistemology: Relativism and engagement in the politics of science. *Social Studies of Science, 26,* 393–418.

Latour, B. (1993). *We have never been modern.* Cambridge, MA: Harvard University Press.

National Science Board. (1996). *Science and engineering indicators—1996.* Washington, DC: U.S. Government Printing Office.

Nelson, R. R. (1990). U.S. technological leadership: Where did it come from, and where did it go? *Research Policy, 19,* 117–132.

Norman, C. (1992). Commission sees NSF's future in its past. *Science, 258,* 1434.

Park, R. L., & Goodenough, U. (1996). The unmaking of American science policy: The end of the scientific era? *Academe, 82,* 12–15.

Price, D. K. (1983). *America's unwritten constitution.* Cambridge, MA: Harvard University Press.

Putnam, R. D. (1992). *Making democracy work: Civic traditions in modern Italy.* Princeton: Princeton University Press.

Putnam, R. D. (1995). Bowling alone: America's declining social capital. *Journal of Democracy, 6,* 65.

Sclove, R. E. (1995). *Democracy and technology.* New York: Guilford Press.

Skocpol, T. (1992). *Protecting soldiers and mothers.* Cambridge, MA: Harvard University Press.

Smith, B. L. R. (1990). *American science policy since World War II.* Washington, DC: Brookings Institution.

Sokal, A. D. (1996). Transgressing the boundaries: Toward a transformative hermeneutics of quantum gravity. *Social Text, 14,* 217–252.

Wagner, P. (1994). *A sociology of modernity: Liberty and discipline.* London: Routledge.

Weeks, L. (1996, February 28). Library of Congress's Freudian Flip. *Washington Post,* Section B, pp. 1, 4.

Weinberg, S. (1995). Night Thoughts of a Quantum Physicist. *Bulletin of the American Academy of Arts and Sciences, 49,* 51–64.

Willke, H., Kruck, C., & Thorn, C. (1995). *Benevolent conspiracies: The role of enabling technologies in the welfare of nations.* Berlin: Walter de Gruyter.

Winner, L. (1986). *The whale and the reactor.* Chicago: University of Chicago Press.

Wynne, B. (1995). Public understanding of science. In S. Jasanoff, G. E. Markle, J. C. Petersen, & T. Pinch (Eds.), *The handbook of science and technology studies* (pp. 361–388). Thousand Oaks, CA: Sage Publications.

CHAPTER 3

FROM NORTHERN TO SOUTHERN EUROPE— POLITICAL MOTIVATIONS BEHIND RECENT DISCOURSE ON THE "PUBLIC UNDERSTANDING OF SCIENCE"

Maria Eduarda Gonçalves

From Cultural Practice to Political Concern

In Western countries, the teaching of science in schools and the diffusion of scientific discoveries beyond the boundaries of the research community began as a strategy of scientists and scientific societies to enable the public to share in their knowledge. Since the 17th century, academies of science have built a remarkable record of pedagogical activity through public lectures in chemistry, physics, and other natural sciences. Following the industrial revolution, in the late 18th and 19th centuries, scientists were called upon to respond to a popular demand for greater knowledge about science in a Europe somehow fascinated by the "merveilles de l'industrie" (Béguet, 1990, p. 10). Around the same period, in the United States of America, professional lecturers traveled throughout the nation speaking on "scientific" topics, often accompanied by elaborate and spectacular demonstrations of scientific phenomena (Massey, 1989, p. 915). As pointed out by Caro (1990), the origins of the popularization of science illustrated the ludic features of the experience of discovery and of explanation that were the source of science's allure (Caro, 1990, p. 24). Scientists' endeavors in this field had a well-intentioned purpose, though it was certainly also linked to the search for credibility and social power by the scientists. In this context, informing the public about developments in science was part of an essentially cultural practice.

Throughout the 20th century, concerns about the public's access to scientific and technical knowledge have spread from the academic to the political and economic realms. This extension has been directly related to the progressive recognition of the role of science and technology in economic development and social welfare. One reason is that researchers themselves advanced the notion that the responsibility of science involves promoting the scientific attitude among large numbers of people. Another reason is that the issue of scientific and technical literacy entered a new political discourse where it became associated with the idea that people living in a complex scientific and technological civilization should possess a certain degree of scientific knowledge, know-how, or both (Durant, 1993). In more advanced countries, concerns about the level of workers' technical skills, its impact on the competitiveness of industry have been particularly recurrent. Some observers have called it *functional literacy,* meaning the ability to respond meaningfully to the technical issues that pervade modern daily lives and the world of political action (Ayala, 1996, p. 2). In this context scientific and technical literacy began to be looked at primarily as a question of educating and training the work force. Only since the mid-1970s has knowledge of science become associated in the political sphere with comprehension of science, meaning the public acceptance of science-based technological change. This acceptance includes consent to support state funding of research and development (R&D), also known as the "public understanding of science." Among the most important roles of governments in science and technological policy were the promotion of scientific and technological literacy through education and other means (OECD, 1987).

As a result of these trends, discourse on the public understanding of science became to some extent instrumental to achieving other goals, of both the political system and the scientific community. By the same token, the purposes underlying the diffusion of scientific knowledge, and even the concepts with which one works in this area, became much more ambiguous than in the past. Shortland (1989, pp. 9–14), for instance, provided a long list of scientific, economic, military, ideological, cultural, intellectual, aesthetic and ethical arguments in favor of popularizing science in this changing world. Fourez (1994, p. 12) argued that scientific and technological literacy should further three things for all citizens: humanistic objectives, a democratic society, and economic growth. He added: "Almost everyone would agree that, without any familiarity with sciences and the technologies, it would be useless to pretend to play a role in the present world" (p. 185; my translation). Such broad approaches to the notion of the public understanding of science approximate the thesis of the well-known 1985 report by the Royal Society of London, according to which "better public understanding of science can be a major element in promoting national prosperity, in raising the quality of public and private decision-making and in enriching the life of the individual" (p. 9).

These statements indicate that the public understanding of science and technology has often been portrayed in somewhat utilitarian terms as science-based knowledge that is required for dealing with practical problems either in professions or daily life. On other occasions, a link has been established between the public understanding of science and the workings of the democratic system: Everyone should be given the opportunity to understand science to the extent required for the sound judgment needed in public decisions on scientific or technical matters. This viewpoint approaches the concept of a right to know, a right of citizenship, a basic freedom of information—the principle held by liberal thinkers to be essential to participation in a democratic society. Hence, in some political and academic circles, the popularization of science as a cultural activity (one that can be seen as a good in its own right) came to be emphasized as much as the evaluation and development of methods for judging whether people are adequately informed or interested in science and technology (a partially instrumental goal).

The interest that public authorities have shown in this issue in recent years can also be explained by their need for social support of science policies, including investment in research and development, especially at a time when the credibility of science is being challenged in advanced societies. Behind the discourse on the public understanding of science is both a will to deepen democratic processes and a search for the public legitimacy of science and public policies on science (Wynne, 1987, p. 5). Felt and Nowotny (1993, p. 286) added that scholarly efforts tend to be made in this field whenever the political and economic support systems of science are under severe strain. They asserted that the reemergence of empirical studies on the public understanding of science is connected with a sense of crisis about the overall societal condition of science, the level of funding and general authority of science, and the esteem in which science is held. This new stage in the relations between the public and science also helped create the atmosphere for the development of what Wynne (1993) called the "self-consciousness research field of the public understanding of science" since the mid-1980s (p. 323).

Under these circumstances, policy action designed to foster the population's scientific literacy and the involvement of scientists themselves in communication with the public about their work can be regarded as prerequisites for reducing the distance and tension between science and society. As Callon, Larédo, and Mustar (1995) state in their book on strategic management of research and technology: "Without a public of concerned citizens willing to support science and technology, research would run the risk of lacking a favorable environment" (p. 12; my translation).

Against this background a number of inquiries and studies have been carried out in recent decades at the initiative of scientific groups and political bodies, alike in order to assess levels of laypeople's scientific and techno-logical knowledge, interest, and attitudes. Examples of such investigation are found in France, where the Ministry for Science led the first major national survey in this area as early as 1972, and in the European Community since the mid-seventies (Boy, 1992, p. 19).

What have been the goals behind such inquiries? To inform policies for science? To justify or legitimate them in the eyes of citizens? In other words, has political discourse on the public understanding of science and specific policy action in this field mostly been part of a strategy of political actors, alone or in alliance with scientists, to justify their policies for science? Or has it paved the way for improved democratic forms of citizen participation in decision-making on scientific and technological issues? Has it been fol-lowed by broadened public access to and use of science? These questions cannot be fully answered in this brief review, but it is reasonable to start by examining the content of the questionnaires and the ways in which the results have been used. In an attempt to interpret the motives behind them, I focus on the inquiries by the Commission of the European Communities (CEC) since the mid-1980s. Consideration will also be given to the generally acknowledged need to explore the results of these surveys further and to complement them with studies on specific social groups (publics) and with contextual studies. In the final section I provide background for specific local treatment of the public understanding of science by delving into the peculiarities of a European country that is not highly industrialized. My assumption is that the following comments will supply arguments that favor the continued study of specific groups of people and various contexts in this field.

Surveys by the European Union and their Political Background

As part of a growing effort to address the issue of the public understanding of science in recent years, the European Union (EU) has launched surveys of the European public opinion (Commission of the European Communities [CEC], 1990; Institut National de Recherche Agricole [INRA] & Report International, 1993). These surveys have been carried out in the framework of Eurobarometer, a special service provided by the Commission of the EC to evaluate European public opinion about various areas of relevance to Union policies and activities. Specific surveys have been designed to assess levels of the European public's knowledge, attitudes, and expectations regarding science. They have also been used to investigate opinions on the directions of EU policies on R&D.

One major justification for such surveys has rested on what can be called the democratic argument: "Within democratic societies, public policies must be rooted in the consensus of a large proportion of the population" (CEC, 1994, p. 331). Theoretically, levels of scientific literacy in the population of a democratic society have two important implications for science policy decisions: (a) effective citizen participation in decisions on scientific or technological matters implies some understanding of the underlying issues, and (b) educational, cultural, and information policy measures should be designed to lower the level of scientific illiteracy and thereby help promote equality among citizens in their understanding of science and technology.

This formulation of the democratic argument underlying the Eurobarometer surveys is especially fitting for questions about attitudes and general convictions relating to scientific developments and R&D policy. However, European surveys have not focused exclusively on attitudes and general convictions. Given the general public's recognized lack of scientific and technological expertise, and in the light of the assumption that this deficit is correlated with rather unfavorable attitudes toward science (see the chapter by Durant et al. in this volume), European surveys have also included questions designed to identify and compare levels of scientific knowledge by the public.

Being aware of the chronic tension between the Council of Ministers and the Commission regarding the EU budget for research, one can understand why the Commission may be interested in acquiring quantitative data on the opinions of the European public. Expecting the public to be generally supportive of EU investment in R&D, the Commission may hope to use the results of the inquiries to justify its aims in this area, which are more ambitious than the Council's. Member states tend to constrain the propensity of the CEC to regularly enlarge the scope of EU R&D policy and to strengthen the means assigned to this policy. In this struggle over jurisdiction and resources, the Commission feels the need to rely upon the support of other political and social forces, namely, the European Parliament and European public opinion. The CEC's concerns about social legitimacy are not independent from the broader environment in which the EU R&D policy is shaped. That environment is marked, at present, by both financial and social constraints. As pointed out by Felt and Nowotny (1993), motivations behind European inquiries into the public understanding of science "reflect the predominant institutional and epistemological insecurity of science" (p. 287). If these statements are acceptable explanations for the decision to launch the Eurobarometer surveys on the public understanding of science, they would confirm the hypothesis that the goal of the surveys is to legitimate the EU's current R&D policy.

In any case, the question remains as to whether such studies and their results have given any impetus to EU policies or endeavors to promote the public understanding of science. If EU decision-makers take the Eurobarometer surveys seriously, the data provided by these instruments could become a basis for policy designed to popularize science and technology and strengthen scientific and technological culture in European countries. More specifically, the CEC's assumption that levels of knowledge are directly related to levels of support for science lead one to expect that survey findings and conclusions based on them will prompt concrete policy action conceived to fill gaps in public knowledge about science and technology. This link may be seen as a logical corollary of the democratic argument.

It has, however, been recognized that data from the Eurobarometer studies are too general to be completely useful. Both the concept of science itself and what the term public is taken to mean for the purposes of the questionnaires are broadly applied without consideration of the particular sciences and technologies involved, the specific sections of the public being surveyed, or the contexts in which those different publics use science and technology (Ziman, 1991, p. 100). Whereas the results of the surveys may yield information about the behavior and sentiments of Europeans toward science at the European level, their "practical bearings are at best unclear" (Durant et al., in this volume, p. 134). In other words, Eurobarometer studies cannot be used for purposes of practical policy.

So-called EU research and development policy has been implemented almost exclusively through transnational cooperative projects that bring together universities, research laboratories, and enterprises from various European countries, primarily in the areas of information and telecommunications technologies, environmental research and technologies, and industrial technologies. Specialized training activities have also been promoted under this framework. In addition, a few initiatives, such as the European Science and Technology Forum, have been launched to stimulate reflection and debate on the social, political, and cultural aspects of science and technology (André, 1995). Among the issues being debated at this forum are science in school and the future of scientific culture in Europe. The fact that a chapter on European attitudes toward science was included in the European Report on Science and Technology Indicators (CEC, 1994, Part IV) also indicates that the public understanding of science is now an integral part of the overall background of European science policy.

Such reflection and informative efforts have not yet been followed by any special line of action by the CEC to encourage the active promotion of scientific literacy through informational, educational, or related activities in Europe. A number of arguments could, however, be made in favor of broadening the scope of EU action in this kind of direction. For instance, levels of participation in European R&D programs and projects, which are the core of EU action in this area, are not separable from the way various social and economic actors perceive, produce, and use—that is, understand—science and technology. There is an apparent gap between research teams and industrial enterprises of industrialized northern Europe and those of less industrialized southern Europe in this regard. For the teams and enterprises of southern Europe, an additional, more immediate obstacle to participation in the EU R&D programs has been the lack of national funding for actors willing to be involved in projects. In southern European countries, it therefore seems perhaps more pressing than elsewhere to dwell on the peculiarities of the social, economic, and politicoadministrative systems in order to better grasp the respective relations to science and the ways in which scientific knowledge is perceived, received, and used within those systems. More specifically, an effort to deepen insights into the understanding that specific social actors have of science might help explain factors hindering a more homogeneous participation in EU programs and projects and might thus offer a basis for new lines of EU policy designed to promote scientific and technological literacy.

Moreover, controversy such as that over mad-cow disease in Europe have clearly shown that public attitudes regarding science are closely related to the transparency of decision-making procedures at both the national and European levels, including the citizens' sense of control over how scientific knowledge is produced and used. The increasing involvement of the EU in the running of complex, highly media-driven, regulatory matters, which often require scientific and technological knowledge or advice, definitively implies that some effort should be made to develop a scientifically better informed public. The public "must be able at least to evaluate the cogency of the arguments advanced by experts and understand the [possible] economic, ecological, or health consequences" (Ayala, 1996, p. 4).

Case studies focusing on the relations between the lay public and knowledge are the tool most likely to offer important insights, as are inquiries into the perceptions, attitudes, and behavior of specific social groups. In various member states, special attention should be given to groups that are directly responsible for either conducting research activities (scientists) or directing or managing them (business managers and policy-makers).

Contextualizing the Analyses: The Special Case of a Southern European Country

It has been pointed out that the results of public opinion surveys on science depend significantly upon the practical relevance of the questions asked (Boyadgieva & Tchalakov, 1993, p. 73). Scholars who have called for further research into the contexts in which different people deal with science or technology argue that more information and analysis is required in order to better understand the complex relations between knowledge levels and attitudes (Wynne, 1991, p. 113). As case studies on the behavior and perceptions of specific local communities have indicated, people's perceptions about science in the abstract (the kind of perceptions that have guided general surveys) are not as pertinent as people's actual experiences and their sense of the relevance or irrelevance of science for the satisfaction of particular needs and interests (Popay & Williams, 1996). This observation suggests that further analyses, including comparative ones, should be carried out in specific regional or local contexts within Europe.

In their interpretation of the results of the 1993 Eurobarometer survey, Durant et al. (in this volume) consider the level of industrialization as one factor that might explain the differences in people's responses from country to country. They argue that people in an industrialized society tend to idealize science as the preferred route to economic and social progress, whereas people in a less developed society (such as Portugal, according to the authors) also tend to regard science as a progressive force but are relatively pessimistic about science's possible contribution to economic and social development. This apparent contradiction might be due to the fact that science and its applications have not proved to be a decisive factor in internal economic and social development in a less industrialized country despite of the existing awareness of science's role in society (which is essentially an abstract perception largely imported through the mass media). Such lack of relevance breeds uncertainty about or disbelief in the practical utility of science.

Comparatively weak science and scientific institutions and underdeveloped relations between the research community and industry cannot be dissociated from practices and attitudes of the major (i.e., economic and political) agents involved. As Lloyd (1996) put it, "the public understanding of science may be a necessary condition for the support of relevant policy, but the political understanding of science in our legislatures is most certainly the keystone of the arch of that condition" (p. 58). Hence my suggestion that comparative surveys be conducted on cognitive as well as attitudinal features of such social groups as the scientific community and political or industrial agents. Such studies might help bring out the contrasts in policy orientation and institutional practices in the field of science and demonstrate their impact on the knowledge levels and attitudes of the general public.

Durant et al. (in this volume) recognize that optimistic and pessimistic attitudes concerning the role of science are closely related to trust in institutions. According to them, when that trust is high, optimism grows; when trust is low, pessimism increases. This general assertion could be usefully tested on science and science-policy institutions. Pessimistic attitudes toward science as an activity or a product, particularly in a less-developed country, may be connected with the low relevance assigned to the science system by political authorities.

It is true that limited trust in political institutions, which permeates a society like the Portuguese, may generally be explained by a recent history of authoritarian and centralized politics (the period of the *Estado Novo* from 1926 to 1974) in which citizens' interests were not at the forefront. However, the public's attitudes toward science cannot be accounted for solely on those grounds. Rather, they appear to be influenced more by people's sense of *policy-makers'* attitudes and behavior toward science.

In modern industrial societies with long liberal and democratic traditions, scientific justifications played a role in building the liberal state, and science provided the basis for technological, economic, and political achievements. In those countries problems today seem to center instead on the decreasing credibility of science, an erosion resulting from improved perception of the risks associated with developments in research and its applications (Ezrahi, 1990, p. 11). By contrast, science in less industrialized countries, including some that are members of the European Union (e.g., Portugal), developed in isolation from the public sphere. That isolation often coincided with an authoritarian political history that reflected little sensitivity to the desires, aspirations, or even capabilities, including scientific capabilities, of society at large. The popularization of science in such countries was mostly confined to a few elites and, since the early 19th century to educational institutions. It had almost no direct involvement of or impact on society and perpetuated levels of illiteracy much higher than the European average.

This state of affairs, together with politician's limited awareness of the economic, social, and cultural role of scientific knowledge, left scientific research as an underbudgeted, socially marginal activity for a long time. Studies done in Portugal on a few recent science-based social controversies have shown, for example, that policy-makers feel at ease publicly disregarding and even denying results of scientific research confirmed by the scientific community. Scientists have often found it difficult to make their points of view heard and accepted by decision-makers and lay persons whenever other, more immediate interests are at stake (Gonçalves, 1996).

In summary, economic agents, the political system, and social organizations in less developed societies make no relevant use of scientific knowledge, there is no real social and economic pressure for investment in R&D, and there is no clear public awareness of risks entailed by research and applications of technology. Given these features of political and scientific culture, it is not hard to understand that there has been little political interest in evaluating the general public's levels of scientific literacy or its attitudes toward science and science policy.

Awareness of the relevance of an issue like the public understanding of science has been largely confined to the scientific community. In the Portuguese context, this restriction of discussion is linked to scientists' struggle for social status in a society where science has been neglected. In the wake of concern about this matter in Europe, inquiries about public opinion of science have been carried out in Portugal almost exclusively at the initiative of scientific associations and research groups rather than by the government or public administration. One example is a 1987 public opinion survey carried out by the *Associação de Ciência e Tecnologia para o Desensolvimento* (*ACTD*), an association formed in 1985 by scientists to pressure policy-makers to raise the status of science policy in the country (Norma, 1987). Two other, somewhat related surveys were launched later. The first one sought to clarify attitudes and expectations within the scientific community itself and was conceived and conducted by a team from the Center for Research and Studies of Sociology (CIES) at the Higher Institute for Business and Labor Sciences (ISCTE) of the University of Lisbon (Jesuíno et al., 1995). The second survey was undertaken jointly by the Portuguese Federation of Scientific Associations and Societies (FEPASC), which was created in 1991, and CIES (FEPASC/CIES, 1996; Gonçalves, Patrício, & Costa, 1996, p. 395). By confronting parliament with an issue (science and technology) alien to its normal business and by comparing the expressed opinions of Portuguese parliamentarians (MPs) with their legislative behavior toward science, the 1996 survey had the purpose of shedding light on the perceptions that the MPs have on science.

Interestingly enough, none of these studies were limited to mere academic issues. To some extent they followed a policy-oriented approach, for they were considered as a means to put pressure on public authorities by making them aware of opinions of relevant actors who were thought to have a positive regard for science. This interpretation may be one reason why such studies encompassed not only the public as a whole but also very selective social groups, namely, the scientific community and MPs. Such inquiries into the understanding of science were certainly not used to justify policy (for policy-makers were not involved), but to inform it by encouraging public and private action in this area, as intended by the researchers who carried out the studies.

When interpreting some of the outcomes of earlier surveys on public opinion of science and technology in Portugal, one may find it helpful to be aware of a central contradiction brought to light by the 1996 survey of MPs. The survey did confirm results of EU inquiries showing that science is, on the whole, highly regarded by the Portuguese people. MPs did state opinions indicating that they believe science promotes both national development and the policy-making process. Moreover, they did express a desire for scientists to play a greater role in the formulation of science policies. The priority that the MPs accorded scientific knowledge and expertise as a guide for political and legislative decision-making processes indicates the high value they placed on science. And in matters of policy formulation, it is striking that the MPs attached greater importance to expert and specialized knowledge than to pressure groups, public opinion, past experience, or foreign legislation. As already pointed out, however, the actual practices of policy-making in Portugal sharply diverge from these stated opinions. Scientists are rarely heard, or even asked to participate in the policy-making process. The political system has thus failed to create the conditions for ensuring their regular and systematic participation.

Why do the opinions that Portuguese parliamentarians have about the role of science differ so greatly from their actual practice? Finding an answer to this question could, hypothetically, help explain why the Portuguese public, too, praises science while showing limited interest in scientific information, scarcely using it, and distrusting its capacity to contribute to the country's development.

The 1996 survey clearly showed major consensus among the MPs on what they take to be the public's growing trust in science. In the opinion of 71.2% of the legislators, trust in science has grown in more industrialized societies over the last decade. This view is at odds with the skepticism that has developed toward science in recent years in technologically advanced states. Is the attitude of the MPs due to relative ignorance of the way science and its impacts are presently perceived in the more industrialized world? Or is it due instead to the idea that the population still expects important contributions from science in a country like Portugal?

Another interesting outcome of the study is how widespread the view is that there is no real science policy in Portugal, with 46.6% of the MPs believing it to be nonexistent. However, most of the respondents in that large grouping were members of opposition parties (60.9% of the MPs of the

Socialist Party and 85.7% of the Communist Party). These results can be compared to those from the survey of the Portuguese scientific community, where a similar question also elicited predominantly negative responses. Of the researchers who were polled, 66.3% believed there to be no science policy in Portugal (Costa, Ávila, & Machado, 1995, p. 170). This finding, too, indicated the lack of confidence that the Portuguese people have in the country's policy-making institutions. It could provide insight into the responses registered by the 1993 Eurobarometer survey on the European public (INRA), & Report International, 1993).

Concluding Remarks

Recent interest in public understanding of science, particularly in ascertaining people's knowledge of and regard for science, reflects underlying motivations that differ depending on the social or political actors and background involved. The CEC has been one such actor that has invested most in surveys on the public understanding of science. It is, therefore, worthwhile clarifying why the CEC has been willing to launch surveys on this topic and what practical use, if any, has been made of the information. Such studies should not result merely in a collection of quantitative data on the public understanding of science, as seems to have been the case thus far. They should also permit analysis of specific policy measures for their ability to raise the level of scientific and technological literacy in member states of the EU. In other words, their function should not be simply to legitimate the powers that be but also to encourage the popularization of science and technology in the EU.

One political argument in favor of an increased EU action in the field of science and technology policy is the close relation between the levels of scientific literacy in member states and the capacity of economic and social actors to participate in EU R&D programs and projects. In light of the arguments and findings presented in this chapter, the EU would be perfectly justified in launching special informational and educational programs for actively promoting the public understanding of the substance, contexts, processes, applications, and impacts of science and technology.

As I have noted, however, the general character of the results of recent EU public opinion surveys prevents them from providing an adequate basis for the sort of concrete policy action suggested here. Case studies focusing on the relation between the lay public and knowledge and inquiries into the perceptions, attitudes, and behavior of specific social groups would be the tools most likely to deepen insight into the public understanding of science and technology. Special attention should be given to those groups that, unlike science "publics" in the passive sense, are directly responsible either for conducting research activities (scientists) or for directing or managing them (business managers, and policy-makers).

In my view it makes sense for the EU to back these kinds of studies and analyses. They might ultimately help to build the momentum and afford the assistance needed for designing a new line of EU activity in the area of science and technology—a line guided by the ideal of sharing scientific knowledge more effectively across European societies and deepening the commitment to the development of science on the part of policy-makers and the general public.

References

André, M. (1995). Thinking and debating about science and technology at European level. *Science and Public Policy, 22,* 205–207.

Ayala, F. (1996). Introductory essay: The case of scientific literacy. In UNESCO (Ed.), *World Science Report 1996* (pp. 1–5). Paris: UNESCO Publishing.

Béguet, B. (1990). La vulgarisation scientifique en France de 1850 à 1914: Contexte, Conceptions, Procédés [The popularization of science in France from 1850–1914: Contexts, conceptions, processes]. In B. Béguet (Ed.), *La Science pour Tous* (pp. 6–29). Paris: Bibliothèque du Conservatoire National des Arts et Métiers.

Boy, D. (1992). Évolution des Attitudes depuis 1972 [The evolution of attitudes since 1972]. In *Colloque Pour La Science, Actes du Colloque des 3 et 4 décembre 1991* (pp. 17–33). Paris: Fondation Électricité de France.

Boyadgieva P., & Tchalakov, I. (1993, Summer/Autumn). À la périphérie de l'Europe [On the periphery of Europe]. *Alliage 16–17,* 67–73.

Callon, M., Larédo, P., & Mustar, P. (1995). *Introduction Générale: La Gestion Stratégique de la Recherche et de la Technologie* [General introduction: The strategic management of research and technology]. Paris: Economica.

Caro, P. (1990). *La vulgarisation scientifique est-elle possible* [Is the popularization of science possible?]. Nancy: Presses Universitaires de Nancy.

Commission of the European Communities (CEC). (1990, January). *Les Européens, la Science et la Technologie, Rapport principal préparé par Faits et Opinions pour la Communauté Économique européenne* [Europeans, science, and technology—Main report prepared by Faits et Opinions for the European Economic Community]. Brussels: CEC.

Commission of the European Communities (CEC). (1994). *The European report on science and technology indicators* (EUR 15897 EN). Luxembourg: CEC.

Costa, A. F., Ávila, P., & Machado, F. L. (1995). Políticas Científicas [Science Policies]. In J. C. Jesuíno (Ed.), *A Comunidade Científica Portuguesa Nos Finais do Século XX* (pp. 163–179). Oeiras: Celta.

Durant, J. (1993). Qu'entendre par culture scientifique? [What does scientific culture mean?] *Alliage, 16–17,* 205–210.

Ezrahi, Y. (1990). *The descent of Icarus: Science and the transformation of contemporary democracy.* Cambridge, MA: Harvard University Press.

Felt, U., & Nowotny, H. (1993). Science meets the public—A new look at an old problem. *Public Understanding of Science, 2,* 285–290.

FEPASC/CIES (Portuguese Federation of Scientific Associations and Societies/Center for Research and Studies of Sociology). (1996). A Ciência na Assembleia da República [Science in the Portuguese Parliament]. Inquiry to the Portuguese parliament. In M. E. Gonçalves (Ed.), *Ciência e Democracia* (pp. 363–384). Lisbon: Bertrand.

Fourez, G. (1994). *Alphabétisation Scientifique et Technique. Essai sur les Finalités de l'Enseignement des Sciences* [Scientific and technology literacy: Essays on the finalities of science education]. Brussels: De Boeck.

Gonçalves, M. E. (1996). Ciência e Política em Portugal. O Caso da Doença das Vacas Loucas [Science and politics in Portugal: The case of 'mad-cow' disease]. In M. E. Gonçalves (Ed.), *Ciência e Democracia* (pp. 121–139). Lisbon: Bertrand.

Gonçalves, M. E., Patrício, M. T., & Costa, A. F. (1996). Political images of science in Portugal. *Public Understanding of Science, 5,* 395–410.

Institut National de Recherche Agricole (INRA), & Report International. (1993, June). *Europeans, science and technology: Public understanding and attitudes* (EUR 15461). Brussels: Commission of the European Communities.

Jesuíno, J. C., Amâncio, L., Ávila, P., Carapinheiro, G., Costa, A. F., Machado, F. L., Patrício, M. T., Stoleroff, A., & Vala, J. (1995). *A Comunidade Científica Portuguesa Nos Finais do Século XX* [The Portuguese scientific community at the end of the 20th Century]. Oeiras: Celta.

Lloyd, I. (1996). Science and public policy—Has the 20th century made any progress? *Science and Public Policy, 23,* 55–58.

Massey, W. E. (1989). Science education in the United States: What the scientific community can do. *Science, 245,* 915–921.

Norma (1987). *Ciência e Opinião Pública Portuguesa, Vol. I: Relatório de Análise* [Science and public opinion in Portugal, Vol. I: Analytic report]. Lisbon: Norma.

OECD (1987, October 9). *The contribution of science and technology to economic growth and social development.* Discussion paper, Paris.

Popay, J., & Williams, G. (1996, June). *Lay knowledge in the public sphere: Barriers to citizen involvement in public health policy.* Paper presented at the International Symposium on Technology and Society, Princeton, NJ.

Royal Society of London (1985). *The public understanding of science.* London: The Royal Society.

Shortland, M. (1989). Promoção da Ciência e da Cultura Científica [Promotion of science and scientific culture]. *Impacte—Ciência e Sociedade, 3,* 7–19.

Wynne, B. (1987, February). From content to process. *EASST Newsletter, 6*(1), 5–9.

Wynne, B. (1991). Knowledges in context. *Science, Technology, and Human Values, 16,* 111–121.

Wynne, B. (1993). Public uptake of science: A case for institutional reflexivity. *Public Understanding of Science, 2,* 321–337.

Ziman, J. (1991). Public understanding of science. *Science, Technology, and Human Values, 16,* 99–105.

PART TWO

COMPARATIVE ANALYSIS: RESULTS OF AND REFLECTIONS ON SURVEYS AS METHODOLOGICAL INSTRUMENTS

INTRODUCTION

Meinolf Dierkes and Claudia von Grote

This part contains the chapters dealing directly with the results, methods, and future possibilities of surveys on the public understanding of science and technology, particularly the Eurobarometer survey. What makes this collection of contributions especially engaging is that their authors approach their common topic—surveys as a research instrument—in very different ways and try to advance it by tackling a variety of methodological and theoretical problems raised by the use of this kind of instrument. Because the Eurobarometer survey is a unique source of comparative data and because it can be combined with similar national surveys, questions about systematic cross-national comparison are stressed in chapters 4 through 6. In the first two chapters (4 and 5), surveys on the public understanding of science and technology are closely examined for patterns within the various member states of the EU (J. Durant et al.) as well as between those and other industrialized nations (J. Miller & R. Pardo) and for patterns that can be used as a heuristic aid for building hypotheses. Chapters 6 and 7 are each concerned with one of the two sides of this research: long-range objectives of expanding the theoretical and empirical foundations of the survey as a research instrument (M. Bauer) and substantive and methodological problems of large-scale surveys of the general public (A. Hamstra).

The contribution by Miller and Pardo (chapter 4) and that by Durant et al. (chapter 5) are secondary analyses. Both are based on the results of the 1992 Eurobarometer but have different points of departure. Durant et al. discuss the data set by developing a hypothetical model for analyzing different cultures of science in Europe. By contrast, Miller and Pardo focus on the methodological development of comparing the data sets from the European Union (1992), Japan (1991), the United States (1995), and Canada (1989) for the essential concepts of their approach. Those central concepts are *scientific literacy* as a precondition for comprehending the arguments in serious controversies involving science and technology, and *the attentive public* as a structural concept for treating the policy views of citizens. Different ways of presenting the material emerge from these two perspectives.

Miller and Pardo discuss the way they measure these concepts, and they present the results of the data analysis they conducted with this new methodological approach, which draws on item-response theory and other resources to improve the accuracy of estimates of civic scientific literacy. Their approach also increases the precision of inquiries into the relation between knowledge and attitudes. They find that the idea of attitudes as a unidimensional process must be abandoned.

Taking a different slant, Durant et al. base their data analysis directly on the assumption made in the so-called deficit model that knowledge and attitude are positively related. They acknowledge that this link is not found throughout the European Union but rather only in the member states that are not yet industrially advanced. The authors therefore develop a two-dimensional model in which the degree of industrialization, as the key variable in the understanding of scientific culture in Europe, is combined with the pattern of the relation between knowledge, interest, and attitudes. With this model the authors formulate predictions about structural features and possible directions in the public understanding of science and technology. These predictions are then tested against the empirical base. Variables of knowledge and interest are thus handled differently than in the chapter by Miller and Pardo. Durant et al. test their hypothesis-based variables against a higher-order concept, the degree of industrialization, rather than against the politicoadministrative units called countries. Given the finding that public attitudes toward science and technology are "chaotic" in postindustrial states, Durant et al. bring in an additional category—trust in the institutions of the industrialized countries—to explain this lack of a uniform pattern. The main problem that Durant et al. see in analyzing data produced by instruments like the Eurobarometer, however, is this instrument's weak theoretical grounding.

Bauer (chapter 6) argues for such an instrument's deeper theoretical anchoring and describes the way in which a broad theoretical framework can be developed. The concept of a general scientific and technological climate as part of a general *Zeitgeist* is introduced by Bauer into the analysis of the public understanding of science and technology. He assumes that the scientific and technological climate is of a long-term nature and that it is more stable than what is captured by the opinions recorded in surveys. He therefore argues that opinion polls need to be supplemented by systematic content analysis of science articles in major newspapers over an extended period. This suggestion contains the interesting idea of studying the public understanding of science and technology by taking a two-track approach. That is, broadly based surveys of the general public as well as content analyses of the media should be used to create European-wide indicators by which to ascertain a country's cultural climate of science and technology, which provides a background for interpreting current information gathered by large-scale surveys. Bauer describes the British Media Monitor Project, presenting it as an example of a system for analyzing press coverage of science for the purpose of developing indicators of a popular-science culture.

Examining the Eurobarometer biotechnology surveys, the author of the final chapter in this part (A. Hamstra) investigates some of the unconsidered interrelationships involved in measurements of public understanding at the more clear-cut level of a specific technology and its application. In keeping with her psychological approach, she discusses the links between knowledge and attitudes in the general context of the relation between public perception and attitude, a concept that also allows for reflection on the way in which the public interacts with science and technology. Hamstra points out the questionable nature of inferences drawn from the measurements of general attitudes toward the acceptance or rejection of specific technologies. Comparing a number of studies, she explores the merits and drawbacks of various measurement approaches. This point of view enables the reader to see connections to similar issues in chapters 4 and 5 and to see how one can follow up on calls to differentiate the measurements.

CHAPTER 4

CIVIC SCIENTIFIC LITERACY AND ATTITUDE TO SCIENCE AND TECHNOLOGY: A COMPARATIVE ANALYSIS OF THE EUROPEAN UNION, THE UNITED STATES, JAPAN, AND CANADA

Jon D. Miller and Rafael Pardo

Civic Scientific Literacy and Attitude to Science and Technology

In the last half of the 20th century, a growing body of research and literature has focused on public perceptions of science and technology among the adult population of developed and developing countries. In the short history of the field generally known as public understanding of science, its theoretical approach and practical objectives have evolved in response to the widening impact of science and technology on the lives of individuals, organizations, and communities. In the first phase, from the end of World War II through the end of the 1950s, there was widespread public awe and admiration for the work of scientists and engineers in producing a series of miracle drugs, jet aircraft, ever improving communications, and a rising standard of living. With the launch of Sputnik I in 1957 and the advent of the space race, a popular standard for intelligence became the rocket scientist.

A second phase began in the 1960s, marked by a series of books and events that emphasized that there were important public policy choices to be made in regard to science and technology. Rachel Carson's *Silent Spring* (1962) revealed important long-term negative effects from the excessive use of DDT and other pesticides. For most of the preceding two decades, the postwar development of pesticides had been one of the miracles of modern science, but in this second phase pesticides were found to have consequences for nature that were unanticipated by both developers and users (Bosso, 1987). Similarly, new drugs such as thalidomide were found to have catastrophic results when taken by pregnant women. In the United States and other major industrial nations, the evidence resulted in a series of new laws requiring an extensive review of new drugs prior to their distribution on the market (Mintz, 1965).

The institutional cynicism that followed the Vietnam War, combined with a growing awareness of actual and potential environmental damage, grew into the environmental movement in the United States and the green movement in Europe that marked the beginning of a third phase. One of the most visible policy disputes during this period occurred over the use of nuclear energy to generate electricity (Freudenburg & Rosa, 1984; Morone & Woodhouse, 1989; Nelkin, 1977). By the beginning of the 1980s, there was broad recognition among government, scientific and technological organizations, and the scientific community generally that the public could veto a program or project if sufficiently aroused. Worldwide access to new computer-based information technologies in the 1990s has increased both the speed and scope of public discourse on science and technology policy issues.

Throughout these three phases, scholars have debated the question of how many citizens understand the basic ideas of modern science and its associated new technologies and whether an understanding of scientific constructs is related to the views of citizens on substantive science and technology issues. Using comprehensive national surveys from the European Union, the United States, Japan, and Canada, we examine the levels of civic scientific literacy, interest in science and technology policy issues, participation in the formulation of science and technology policy, and substantive attitudes toward one major public policy issue, public funding for basic scientific research. The analysis will conclude with a discussion of the implications of our findings for democratic government in the 21st century.

The Concept of Civic Scientific Literacy

To understand the concept of civic scientific literacy, it is necessary to begin with an understanding of the concept of *literacy* itself. The basic idea of literacy is to define a minimum level of reading and writing skills that an individual must have to participate in written communication. Literacy is most often presented as a dichotomy—literate versus illiterate—precisely because it is a threshold measure. The focus on a threshold level of knowledge is inherent in the concept of literacy.

Historically, an individual was thought of as literate if he or she could read and write their own name. In recent decades, there has been a redefinition of basic literacy skills to include the ability to read a bus schedule, a loan agreement, or the instructions on a bottle of medicine. Adult educators often use the term "functional literacy" to refer to this new definition of the minimal skills needed to function in a contemporary industrial society (Cook,

1977; Harman, 1970; Kaestle, 1985; Resnick & Resnick, 1977). The social science and educational literature indicates that about a quarter of Americans are not "functionally literate," and there is good reason to expect that roughly this proportion applies in most mature industrial nations and a slightly higher rate in emerging industrial nations (Ahmann, 1975; Cevero, 1985; Guthrie & Kirsch, 1984; Northcutt, 1975).

In this context, civic scientific literacy is conceptualized as the level of understanding of science and technology needed to function as citizens in a modern industrial society. This conceptualization of scientific literacy does not imply an ideal level of understanding but rather a minimal threshold level. It is neither a measure of job-related skills nor an index of economic competitiveness in a global economy.

There is broad agreement that civic scientific literacy includes at least two basic dimensions: a basic vocabulary of scientific terms and concepts and an understanding of the process or methods of science for testing models of reality (Durant, Evans, & Thomas, 1989, 1992; Evans & Durant, 1995; Miller, 1983b, 1987, 1989, 1995). Although the literature includes several minor variations in operationalizing these dimensions, it is widely recognized that a significant portion of citizens of modern democratic nations need to understand the basic approach of science and be able to utilize a set of basic constructs to make sense of contemporary public policy arguments involving science and technology.

In Miller's original model (Miller, 1983b), he conceptualized a third dimension that reflected an awareness of the impact of science and technology on individuals and on society. Although this dimension has been operationalized within U.S. data sets, it is more difficult to construct accurate cross-national measures of this dimension because science and technology may be experienced differently, depending on the emergence of public policy issues in a given country. For example, countries that received the largest amounts of fallout from Chernobyl may be more aware of the potential impact of radiation than people in countries with significantly less exposure. Durant et al. (1992) have suggested the measurement of an additional dimension they label "understanding the institutional structures of science," which has not been operationalized to date. This dimension includes the arrangements and organizational procedures by means of which scientific knowledge is produced and validated. Future research should attempt to develop reliable measures for addressing this institutional dimension.

The primary point of difference in the literature has been whether these two dimensions (or three in analyses focusing on a single nation) should be combined into a single index of scientific literacy. Miller (1983a, 1983b, 1987, 1989, 1995) has argued that it is at the point of a policy controversy that scientific literacy becomes functional and that policy debates at this level require a combination of a sound vocabulary of scientific constructs and an understanding of the nature of scientific inquiry. Nelkin's (1977) analysis of the nuclear power controversy in Sweden illustrated the higher threshold of knowledge needed for effective participation in the formulation of public policy.

Durant et al. (1992) acknowledged the clarity of a threshold measure but elected to utilize scalar measures to avoid stigmatization and to support the use of correlational analysis methods. They did not address the question of the level of functional scientific literacy needed in the context of a public policy controversy. Further, the development of Jöreskog and Sörbom's (1993) structural equation techniques allowed the effective integration of dichotomous, ordinal, and interval measures into a common analytic metric.

The Measurement of Civic Scientific Literacy

In developing a measure of civic scientific literacy, it is important to construct a measure that will be useful over a period of years, providing a time-series indicator. If an indicator is revised periodically, it is often impossible to totally separate the variation attributable to measurement changes from real change over time. The current debate over the composition of consumer price indices in the United States and other major industrial nations is a relevant reminder of the importance of stable indicators over periods of time.

The durability problem can be seen in the early efforts to develop measures of the public understanding of science in the United States. In 1957, the National Association of Science Writers (NASW) commissioned a national survey of public understanding of and attitudes toward science and technology (Davis, 1958). Because the interviewing for the 1957 study was completed only a few months before the launch of Sputnik I, it is the only measure of public understanding and attitudes prior to the beginning of the space race. Unfortunately, the four major items of substantive knowledge were (a) radioactive fallout, (b) fluoridation in drinking water, and (c) polio vaccine, and (d) space satellites. Twenty years later, at least three of these terms were no longer central to the measurement of public understanding, and it is impossible to find current terms that are *equally relevant*.

Recognizing this problem, Miller attempted to identify a set of basic constructs, such as atomic structure or DNA, that are the intellectual foundation for reading and understanding contemporary issues but that will have a longer durability than specific terms, such as the fallout of strontium 90 from atmospheric testing. In the late 1970s and the early 1980s, when the National Science Foundation began to support comprehensive national surveys of public understanding and attitudes in the United States, there was little experience beyond the 1957 NASW study in the measurement of adult understanding of scientific concepts. The first U.S. studies (National Science Board [NSB], 1981, 1983, 1986) relied heavily on each respondent's self-assessment of his or her level of understanding of various terms and concepts, building on a survey research literature that suggested that when respondents are offered a trichotomous set of choices (i.e., do you have a clear understanding of [construct A], a general sense of [construct A], or not much understanding of [concept A]?) individuals selecting the clear-understanding choice would be very likely to understand the concept, while individuals who were unsure about the concept or who did not understand it might select the middle or lower category (Converse & Schuman, 1984; Dillman, 1978; Labaw, 1980; Oppenheim, 1966; Sudman & Bradburn, 1982). The basic idea was that respondents' inflation of their knowledge would occur primarily between the little-understanding and general-sense categories. This approach, which is still used in national studies in Japan and some other countries, can provide useful estimates but at a lower level of precision than that provided by direct substantive inquiries.

In a 1988 collaboration between Miller in the United States and Thomas and Durant in the United Kingdom, an expanded set of knowledge items was developed that asked respondents direct questions about scientific concepts. In the 1988 studies, a combination of open-ended and closed-ended items were constructed that provided significantly better estimates of public understanding than had been collected in any previous national study. From this collaboration, a core set of knowledge items emerged that have been used in studies in Canada, China, the European Union, Japan, Korea, New Zealand, and Spain. To a large extent, these core items have provided a durable set of measures of a vocabulary of scientific constructs, with minor additions and deletions over the last decade.

The understanding of basic scientific constructs. In the context of a search for durable measures, it is useful to begin with an examination of the development of measures of the public understanding of basic scientific constructs. Many of the measures used over the last decade to measure construct understanding emerged from the 1988 U.K.–U.S. collaboration, which basically produced a set of open-ended items, several multipart questions, and a closed-ended true-false quiz (Miller, 1989, 1995; NSB, 1988, 1990). It may be useful to look briefly at some of these core items.

One of the core items that emerged from the 1988 U.K.–U.S. collaboration was an open-ended question about DNA. Typical of a series of open-ended questions used in later U.S. studies, the question began with a closed-ended inquiry: "When you read the term DNA in a newspaper or magazine, do you have a clear understanding of its meaning, a general sense of its meaning, or not much understanding of its meaning?" Respondents who indicated that they had either a clear understanding or a general sense of the meaning of DNA were then asked: "Please tell me, in your own words, what does DNA mean?" The interviewers, regardless of whether the interview was conducted in-person or over the telephone, were instructed to record the response verbatim, and these text responses were subsequently coded independently by teams of individuals knowledgeable about the definition and meaning of DNA. Standard double-blind coding procedures were employed (Hughes & Garrett, 1990; Montgomery & Crittenden, 1977; Perreault & Leigh, 1989), and in 1988 full sets of text responses were coded by both American and British coders to assure cross-national comparability. The results of this work demonstrated that double-blind coding practices could produce highly reliable data and that there were few cross-national variations in coding judgments between the United Kingdom and the United States. It is recognized that translation and coding across several languages may introduce some additional problems, but we believe that the benefits of cross-national use of open-ended questions outweigh its difficulties. In subsequent U.S. studies, similar open-ended questions have been employed to measure the understanding of basic concepts such as molecule, radiation, acid rain, computer software, and the thinning of the ozone layer around the Earth (NSB, 1986, 1988, 1990, 1992, 1994, 1996).

In addition to these open-ended items, a set of multipart items was first developed in the 1988 U.K.–U.S. collaboration. A two-part question about the movement of the Earth and the Sun has been widely cited in the popular press. In this question, each respondent is asked whether "the Earth goes around the Sun, or the Sun goes around the Earth?" Those respondents who indicate that the Earth goes around the Sun are asked whether the Earth goes around the Sun "once a day, once of month, or once a year?" In 1988, approximately 47% of American respondents and 33% of British respondents were able to report that the Earth moves around the Sun once each year. The percentage of Americans able to answer this question correctly has remained stable since 1988.

Given the difficulty of asking too many open-ended and hard questions to respondents, especially in a telephone setting where the respondent can terminate the interview by hanging up the telephone, it is important to use some less stressful forms of inquiry. In the 1988 U.K.–U.S. study, a series of items were constructed for use in a true–false format, with an invitation for a respondent who was unsure to indicate his or her uncertainty and continue to the next question. Examples of items in this true–false quiz are:

- Lasers work by focusing sound waves.
- All radioactivity is man-made.
- The earliest human beings lived at the same time as the dinosaurs.
- The center of the Earth is very hot.
- Antibiotics kill viruses as well as bacteria.
- The continents on which we live have been moving their location for millions of years and will continue to move in the future.
- Radioactive milk can be made safe by boiling it.

Finally, a few items were asked as direct inquiries. For example, respondents in the 1998 study and subsequent studies have been asked the question: "Which moves faster, light or sound?"

By using various sets of these construct-understanding questions, researchers in Canada and Japan have been able to collect comparable construct-knowledge measures. In 1989, a national study conducted in Canada included a substantial number of knowledge items identical to those used in the 1988 U.K.–U.S. studies, including an open-ended question about DNA. In 1991, the National Institute of Science and Technology Policy in Japan sponsored a similar study that included a smaller core set of items identical to, or comparable to, items used previously in the European or U.S. studies. Although the exact items have varied from study to study (for a detailed discussion of the selection of items and the comparability of items in the four studies included in this analysis, see Miller, Pardo, & Niwa, 1997), the essential point is that each of these sets of items should be viewed as a sample of constructs from a universe of perhaps a hundred or more constructs that are important to civic scientific literacy. The range of constructs developed by Project 2061 provides a useful approximation of the range of substantive concepts that might constitute this universe of relevant constructs (Rutherford & Algren, 1989).

The process of constructing reliable and comparable measures of construct vocabulary can be understood by examining the construction of these indices for the 1992 Eurobarometer study and the 1995 U.S. study. The two studies had a common core of construct knowledge items, but the U.S. study included more open-ended items, and some items were not asked in both studies (see Table 4-1). An examination of the percentage correct on the nine common items suggests that there is little difference between the European Union and the United States on this dimension. American adults were able to answer an average of 5.1 questions correctly, and citizens of countries in the European Union answered an average of 4.9 items correctly. It is possible to limit cross-national comparisons to only those items asked in a comparable manner in each study, but this approach does not utilize the full array of information available from each study and may eliminate the possibility of cross-national comparison when there are minor variations in wording or translation.

Table 4-1. Common Vocabulary Construct Items

Correct response expected	Percentage of sample responding correctly	
	European Union 1992	United States 1995
Provide a correct open-ended definition of a molecule	–	9
Provide a correct open-ended definition of DNA	–	21
Disagree that "antibiotics kills viruses as well as bacteria"	27	40
Disagree that "lasers work by focusing sound waves"	36	40
Agree that "electrons are smaller than atoms"	41	44
Indicate through a pair of closed-ended questions that the Earth goes around the Sun once each year	51	47
Disagree that "the earliest humans lived at the same time as the dinosaurs"	49	48
Disagree that "all radioactivity is man-made"	53	72
Indicate that light travels faster than sound	–	75
Disagree that "radioactive milk can be made safe by boiling it"	66	61
Agree that "the continents on which we live have been moving their location for millions of years and will continue to move in the future"	82	78
Agree that "the center of the Earth is very hot"	86	78

Note. A dash means that the item was not asked.

To facilitate cross-national comparisons, it was necessary to create summary measures across the four studies. Methods of multiple group item-response-theory (IRT), as implemented in the BILOG-MG program, provide a means for computing item values and test scores that take into account the relative difficulty of the items and the different composition of each test (Bock & Zimowski, 1977). The program places the items from all tests on a common scale by jointly estimating the item parameters and the latent distribution of each group, or nation, and uses the maximum marginal-likelihood method. This method is capable of providing reliable results for tests or scales with fewer than 10 items (Bock & Aitken, 1981).

Based on the assumption that all respondents taking a test could be arrayed in an order reflecting their knowledge of the area being tested, the concept of IRT is that the responses to a knowledge item will form an item-response curve (see Figure 4-1), which is a probability distribution of responses to each item, usually of a logistic nature (although it can be a normal distribution). The item-response curve indicates that few individuals with a low level of knowledge of the subject will be able to answer the hypothetical question in Figure 4-1 and that most of the respondents with a high level of knowledge will be able to answer it.

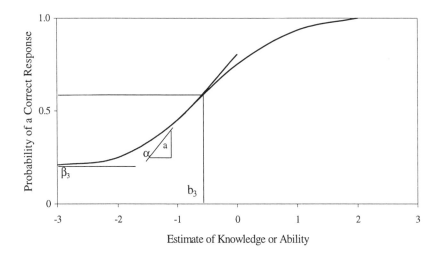

Figure 4-1. Item response curve.

For each item in the four studies, the BILOG-MG program calculates three separate IRT parameters (a threshold parameter, a slope parameter, and a guessing parameter) to discriminate among subjects with different degrees of knowledge. Curves to the left of the figure indicate items of a lower difficulty level than items located on the right, and items with steeper curves represent greater discrimination capacity than items with flatter curves (see Table 4-2). The threshold parameter is a measure of item difficulty, with higher values meaning that fewer respondents were able to answer it correctly. The slope parameter estimates the measurement efficiency of the item. The guessing parameter provides a correction for guesses in closed-ended questions and for coding error in open-ended questions. In the hypothetical item-response curve shown in Figure 4-1, the guessing parameter raises the base from the level of zero correct to a level that would be obtained

Table 4-2. IRT Parameters for Basic Scientific Construct Items

Correct response expected	Slope parameter	Threshold parameter	Guessing parameter
Provide a correct open-ended definition of DANN	1.006	1.191	0.020
Provide a correct open-ended definition of a molecule	1.179	1.902	0.000
Indicate that light travels faster than sound	0.775	–0.872	0.234
Indicate that the Earth goes around the Sun once each year through a pair of closed-ended questions	0.600	0.066	0.077
Disagree that "lasers work by focusing sound waves"	0.893	0.435	0.018
Disagree that "all radioactivity is man-made"	1.044	–0.185	0.117
Agree that "electrons are smaller than atoms"	0.535	0.312	0.000
Disagree that "the earliest humans lived at the same time as the dinosaurs"	0.536	–0.074	0.000
Agree that "the continents on which we live have been moving their location for millions of years and will continue to move in the future"	0.750	–1.636	0.000
Disagree that "radioactive milk can be made safe by boiling it"	0.853	–0.499	0.169
Agree that "the center of the Earth is very hot"	0.863	–1.887	0.000
Disagree that "antibiotics kill viruses as well as bacteria"	0.534	1.158	0.000

solely by guessing with no substantive knowledge of the domain. All of the item parameters are estimated in a standardized form, with a mean of zero and a standard deviation of 1.0.

The responses from all adults participating in the studies were used for computing item parameters for this set of basic scientific constructs. This approach produced one set of item parameters that applied to all items in the four studies. The common, or linked, questions provide a means for placing on the same scale the parameters for items asked in some countries but not in others. It is possible to compute comparable scores from each set of items because all of the item parameters are placed on a single metric.

The computation of individual scores by BILOG-MG utilizes a standardized metric, with a mean of zero and a standard deviation of 1.0 in the combined population of respondents. This standardized metric, however, may be confusing because approximately half of the respondents would have a negative score. To obtain a more comprehensible metric, the mean for the combined pool of respondents in the four studies is set to a value of 50, with a standard deviation of 20. In practice, this means that for all respondents within 2.5 standard deviations of this common mean the score varies between zero and 100. With this metric, the mean construct vocabulary score was 54.5 for the United States, 49.3 for the European Union, 46.3 for Canada, and 36.1 for Japan. These results are consistent with the mean percentage correct on the nine common items between the European Union and the United States, but they provide more precise individual scoring, allowing comparisons among nations with overlapping, but not identical, sets of knowledge items (Miller, 1996).

It is useful to estimate the minimum score above which an individual is likely to understand a public policy dispute over a scientific or technological issue. In previous estimates of civic scientific literacy, Miller (1987, 1989, 1995, 1996) used a threshold level of 67 or more, reflecting the ability of a respondent to score two-thirds of the possible points on the construct vocabulary index. When this standard is applied to the 1995 U.S. data, 27.2% of Americans score at or above the 67-point level compared to 17.4% of Canadians and 20.2% of Europeans. In a separate analysis of 14 industrial nations, Miller (1996) found the proportion of adults meeting this standard within the European Union ranged from 27.2% in the Netherlands to 6.6% in Greece. This result suggests that approximately three of four adults in the United States and four of five adults in Canada and Europe would be unable to read and understand news or other information that utilized basic scientific constructs such as DNA, molecule, or radiation.

The application of this standard to the data from Japan caused problems. The 1991 Japanese survey included a smaller set of knowledge items than the other studies, and the items tended to have lower threshold values—they were ranked easier by IRT scoring. As a result, a Japanese respondent who answered all five of the items in the vocabulary scale correctly would not have achieved 67 points. The IRT parameters for each question can be compared with the difficulty score assigned to various dives in Olympic diving competitions. A relatively easy dive might have a difficulty value of 2.9, whereas a very difficult dive might have a value of 4.1. Each Olympic judge scores each dive on a zero to 10 scale, and this score is then multiplied by the difficulty score for the dive. In IRT scoring, a respondent gets the item either right or wrong, which might be thought of as having the values of one and zero, respectively, and each correct response is multiplied by its IRT parameters to compute a total score.

The five items with adequate characteristics to be included in the construct vocabulary dimension for Japan had sufficiently low thresholds that their collective total was still markedly lower than the items used in the collective sets of items used in other countries. The level of difficulty, not the number of items included, created the scoring problem. Even nine or more items of this relatively low difficulty level would have produced the same result. This is analogous to a diver who makes five excellent, but easy, dives. Even though the diver might have done well on more difficult dives, no evidence supports or opposes that possibility.

The two choices for this analysis were to drop the Japanese study results from the comparison or to lower the minimum standard because the composition of the 1991 interview questionnaire made it impossible for any of its respondents to meet the standard of 67 points. A review of Japanese adult scores showed a break in the data just below 60. Rather than drop the Japanese study from the comparison, the minimum standard was reduced from 67 to 60 to estimate the percentage of Japanese adults having a sufficient vocabulary of basic scientific constructs to read and follow science and technology policy issues in the news. Approximately 25% of Japanese adults met this standard.

Understanding the nature of scientific inquiry. The second major dimension of civic scientific literacy requires that an individual display a minimal understanding of the empirical basis of scientific inquiry, ideally understanding science as theory-building and testing, but minimally as the empirical testing of propositions (Kuhn, 1962). The idea that scientific ideas are subjected to empirical scrutiny with the possibility of being falsified is an important component of understanding the nature of scientific inquiry (Popper, 1969).

In his original U.S. studies, Miller (1983b, 1987, 1989, 1995) utilized a combination of a single open-ended question and a closed-ended question about astrology to identify respondents who held at least a minimal understanding of the process of scientific inquiry. Following the two-part approach described above, respondents were asked whether they have a clear understanding, a general sense, or not much understanding of what it means to study something scientifically. Those individuals who reported that they had a clear understanding or a general sense of it were then asked to describe, in their own words, what it means to study something scientifically. The responses were collected verbatim and subsequently coded by teams of three or more independent coders. In the 1988 U.K.–U.S. study, teams of American and British coders coded all of the responses from both countries, and the final results had an intercoder reliability coefficient above .90.

One of the common responses to the open-ended question about the meaning of scientific study was that a scientific study involved "an experiment." Often, this was the only response provided, and it was coded as correct, but Miller and others wanted an expanded measure of the meaning of experimentation. In the 1993 Biomedical Literacy Study, Miller and Pifer (1995) were able to introduce a new question concerning experimentation:

> Now, please think of this situation. Two scientists want to know if a certain drug is effective against high blood pressure. The first scientist wants to give the drug to 1,000 people with high blood pressure and see how many experience lower blood pressure levels. The second scientist wants to give the drug to 500 people with high blood pressure, and not give the drug to another 500 people with high blood pressure, and see how many in both groups experience lower blood pressure levels. Which is the better way to test this drug?

> Why it is better to test the drug this way?

All respondents were asked the follow-up probe, regardless of which group they selected. This decision proved to be useful in assessing the level of understanding. Whereas most of the 17% of American adults who selected the single-group design in 1995 did not understand the rationale for a control group, a small number of respondents explained that they understood the logic of control groups and placebos but that they could not ethically withhold medicine from a sick person. The ethical argument notwithstanding, it is clear from this response that this small group of respondents, who represented approximately 1% of the total sample, did have an adequate level of understanding of experimental logic, and they were coded as understanding the nature of scientific inquiry.

Among the 69% of individuals who selected the two-group design in 1995, the open-ended probe found substantial misunderstanding of the rationale for experimental design. A majority of this group, representing approximately 40% of the total population, indicated that they selected the two-group design so that if the drug "killed a lot of people," it would claim fewer victims because it would have been administered to fewer subjects. This is hardly the understanding of experimental logic that one would infer from the selection of the two-group design and illustrates one of the hazards of closed-ended questions. Approximately 12% of American adults selected the two-group design and were able to explain the logic of control groups. An additional 14% of Americans interviewed in the 1995 study selected the two-group design and provided a general rationale that included a "comparison" between the two groups but lacked the language or logic of control groups.

A closed-ended question that assessed each respondent's understanding of probability was used as the third part of the typology to certify a minimal level of understanding of the nature of scientific inquiry. The question posed a situation in which a doctor "tells a couple that their genetic makeup means that they've got a one-in-four chance of having a child with an inherited illness." Each respondent was asked to indicate whether each of four statements was a correct or incorrect interpretation of the meaning of "a one-in-four chance":

(a) If they have only three children, none will have the illness.
(b) If their first child has the illness, the next three will not.
(c) Each of the couple's children has the same risk of suffering from the illness.
(d) If their first three children are healthy, the fourth will have the illness.

Responses were expected to select the (c) response as correct and label the other three choices as incorrect. Approximately 54% of American adults were able to demonstrate an understanding of probability in 1995.

To be classified as having a minimal understanding of the nature of scientific inquiry in the 1995 U.S. study, a respondent was required to meet two criteria. First, it was necessary to demonstrate an understanding of the nature of scientific study by either describing the purpose of scientific study as theory-building and testing or by demonstrating a correct understanding of experimental design or procedure. Second, it was necessary to demonstrate a correct understanding of probability. A total of 21% of Americans met this standard in 1995.

The 1992 Eurobarometer included no open-ended items, but three sets of closed-ended items that were related to an understanding of the nature of scientific inquiry loaded on a second factor in a confirmatory factor analysis in a pattern similar to that described above for the United States. Given the importance of the two-dimension hypothesis to this analysis, it is important to look briefly at each of these three sets of questions. First, the total 1992 Eurobarometer sample was randomly split into two groups. Then respondents were presented with a pair of closed-ended questions asking them to think about either a medical example or a machine-tool example and to determine how they would obtain information to assess the effectiveness of a drug or the likely durability of a metal. Each question offered the respondent three alternatives for procuring the information: (a) ask the opinion of an expert in the field, (b) use your own scientific knowledge, and (c) do an experiment. The experimental choice was coded as the correct choice. Approximately 38% of European adults provided a correct response.

An additional question was asked in the 1992 Eurobarometer to assess the understanding of experiments. It used the first part of a question written for the U.S. Biomedical Literacy Study. Each respondent in the 1992 Eurobarometer was asked:

> Let us imagine that two scientists want to know if a certain drug is effective against high blood pressure. The first scientist wants to give the drug to 1,000

people with high blood pressure and see how many of them experience lower blood pressure levels. The second scientist wants to give this drug to 500 people with high blood pressure, and not give this drug to another 500 people with high blood pressure, and see how many in both groups experience lower blood pressure levels. In your opinion, which is the better way?

Approximately 65% of European respondents in 1992 selected the two-group model. The 1992 Eurobarometer did not use the follow-up probe employed in the United States in 1993 and 1995, making it likely that this response overstates the real level of public understanding of the logic of experimentation.

Although neither of these questions utilized an open-ended probe, the combination of the two items into a single indicator improves the quality of the measure. All respondents who selected the experimental option in the first closed-ended question and who selected the two-group model in the second question were classified as having at least a minimally acceptable level of understanding of experimentation. Approximately 28% of European adults qualified as knowledgeable about experimentation in the 1992 Euro-barometer.

Second, a split-ballot approach was employed with a question about the scientific or nonscientific basis of astrology. All respondents in the 1992 Eurobarometer were asked to rate *how scientific* a set of disciplines or activities were, using a scale that ranged from 1 for *not at all scientific* to 5 for *very scientific*. The list included biology, astronomy, history, physics, astrology, economics, medicine, and psychology. A random half of the respondents were given an additional sentence of explanation for each of the disciplines. For example, astronomy was defined as "the study of the heavenly bodies" and astrology was defined as "the study of occult influence of stars, planets, etc. on human affairs." Nearly 40% of European adults interviewed in the 1992 Eurobarometer study indicated that astrology is not at all scientific, but a majority of European respondents thought that there was at least some scientific content in astrology.

Third, a closed-ended question assessed each respondent's understanding of probability. Parallel to the question used in the 1995 U.S. study, this 1992 Eurobarometer question posed a situation in which a doctor "tells a couple that their genetic makeup means that they've got a one-in-four chance of having a child with an inherited illness." Each respondent was then shown a card with the same four choices used in the United States and described above. Approximately 71% of European adults selected the correct equal-probability choice in the 1992 Eurobarometer study.

To estimate the proportion of Europeans with an understanding of the nature of scientific inquiry, a simple typology was constructed on the basis of the framework employed in the analysis of the U.S. data set. All respondents who demonstrated a minimally acceptable level of understanding of experimentation, who recognized that astrology is not at all scientific, and who provided a correct response to the probability question were coded at understanding the nature of scientific inquiry. Approximately 12% of European adults in the 1992 Eurobarometer study met this standard.

The 1989 Canadian study included an open-ended question concerning the meaning of scientific study, identical to the question used in previous and subsequent U.S. surveys. The coding utilized categories that we were able to collapse into a metric comparable to that used in the United States. In addition, the 1989 Canadian study included the question on the scientific basis of astrology that Miller had previously used in his estimates of civic scientific literacy in the United States. As with the prior U.S. metric, respondents who were able to provide an acceptable explanation of scientific study and who understood astrology to be not at all scientific were classified as having a minimally acceptable understanding of the nature of scientific inquiry. Approximately 8% of Canadians met this standard in the 1989 study.

The 1991 Japanese study included two items that represented the level of understanding of the nature of scientific inquiry. Given the importance of the two-dimension hypothesis to this analysis, it is necessary to examine each of these items. First, each respondent in the 1991 Japanese study was asked to classify his or her own understanding of the meaning of scientific study. The question used for this purpose was the same as the one posed in Canada, the European Union, and the United States. The 1991 Japanese study did not include an open-ended probe. When Japanese adults were asked how well they understood what it means to study something scientifically, 4% of them claimed a clear understanding and 36% said that they "understood," but at a level below "clearly understand." Nearly half of Japanese adults said that they hardly understood, and 12% said that they did not understand it at all. Though it would have been desirable to have a confirmatory open-ended probe, the literature on Japanese culture suggests that most respondents would be reluctant to exaggerate their level of understanding. We therefore utilize this self-reported response as one indicator of the understanding of scientific study among Japanese adults.

Second, each respondent was asked a closed-ended question about a hypothetical drug-testing issue. A question similar to one in the 1989 and 1992 Eurobarometers was used to describe a situation in which medical researchers want to determine the efficacy of a given drug. Each respondent was asked to determine if the best information would be obtained by asking the patient, using the biochemical knowledge of the physicians, or conducting an experiment. The experiment response was coded as correct, and 38% of Japanese adults in the 1991 study provided a correct response.

To provide a single indicator of the understanding of the logic of an experiment, all respondents who reported that they *clearly understood* or *understood* the meaning of a scientific study, and those respondents who selected the experiment option in the drug assessment question, were coded as having at least a minimal understanding of experimentation. Approximately 18% of Japanese adults qualified in the 1991 study.

The second item that loaded on the understanding of scientific inquiry factor was an item concerning probability. The designers of the 1991 Japanese study took the probability question on inherited illness that had been used in Europe and the United States and changed it into a question about plant genetics, using an example based on red and white flowers. The question context was changed because the study design committee felt that Japanese students are more likely to encounter probability concepts in the context of plant genetics than drug testing and that the question would make more sense to younger Japanese respondents who had finished school in the last two decades. The question used four probes that were structured like the probability question in the U.S. studies. In the 1991 study 20% of Japanese adults were able to provide the equal-probability response and reject the three incorrect responses.

Following the same procedure utilized in U.S., Canadian, and EU analyses, researchers constructed a simple typology to categorize the level of understanding of the nature of scientific inquiry. Japanese respondents who qualified as understanding both experimentation and probability were classified as understanding the nature of scientific inquiry, and only 5% of Japanese adults met this standard in 1991. This is a surprisingly low result because 18% understood experimentation and 20% understood probability, but the polychloric correlation coefficient for the relation between the two variables is .21. From the limited Japanese data set, it is not clear whether this second dimension would have been estimated differently if there had been additional process-oriented questions in the Japanese study or if open-ended items had been used.

The Computation of Results

Given the two-dimensional model described above, how might these dimensions be used to provide a single estimator of the level of civic scientific literacy? Conceptually, individuals who demonstrate a high level of understanding on both dimensions are expected to be the most capable of acquiring and comprehending information about a controversy over science or technology policy, and these individuals will be referred to as being *well informed*, or *civic scientifically literate*. At the same time, individuals who demonstrate either an adequate vocabulary of scientific constructs or who display an acceptable level of understanding of the nature of scientific inquiry are expected to be more capable of receiving and utilizing information about a science or technology policy dispute than citizens who understand neither dimension. This second group will be referred to as *moderately well informed* or *partially civic scientifically literate*. In the 1995 study, 12% of American adults qualified as well informed, or civic scientifically literate, and approximately 25% as moderately well informed (see Table 4-3).

By means of the same procedure, all European respondents who earned a score of 67 or more on the Index of Scientific Construct Knowledge and who demonstrated at least a minimally acceptable level of understanding of the nature of scientific inquiry were classified as well informed or civic scientifically literate. Individuals who qualified on one dimension but not the other were classified as moderately well informed. On the basis of the 1992 Eurobarometer, 5% of European adults were well informed or civic scientifically literate, while an additional 22% were moderately well informed.

By means of the same procedures, all Canadian respondents who scored 67 or higher on the Index of Construct Vocabulary and who were able to demonstrate a minimal understanding of the nature of scientific inquiry were classified as being civic scientifically literate. On the basis of the 1989 study, 4% of Canadian adults qualified as very well informed or civic scientifically literate, and an additional 17% were classified as moderately or partially civic scientifically literate.

As noted above, the composition of the 1991 Japanese survey poses some difficulties in estimating the proportion of adults qualifying as civic scientifically literate. Using the procedures outlined above, all Japanese adults with a score of 60 or higher on the Index of Construct Vocabulary and with a minimal understanding of the nature of scientific inquiry were classified as civic scientifically literate. Approximately 3% of Japanese adults were civic scientifically literate on the basis of the 1991 study, and an additional 22% qualified as moderately well informed.

Table 4-3. Percentage of Adults Estimated to be Scientifically Literate

Variable	European Union	Japan	United States	Canada
All adults aged 18 and over	5/22[a] $N = 12{,}147^{b}$	3/22 $N = 1{,}458^{b}$	12/25 $N = 2{,}007^{b}$	4/17 $N = 2{,}000^{b}$
Level of education				
Less than secondary school	1/10 $n = 3{,}324$	2/11 $n = 433$	1/8 $n = 387$	1/9 $n = 1{,}143$
Secondary school graduate	4/22 $n = 6{,}103$	2/25 $n = 701$	8/28 $n = 1{,}228$	4/23 $n = 651$
University graduate	11/37 $n = 2{,}712$	7/30 $n = 323$	35/33 $n = 392$	21/40 $n = 206$
Gender				
Female	3/17 $n = 6{,}372$	1/15 $n = 746$	8/20 $n = 1{,}053$	2/11 $n = 1{,}024$
Male	7/27 $n = 5{,}775$	6/29 $n = 711$	16/30 $n = 953$	6/23 $n = 976$
Age (in years)				
18 to 29	6/27 $n = 3{,}028$	5/26 $n = 330$	12/32 $n = 479$	5/20 $n = 582$
30 to 39	7/26 $n = 2{,}317$	6/24 $n = 252$	15/30 $n = 479$	6/25 $n = 445$
40 to 49	6/24 $n = 1{,}837$	3/18 $n = 302$	17/23 $n = 383$	4/17 $n = 339$
50 to 64	3/20 $n = 2{,}734$	1/26 $n = 243$	8/20 $n = 340$	2/9 $n = 364$
65 or older	2/11 $n = 2{,}231$	2/18 $n = 331$	3/13 $n = 321$	1/8 $n = 245$

Note. [a] % Well informed/% Moderately well informed. [b] The total N may vary because of the rounding of weighted case files.

Public Participation in the Formulation of Science Policy

The implications of these low rates of civic scientific literacy for democratic government are serious in any political system. It is important to examine the processes of public participation in the formulation of national policy in specialized areas such as science and technology. To assess these results in context, it is necessary to turn to a discussion of the growing impact of political and issue specialization in modern industrial societies in recent decades.

Political Specialization

For most individuals living in modern industrial societies, the range of demands on time and the array of choices for work, transportation, entertainment, and enlightenment are enormous. Although the legal work week has continued to decrease toward 40 hours per week in most industrial countries in recent decades, the effective work week for many professional and technical occupations has been increasing (Bosch, Dawkings, & Michon, 1993; Christopherson, 1991). The number of two-job families has been growing steadily for at least three decades in the United States and for the last 15 years in most European and Asian countries.

In this marketplace for the individual's time, politics and public affairs are but one competitor among many. Each citizen must decide how much time, energy, and resources to devote to becoming and remaining informed about politics and to overt acts of participation. The evidence suggests that the market share for politics, as measured by the proportion of adults who take the time to vote in national and local elections, has been declining in many countries (Kaase & Newton, 1995; Verba, Schlozman, & Brady, 1995). This decision to follow or not follow political affairs is referred to as *political specialization.*

Among those citizens who decide to devote some of their time and energy to public policy issues, there is a second level of specialization involving the selection of issues about which to become and stay informed. The range of issues at the national level alone is far too broad for any individual to maintain currency, and, when state, provincial, and local issues are included, the full range of potential public policy issues is too vast for any individual to master. Inevitably, all citizens who follow political affairs must focus their attention on a significantly smaller set of issues, and previous studies suggest that few citizens follow more than two or three major issue areas (Almond, 1950; Miller, 1983a; Popkin, 1994; Rosenau, 1974). This focusing on a limited range of issues is referred to as *issue specialization.*

The data sets from the European Union, the United States, Japan, and Canada offer an opportunity to test the specialization hypothesis empirically. In all four studies, individuals were asked to report their level of interest in selected public policy issues, including several issues involving science and technology. Although there were minor differences in wording and format, all of the questions included a *very interested* choice as the highest level of self-reported interest in each issue. To provide a simple descriptive indicator of interest in selected issues, an index was constructed that scored 100 points for being very interested, 50 points for being moderately interested, and no points for having reported that one is not at all interested.[1] The mean interest index score ranged from zero to 100 for any national or demographic grouping.

Whereas the general level of political interest and participation appears to be declining in most industrial nations, scientific and technological issues appear to compete relatively well for the attention of citizens in Canada, the European Union, and the United States, and only slightly less well among Japanese adults. For all four national groupings, the index of interest in environmental issues was in the low to mid-70s range, and approximately two thirds of adults in the European Union, the United States, and Canada expressed a high level of interest in new scientific discoveries. The proportion of citizens in Canada, Japan, and the United States who reported that they were interested in environmental and health issues was higher than that of citizens who indicated they were interested in economic issues (see Table 4-4). This level of interest is reflected in the continuing high numbers of people who watch science television shows and read science magazines and in the abiding popularity of health and environmental reporting in newspapers.

Although a high level of interest in an issue is necessary for effective citizen participation, it is not sufficient. Almond (1950), in a landmark study of public participation in the formulation of foreign policy in the United States, argued that it is also necessary for citizens to feel that they are reasonably well informed and to be continuing consumers of relevant news and information. Citizens who have a high level of interest in an issue, who think that they are reasonably well informed about it, and who follow that issue in the news are significantly more likely to make a voting decision based on that issue, to write or contact a legislator or government officer about that issue, or to engage in overt political activities in pursuit of a particular policy than other citizens (Rosenau, 1974).

To assess the degree of self-perceived knowledgeability, national studies in Canada, Europe, Japan, and the United States have asked respondents to categorize themselves as being very well informed, moderately well informed, or poorly informed about the same set of public policy issues used to assess issue interest. In every country, the proportion of adults who are

---

[1] Unlike the Canadian, EU, and U.S. studies, however, the 1991 Japanese study offered four levels of interest: *very interested, moderately interested, hardly interested,* and *not at all interested.* This differentiation of categories has the potential to complicate comparisons across countries, but a careful analysis of alternative coding rules found that the assignment of values of 100, 67, and 33 for the three highest levels of interest provided an acceptable estimate of the level of issue interest in Japan.

interested in a public policy issue is far higher than the proportion who report that they are well informed about that issue. Though virtually all of the empirical research on the impact of this self-perceived level of knowledge has focused on the United States, it appears that the perception of not being well informed deters individuals from engaging in overt efforts to influence public policy by writing a letter to or personally contacting a public official. Ressmeyer (1994) suggests that the actual level of knowledgeability also contributes to the likelihood of letter-writing or contacting, especially during noncrisis periods.

Table 4-4. Mean Scores on the Index of Issue Interest Scores

Issue area	European Union	Japan	United States	Canada
New scientific discoveries	61	50	66	63
New inventions and technologies	59	53	66	58
New medical discoveries	68	65	82	77
Environmental pollution	75	71	73	74
Space exploration	–	45	49	48
Energy/nuclear power	–	59	54	–
Computers and related technologies	–	–	–	43
Economic policy	–	65	68	52
Education/local schools	–	62	72	–
Agriculture	–	56	47	41
Military/defense issues	–	56	60	–
Foreign & international policy	–	55	47	–
Politics	55	–	–	50
Sports news	48	–	–	42
Taxes	–	71	–	–
Land use issues	–	65	–	–
Senior citizen issues	–	74	–	–

Note. A dash means that the item was not asked.

In this stratified model, those citizens with a high level of interest in a given issue and a sense of being well informed about that issue are referred to as the *attentive public* for that issue. There is an attentive public for almost every issue, and most citizens who follow public policy issues at all tend to become attentive to two or three issue areas. Citizens who are attentive to a particular policy area tend to read, or stay informed, about that issue area on a continuing basis. There is some evidence that attentive citizens hold more elaborate schema concerning any given policy area and are better able to receive and process new information about it than is the case with citizens who are not attentive to that policy area. It is a rare citizen who qualifies as attentive to four or more issue areas.

The Attentive Public for Science and Technology Policy

Miller (1983a, 1983b) and others have utilized the Almond model to define and describe an attentive public for science and technology policy in the United States. Beginning in 1979, the Miller and Prewitt measure of attentiveness to science and technology policy was based on respondent reports of interest in and knowledgeability about issues surrounding new scientific discoveries and issues surrounding the use of new inventions and technologies and on a measure of persistence in the consumption of relevant information. Fortunately, identical or comparable items have been asked in the Canadian, Eurobarometer, and Japanese studies, so an assessment of this construct in different political systems is possible.

Approximately one in ten adults in Canada, the European Union, and the United States qualified as attentive to science and technology policy, compared to seven in ten in Japan (see Table 4-5). Although science and technology policy is not a highly salient issue in any of these four political systems, science and technology policy decisions may be influenced by interested, informed, and concerned citizens and groups in virtually any political system. The likelihood of being able to influence policy, however, may be higher in competitive multiparty systems than in *de facto* one-party systems. To the extent that citizens perceive this statement to be true, there is less incentive to become and remain attentive to science and technology policy issues. Given the nature of our present data sets, we cannot subject this hypothesis to a rigorous falsification test, but we believe that the evidence argues for the further exploration of this hypothesis as one part of an explanation for this difference between Japan and Canada, the European Union, and the United States.

Table 4-5. Percentage of Adults Attentive to (Attn) or Interested in (Int) Science and Technology Policy

Variable	European Union Attn	European Union Int	Japan Attn	Japan Int	United States Attn	United States Int	Canada Attn	Canada Int
All adults	10	33	2	17	10	47	11	40
Education								
Less than high school	5	25	1	9	4	37	9	37
High school graduate	9	33	2	18	8	48	11	45
Baccalaureate	18	40	4	26	21	53	19	46
Gamma	.32		.36		.37		.22	
Gender								
Female	7	30	<1	11	8	45	7	47
Male	13	36	3	24	12	49	14	44
Gamma	−.23		−.48		−.17		−.28	
Age (in years)								
18 through 29	13	35	2	19	7	52	8	38
30 through 39	10	36	2	21	12	48	14	41
40 through 49	10	35	2	18	11	47	10	46
50 through 64	9	32	3	20	9	47	11	46
65 or older	8	25	1	11	10	40	10	28
Gamma	−.13		−.13		−.05		.02	
Civic scientific literacy								
Well informed	18	45	5	45	29	55	26	42
Moderately well informed	14	39	6	31	14	51	16	44
Not well informed	7	27	1	14	7	45	8	40
Gamma	.36		.52		.36		.27	
N	1,226	3,971	27	252	195	946	209	809

The Structure and Distribution of Attentiveness

The aggregate proportion of citizens attentive to science and technology policy in each of these four sociopolitical systems is important. In Europe, one in ten adults was attentive to science and technology policy. Better educated citizens were more likely to qualify as attentive than less well-educated citizens, and civic scientifically literate Europeans were more likely to be attentive to science and technology policy than other citizens. The ordinal measure of association *gamma* (a proportional reduction of error in error statistic) for the relationship between the level of education and attentiveness to science and technology policy was .32, indicating that this bivariate relationship accounts for approximately 32% of the total variation between the two groups (Costner, 1965; Goodman & Kruskal, 1954). Civic scientifically literate Europeans were more likely to be attentive to science and technology policy than other citizens (gamma = .36). Younger citizens of EU member states were slightly more likely to be attentive to science and technology issues than older ones.

In the United States, which has a weak party system and strong legislative committees, the results from the 1995 study indicate that approximately 10% of American adults were attentive to science and technology policy and that an additional 47% were interested in science and technology issues (see Table 4-5). Americans with more years of formal schooling were significantly more likely to be attentive to science and technology policy, with one in five college graduates qualifying as attentive and an additional 53% reporting a high level of interest in these issues (gamma = .32). The level of civic scientific literacy was positively related to attentiveness in the U.S. data, with 29% of civic scientifically literate Americans qualifying as attentive to science and technology policy. American men were more likely to be attentive than American women (gamma = –.17), and there was no clear pattern of attentiveness related to age.

In Japan, 7% of adults qualified as attentive to science and technology policy. An additional 12% met the criteria for the interested public for science and technology policy. This lower level of attentiveness appears to be a reflection of Japanese political and social systems rather than a measurement problem. As discussed above, a modified scale was used in response to the four-level response in the 1991 study. For similar reasons, Japanese respondents who indicated that they were very knowledgeable about new scientific discoveries or new inventions and technologies were classified as meeting the criterion for feeling adequately informed about an issue. There may be a methodological bias resulting in a slight overestimate of the level of attentiveness to science and technology policy among Japanese respondents.

In Canada, the pattern of attentiveness to science and technology policy was similar to that found for the United States. Approximately 19% of college graduates were attentive to science and technology policy (gamma = .22). In the 1989 study, 25% of Canadians who qualified as civic scientifically literate were attentive to science and technology policy (gamma = .27). Canadian men were twice as likely as Canadian women to be attentive to science and technology policy (gamma = −.28). Although there was no association between age and attentiveness in Canada, it is interesting to note that the lowest rates of attentiveness to science and technology policy in both Canada and the United States were among citizens from 18 to 29 years old.

The Role and Importance of Attentive Publics

How, then, does this specialization process affect the formulation of public policy and the resolution of disputes involving scientific and technological issues? How, and under what conditions, can citizens who are attentive to science and technology policy issues influence public policy? To a large extent, the formulation of science and technology policy and the resolution of specific scientific and technical disputes has been moved almost totally out of the electoral process. Virtually no candidate for national office wins or loses a race for office on an issue of science or technology policy. Although elected officers do play an essential role in the process, they are rarely elected on the basis of a commitment to support or oppose a specific science or technology policy.

Focusing on the federal government of the United States (see Figure 4-2), Almond (1950) outlined a pyramidal structure that illustrates the types of public participation in the policy formulation process that is likely to occur under conditions of issue specialization. In this stratified model, policy-makers sit at the pinnacle of the system and have the power to make binding decisions on a given policy matter. In the United States, this group would include a mix of executive, legislative, and judicial officers, and, in the case of science and technology policy, the officers would be primarily at the federal level. In unitary parliamentary systems common in Canada, Europe, and Japan, the decision-makers would include the prime minister, relevant cabinet officers, and relevant committee chairs and members if there is a substantive legislative committee structure. In the European Union, the policy-makers would include a combination of leaders from the Commission and the European Parliament, including the leadership of committees with responsibility for science and technology.

The second level of the system is a group of nongovernmental policy leaders. In the case of science and technology policy, this group would include leading scientists and engineers; leaders of major corporations active in science and engineering; the officers and leaders of scientific and professional societies; the presidents and relevant deans of major research universities; the members of the national academies of science or engineering; and various business, academic, and other leaders interested in science and technology matters. Rosenau (1961, 1963, 1974) and others have noted that there

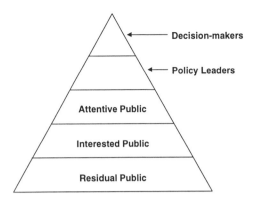

Figure 4-2. A stratified model of policy formulation

is some movement of elites into decision-making posts and of decision-makers into the leadership group from time to time. When there is a high level of concurrence between the decision-makers and the leadership group, policy is generally made, and there is no wider public participation in the policy process.

On some issues, however, there may be a division of views among the policy leadership group itself, or between the policy leaders and the decision-makers. In those circumstances, there may be appeals to the attentive public to join in the policy process and to seek to influence policy-makers by direct contact and by persuasion. Although the processes through which the leadership group mobilizes attentives in various policy areas are only now being studied, it appears that there is a flow of appeals from leaders to attentives through professional organizations, specialized magazines and journals, and employment-related institutions. The contacting of public officials (policy-makers) by attentives appears to take place in traditional modes: letters, telephone calls, and personal visits.

The attentive public does not formulate the national policy agenda or play a significant role in the daily negotiation of public policy relevant to science and technology. It is mainly when there is a conflict in the system, when the leaders or the decision-makers cannot find a solution, that the system turns to the attentive publics to resolve the issue. Almond (1983) compared the role of the attentive public to that of reserve units in the military. Attentive citizens read about the issues that interest them and talk with friends and colleagues who share similar interests. When a science or technology dispute arises, individuals who are attentive may be persuaded to engage in direct efforts to influence decision-makers. Like wars, these conflicts do not happen often, but they are likely to concern the most important issues in the science and technology arena, and certainly the most difficult and controversial ones.

When an issue or controversy cannot be resolved at the leadership level, it is essential that there be a sufficient number of citizens who are attentive to that area and who are able to comprehend the debates among the leaders about the issue. Although it is essential that attentive citizens think of themselves as being sufficiently well informed to enter into the policy debate via letters or direct contacting, it is important that these attentives also be civic scientifically literate, that is, that they know enough about science, mathematics, and technology to follow and evaluate the major competing arguments about an issue.

Public Attitudes Toward Science and Technology

The social psychology and cognitive science literature indicates that most individuals, when faced with a daily barrage of complex information from media, friends, and colleagues, construct *schemas* to filter and manage the receipt and organization of information (Lau & Sears, 1986; Milburn, 1991; Minsky, 1986; Pick, van den Broek, & Knill, 1992; Schank, 1977). A schema is a cohesive set of expectations about objects, situations, or sequences of actions. Minsky argues that people do not organize information in an atomistic way and that ordinary human reasoning does not follow the principles of classical deductive logic (particularly the property called monotonicity) (Minsky 1974, 1986) but processes fairly large blocks or "chunks" of information he calls *frames*. Frames about certain objects, events, or situations are "a sort of skeleton, somewhat like an application form with many blanks or slots to be filled" (Minsky 1986, p. 245), each frame having a set of slots that can accommodate individual pieces of relevant information. The slots have certain values by default:

> [D]efault assignments are of huge significance because they help us represent our previous experience. We use them for reasoning, recognizing, generalizing, predicting what may happen next, and knowing what we ought to try when expectations aren't met. Our frames affect our every thought and everything we do. (Minsky, 1986, p. 245)

Scripts, as originally proposed by Schank (1977), are another knowledge representation scheme used in cognitive science for making sense of the way people deal with and process information about *typical* objects and situations. Instead of just capturing objects, however, they represent sequences of events people expect to take place in familiar situations.

Setting aside issues concerning how much, if any, of the ability to develop effective schemas is an inherited trait, we believe that it is important to recognize the central role that schemas play in each individual's efforts to receive, organize, and make sense of the wide array of new and complex information offered daily in the broadcast and print media of modern industrial and scientific societies. Although we expect that some individuals have much more developed schemas about science and technology than others, as reflected in the variations in interest and civic scientific literacy, we expect that most people in a modern society have some schema or schemas for scientific and technical matters. The combined national surveys from the European Union, the United States, Japan, and Canada provide an opportunity to examine empirically the existence and composition of schemas relevant to science and technology issues.

A series of confirmatory factor analyses demonstrated that a set of six of items used in the 1992 Eurobarometer formed a unidimensional factor that appears to tap a general positive attitude toward science and technology (see Table 4-6). Three other items from the 1992 Eurobarometer form a second dimension that reflects concerns or reservations about the impact of science and technology on individuals and on society. The two factors are nearly independent, with a correlation of $-.11$. This degree of independence suggests that many European adults simultaneously recognized the positive potential of science and technology for themselves and their children and expressed personal fears or concerns about the real or potential negative impacts of science and technology on traditional practices and values.

A parallel set of confirmatory factor analyses of the 1995 *Science and Engineering Indicators* data identified a two-factor structure similar to the pattern found in the European data (see Table 4-7). Substantively, the composition of the two factors found in the U.S. data is consistent with the results from the 1992 Eurobarometer. However, these two factors have a strong negative correlation of -64. This relationship means that an individual with a high score on the first dimension is very likely to have a low score on the second factor, and vice versa. Americans who have positive attitudes toward the benefits of science and technology are less likely to have reservations or concerns about possible impacts, whereas individuals who have strong concerns about science and technology are less likely to recognize contemporary benefits or have optimistic views of future contributions from science and technology. This pattern points to an attitude structure more polarized than that found in Europe.

Table 4-6. Confirmatory Factor Analysis of Attitudes, European Union, 1992

I would like to read you some statements that people have made about science, technology, or the environment. For each statement, please tell me how much you agree or disagree (SHOW CARD).	Dimension		
	Promise of science	Reservation about science	Proportion of variance explained
Thanks to science and technology, there will be more opportunities for the future generations.	.61	–	.37
Science and technology are making our lives healthier, easier, and more comfortable.	.58	–	.33
The benefits of science are greater than any harmful effects it may have.	.57	–	.32
Most scientists want to work on things that will make life better for the average person.	.51	–	.26
Scientific and technological progress will help to cure illnesses such as AIDS and cancer.	.51	–	.26
The application of science and new technology will make work more interesting.	.51	–	.26
Science makes our way of life change too fast.	–	.67	.45
We depend too much on science and not enough on faith.	–	.47	.22
Because of their knowledge, scientific researchers have a power that makes them dangerous.	–	.43	.19

Note. $\chi^2 = \dfrac{46.9}{16}$ degrees of freedom. Root Mean Square Error of Approximation (RMSEA) = .02. Upper limit of the 90% confidence interval of RMSEA = .024. Factor 1 and Factor 2 are correlated at –.11. $N = 6,122$.

Table 4-7. Confirmatory Factor Analysis of Attitudes, United States, 1995

I'm going to read you some statements such as those you might find in a newspaper or magazine article. For each statement, please tell me if you generally agree or generally disagree. If you feel especially strongly about a statement, please tell me that you strongly agree or strongly disagree.	Dimension		
	Promise of science	Reservation about science	Proportion of variance explained
Because of science and technology, there will be more opportunities for the next generation.	.68	–	.47
Science and technology are making our lives healthier, easier, and more comfortable.	.62	–	.38
Most scientists want to work on things that will make life better for the average person.	.54	–	.29
With the application of science and new technology, work will become more interesting.	.53	–	.28
Science makes our way of life change too fast.	–	.66	.44
On balance, the benefits of scientific research have outweighed the harmful results.	–	–.60	.36
We depend too much on science and not enough on faith.	–	.51	.26
It is not important for me to know about science in my daily life.	–	.41	.17

Note. $\chi^2 = \dfrac{30.7}{15}$ degrees of freedom. Root Mean Square Error of Approximation (RMSEA) = .02. Upper limit of the 90% confidence interval of RMSEA = .034. Factor 1 and Factor 2 are correlated at –.64. $N = 2,006$.

Using the techniques described above, a series of confirmatory factor analyses of the 1991 Japanese data identified a two-factor structure similar to that found in Europe and the United States (see Table 4-8). Although the wording in three of the four items on the first factor is slightly different from that used in the European and North American studies, the context is similar. The commonality of the factor is a respect for the intentions of scientists and a sense that science will provide useful results and products for society. One of the strengths of structural analysis is that it allows one to go beyond the issue of whether the wording of items was exactly the same and to look at attitudinal dimensions that may be similar regardless of minor variations in wording. Following the pattern found in Europe, these two factors were weakly and negatively correlated in the 1991 Japanese data, with a correlation of –.22. This relationship means that an individual with a high score on the promise of science and technology schema has an almost equal probability of having a low score on the concern about science and technology schema, and vice versa.

Table 4-8. Confirmatory Factor Analysis of Attitudes, Japan, 1991

For each statement, please tell me if you strongly agree, agree, disagree, or strongly disagree.	Dimension		
	Promise of science	Reservation about science	Proportion of variance explained
With the application of science and new technology, work will become more interesting.	.67	–	.45
Computers and factory automation will create more jobs than they will eliminate.	.45	–	.21
Scientists are seeking to benefit human beings.	.44	–	.19
The benefits of scientific research have outweighed any harms.	.36	–.24	.15
Science makes our way of life change too fast.	–	.76	.58
We depend too much on science and not enough on faith.	–	.64	.42
Because of their knowledge, scientific researchers have a power that makes them dangerous.	–	.26	.07

Note. $\chi^2 = \dfrac{13.5}{9}$ degrees of freedom. Root Mean Square Error of Approximation (RMSEA) = .02. Upper limit of the 90% confidence interval of RMSEA = .038. Factor 1 and Factor 2 are correlated at .22. $N = 1,427$.

Relative to the studies in the European Union, the United States, and Japan, the 1989 Canadian study included markedly fewer attitude items. A series of confirmatory factor analyses of the Canadian data found four items that loaded on a dimension reflecting the concerns and reservations similar to the preceding analyses and two items that loaded on a dimension reflecting the promise of science and technology (see Table 4-9). The two factors were correlated negatively at the –.59 level, reflecting a polarity of views similar to that found in the United States.

Table 4-9. Confirmatory Factor Analysis of Attitudes, Canada, 1989

For each statement, please tell me if you strongly agree, agree, disagree, or strongly disagree.	Dimension		
	Promise of science	Reserva- tion about science	Proportion of variance explained
On balance, the benefits of scientific research have outweighed the harm- ful results.	.68	–	.46
Science and technology are making our lives healthier, easier, and more com- fortable.	.27	–	.07
Science makes our way of life change too fast.	–	.65	.42
We depend too much on science and not enough on faith.	–	.64	.41
It is not important for me to know about science in my daily life.	–	.40	.16
Because of their knowledge, scientists have a power that makes them dan- gerous.	–	.38	.14

Note. $\chi^2 = \dfrac{5.3}{7}$ degrees of freedom. Root Mean Square Error of Approximation (RMSEA) = .00. Upper limit of the 90% confidence interval of RMSEA = .023. Factor 1 and Factor 2 are correlated at –.59. $N = 2,000$.

Functionally, we interpret the factor patterns found in the preceding analyses to support the view that most individuals hold two primary schemas toward science and technology. We characterize the first factor found in each political system as representing the *promise of science and technology*. All of these items reflect the judgment that science and technology have already improved the quality of life or a positive assessment of the likelihood of future benefits. We characterize the second dimension as representing *reservation about science and technology*. These items express concerns about the speed of change in modern life and a sense that science may, at times, pose conflicts with traditional values or belief systems. Even though the exact wording of some items differed slightly from study to study, each of these factors represent general beliefs about science and technology that will serve as filters and organizers for more specific policy preferences.

In the context of the broader cognitive science literature on the formation and use of schemas, we expect to find a full array of possible combinations of these two schemas. Some individuals may have a strong positive schema and a weaker negative schema, leading them to react positively to a wide spectrum of science news. Alternatively, some individuals may have a weak positive schema and a stronger negative schema, leading them to be doubtful or negative about scientific news. It is also possible for an individual to have a strong positive schema and a strong negative schema, recognizing both the substantial promise of science and technology and opportunities for substantial harm from scientific and, especially, technological activities. Given the low salience of science and technology to many adults, it is likely that many people will have weak positive and weak negative schemas, reflecting the absence of direct experience or information about science and technology. Given the vast array of science and technology impacts and their interactions with virtually every aspect of society and nature today, apart from the very different nature of these impacts, most people should not be expected to have clear cut images of where the domain of science and technology begins and ends and where other social domains or variables in which science plays a major role start (the work context, the health context, the political context). Accordingly, people's attitudes toward an object like science, which for individuals outside the scientific community appears not to have sharp boundaries with other objects (economic and political institutions), may exhibit many possible combinations of weak positive and negative schemas. Science, like many other complex social objects, appears for most people— again, not for the scientist or the civic scientifically literate—as a "fuzzy set" in the technical sense of fuzzy set theory (Zadeh, 1987).

To provide a common metric for comparison, a factor score was computed for each dimension (schema) for each of the four data sets, and this score was then converted to a zero-to-100 metric. In this conversion a score of zero was assigned for the factor score reflecting the lowest level of agreement with a dimension and a score of 100 for the highest possible level of agreement with a dimension. Measured in this way, the mean score on the Index of Scientific Promise was 72 in Canada, 69 in the European Union, 68 in the United States, and 55 in Japan (see Table 4-10). The markedly lower level of public belief in the promise of science in Japan is not related to any of the measurement issues discussed previously. It appears to be a reflection of a lower level of scientific optimism, which is consistent with an apparently lower level of optimism and satisfaction on the part of Japanese adults in many other areas of social life (Ladd & Bowman, 1996).

Table 4-10. Mean Scores on the Index of Scientific Promise

Variable	European Union	Japan	United States	Canada
All adults	69	55	68	72
Level of formal education				
Less than completed secondary	68	54	63	68
Completed secondary	69	55	68	75
Post secondary	71	56	71	84
Gender				
Female	68	54	67	68
Male	70	55	69	76
Age (in years)				
18 through 29	69	53	67	70
30 through 39	69	53	69	74
40 through 49	70	54	69	73
50 through 64	71	56	69	75
65 or older	68	57	66	69
Civic scientific literacy				
Well informed	70	64	72	84
Moderately well informed	69	58	69	80
Not well informed	69	54	67	69
Attentiveness to science and technology policy				
Attentive public	74	56	74	79
Interested public	72	59	69	74
Residual public	67	54	65	69
N	6,122	1,457	2,006	2,000

Citizens who qualified as civic scientifically literate were significantly more likely to score higher on the promise-of-science-and-technology schema in Canada, Japan, and the United States, than were citizens who did not qualify as civic scientifically literate, but there was no significant difference in the level of belief in the promise of science by level of civic scientific literacy among respondents in the European Union. Only in Canada was the level of formal educational attainment associated with belief in the promise of science and technology.

In Canada, the European Union, and the United States, citizens who were attentive to science and technology policy issues held more positive views of the promise of science and technology than nonattentive citizens. In Japan, those individuals who were interested in science and technology policy issues but who were not willing to classify themselves as very well informed expressed a slightly more optimistic view toward the promise of science than citizens qualified as attentive.

Computed with the previously explained procedures and metric, the mean score on the Index of Scientific Reservation was 58 in the European Union, 56 in Canada and Japan, and 39 in the United States (see Table 4-11).

Table 4-11. Mean Scores on the Index of Scientific Reservation

Variable	European Union	Japan	United States	Canada
All adults	58	56	39	56
Level of formal education				
Less than completed secondary	64	62	51	60
Completed secondary	57	55	39	52
Post secondary	53	50	27	40
Gender				
Female	60	57	40	58
Male	57	55	38	53
Age (in years)				
18 through 29	53	54	30	45
30 through 39	55	52	38	54
40 through 49	58	56	36	58
50 through 64	62	58	39	60
65 or older	64	63	45	61
Civic scientific literacy				
Well informed	46	45	24	39
Moderately well informed	55	55	30	45
Not well informed	62	56	42	59
Attentiveness to science and technology policy				
Attentive public	57	54	30	45
Interested public	57	52	38	54
Residual public	60	57	42	59
N	6,122	1,457	2,006	2,000

In contrast to the distribution of positive schema, this pattern suggests that the citizens of Canada, member states of the European Union, and Japan hold moderately high levels of reservation about the present and potential negative consequences of science and technology. The significantly lower level of reservation found in the United States is consistent with a wide body of literature pointing to nearly five decades of high optimism for and low concern about science and technology among Americans (Hughes, 1989; Miller, 1983a).

In all four political systems, individuals with the lowest levels of formal education expressed the highest level of reservation about science and technology. Citizens with higher levels of civic scientific literacy reported significantly lower levels of reservations about science and technology than citizens not well informed about science. As with the Index of Scientific Promise, these relationships were strongest in the United States.

In Canada and the United States, citizens who were attentive to science and technology policy reported significantly lower levels of reservation about the impact of science and technology than nonattentive citizens. In the European Union and Japan, attentive and interested citizens were only slightly less concerned about the impact of science and technology than the residual public.

In all four political societies, women were slightly more likely to hold reservations about science and technology than men. The margins of difference were small, and it is likely that most of these observed differences are accounted for by differences in educational attainment.

The Linkage Between Schema and Specific Policy Preferences

Most people appear to hold simultaneously a positive schema toward the achievements and promise of science and technology and a negative schema toward present or prospective harmful results from science or technology. Although the calculation of a mean or ratio of the strength of these two schemas is a simple task, we believe that a mean or ratio of this kind would mask the actual dynamic of the process at the individual level. In practice, science and technology are not one uniform attitude object but rather an assortment of specific activities, some of which may activate primarily one schema over the other. For example, a news report about a new medical test result may activate the promise of science schema and generate optimism, whereas a report of the test of a nuclear weapon may activate reservations about science and technology.

To explore the role of schema in the formulation of more specific policy attitudes, it is useful to take a cross-national look at the responses to the statement[2] that "[government] should provide support for basic scientific research even if it produces no immediate benefits." An overwhelming majority of adults in all four political systems supported the view that their national government should provide support for basic scientific research, with the level of agreement being highest in Canada and Japan, with 88% and 86% of adults, respectively, indicating that they would like for the government to provide, or continue to provide, support for basic scientific research (see Table 4-12). That same attitude toward government support for basic research was expressed by 80% of the EU respondents and 78% of the Americans.

In all four societies the level of formal education and the level of civic scientific literacy were positively associated with support for government funding of basic scientific research. Likewise in all four political systems, citizens who were not interested in science and technology policy issues were the least supportive of basic science funding. Men were slightly more likely to support government spending for basic scientific research than were women.

To explore the relative impact of the science-promise and science-reservation schemas in the development and maintenance of a specific policy preference in favor of government support for basic research, a set of structural equation models[3] were used to predict attitude toward government sup-

---

[2] In the EU, Japanese, and American studies, the reference to the government was the only part of the statement that varied. In the U.S. study, the term "Federal Government" was inserted into the question, whereas a more generic reference to the national government was employed in the European Union and Japan. In all three of these studies, the respondents were asked if they strongly agreed, agreed, disagreed, or strongly disagreed with the statement. *Don't know* or *not sure* responses were coded into a middle category, producing a five-category ordinal variable. In the 1989 Canadian study, this question was not asked, but the Canadians were asked if they thought that their national [federal] government was spending too little, too much, or about the right amount on basic scientific research, producing a three-category ordinal variable. The descriptive results reported in Table 16 include the strongly agree and agree responses in European, Japanese, and American studies and the *about right* and *too little* responses in the Canadian studies. In the structural equation analyses, all five responses categories were retained in the EU, Japanese, and U.S. data, and all three categories were utilized in the Canadian data.

[3] In general terms, a structural equation model is a set of regression equations that provide the best estimate for a set of relationships among several independent variables and one or more dependent variables. For all of the structural analyses presented in this report, we used the program LISREL, which allows the simultaneous examination of structural relationships and the modeling of measurement errors. For

port for basic scientific research (see Figures 4-3, 4-4, 4-5, and 4-6). In these models, gender and age are treated as background variables that may influence all of the variables to the right in each figure. All of the variables to the left of the spending attitude may predict the level of the outcome variable. The objective of the analysis is to identify the minimal set of paths that provide the best prediction of attitude toward support for basic science funding.

Table 4-12. Percentage of Sample Agreeing that Government Should Support Basic Scientific Research

Variable	European Union	Japan	United States	Canada
All adults	80	86	78	88
Level of formal education				
Less than completed secondary	67	81	67	85
Completed secondary	83	86	79	89
Post secondary	89	93	87	98
Gender				
Female	77	83	77	84
Male	83	90	79	91
Age (in years)				
18 through 29	78	86	86	84
30 through 39	85	86	84	90
40 through 49	84	88	78	89
50 through 64	80	88	72	87
65 or older	71	84	65	92
Civic scientific literacy				
Well informed	91	96	90	98
Moderately well informed	87	94	87	93
Not well informed	74	85	75	86
Attentiveness to science and technology policy				
Attentive public	91	89	83	92
Interested public	89	96	85	90
Residual public	73	84	70	84
N	6,122	1,457	2,006	2,000

a more comprehensive discussion of structural equation models, see Hayduk (1987) and Jöreskog & Sörbom (1993).

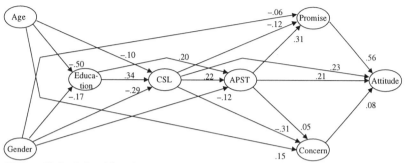

CSL = Civic Scientific Literacy
APST = Attentive Public for Science and Technology

Figure 4-3. A path model to predict attitudes toward government spending for basic scientific research, European Union, 1992.

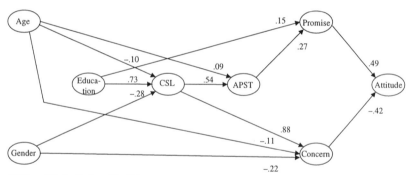

CSL = Civic Scientific Literacy
APST = Attentive Public for Science and Technology

Figure 4-4. A path model to predict attitudes toward government spending for basic scientific research, United States, 1995.

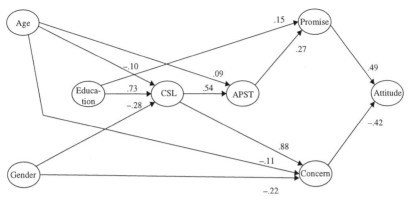

CSL = Civic Scientific Literacy
APST = Attentive Public for Science and Technology

Figure 4-5. A path model to predict attitudes toward government spending for basic scientific research, Japan, 1991.

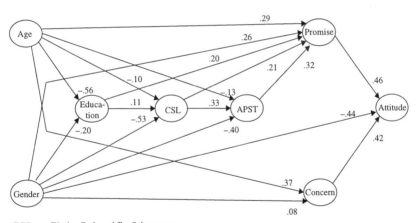

CSL = Civic Scientific Literacy
APST = Attentive Public for Science and Technology

Figure 4-6. A path model to predict attitudes toward government spending for basic scientific research, Canada, 1989.

In these models, the order and placement of the variables reflect a chronological or logical order, with time and logical order flowing from left to right. A path coefficient represents the relationship between two variables, holding constant all other variables that are logically or chronologically earlier (to the left) of the predicted variable (the variable receiving the path). The absence of a path means that there was no significant relationship between the two variables, holding constant other variables as described above. The total effect of any variable on the outcome variable—attitude toward government spending for basic scientific research—may be determined by multiplying all of the path coefficients that connect the two variables and summing the products. The path coefficients and their total effects may be compared across the national models because these models were analyzed as four groups with a pooled variance and a common metric (Zimowski, Muraki, Mislevy, & Bock, 1996).

The intervening relationships between the two schemas and attitude toward government spending for basic scientific research illustrate important differences in the four societies. In Europe, a strong-promise-of-science schema was the strongest predictor of citizen approval of government spending for basic scientific research (see Figure 4-3 and Table 4-13). It appears that European adults favor government spending for research as a means of improving the quality of life. The level of reservation about the impact of science and technology is significantly higher among European adults than U.S. adults, but this concern is virtually unrelated to attitude toward government spending for scientific research. One might see this combination of attitudes as similar to an individual who has a moderate level of anxiety about flying in an airplane but who recognizes the speed and convenience of air transportation and continues to travel by air.

In the United States, the scientific-promise and scientific-reservation schemas display a clear, but differentiated, relationship with citizens' policy attitude on government support for basic scientific research. Reflecting the strong negative correlation between the promise and reservation schemas in the U.S. data, this model indicates that citizens who had a strong belief in the promise of science and technology and relatively little reservation about the impact of science and technology were significantly more likely to approve of government spending for basic scientific research than were citizens who had a weak belief in that promise (see Figure 4-4 and Table 4-13). This model suggests that these two schemas operate separately but not independently.

The role of the two schema in Japan illustrates the special place of science and technology in modern Japan. Both the promise-of-science schema and the reservations about science schema are positively related to approval of government spending for scientific research. It will be recalled that the mean Japanese score on the Index of Scientific Promise was 55 and that the mean score on the Index of Scientific Reservations was 56 but that 86% of Japanese adults voiced approval for government spending for scientific research (see Figure 4-5 and Table 4-13). In the context of the earlier finding that science and technology policy is a low-salience issue for many Japanese adults, this pattern suggests that Japanese adults do not have strong crystallized feelings about either the promise or the dangers of science and technology but that they may attribute some responsibility to science and technology for the strong growth of Japanese industry in recent decades. Support for scientific research may have become a social and political expectation in Japan in the postwar period, but without a clear cognitive foundation of associated schemas. The positive total effect (.20) between age and approval of government spending for scientific research in Japan supports this interpretation.

Table 4-13. The Estimated Total Effects of Different Variables on the Prediction of Attitude toward Support for Basic Research

Variable	Estimated total effect			
	European Union	Japan	United States	Canada
Age	−.09	.20	.01	−.06
Gender (female is positive)	−.17	−.54	−.03	−.45
Education	.15	.10	.39	.09
Civic scientific literacy	.22	.10	.44	.13
Attentiveness to science & technology policy	.38	.15	.13	.18
Science promise schema	.56	.46	.49	.00
Science reservation schema	.08	.42	−.42	−.25
Multiple R^2	.49	.51	.63	.30
N	6,122	1,451	2,006	2,000

Note. $\chi^2 = \dfrac{42.4}{43}$ degrees of freedom. Root Mean Square Error of Approximation (RMSEA) = .00. Upper limit of the 90% confidence interval of RMSEA = .011.

In Canada, the best predictor of approval of government spending for scientific research was a rejection of concern about science and technology (see Figure 4-6 and Table 4-13). It is important to remember that Canadian adults had a higher mean score (72) on the Index of Scientific Promise than EU, Japanese, or U.S. adults did and that 88% of Canadians approved of government spending for scientific research. Given these parameters, the model had limited variance to predict, and the most important factor that distinguished approval from disapproval was a low level of reservation or concern about the impact of science and technology. In this context, the level of belief in the promise of science was so pervasive among Canadian adults that it was not a good predictor of the spending attitude.

Beyond the role of the two schemas, these models provide several additional insights into the social and political context of a specific policy preference. The combined effects of formal education and the development of civic scientific literacy continue to have substantial influence in the United States but significantly less influence in the other three societies. This result appears to reflect some of the differences in access to higher education generally and to university-level science instruction in particular in Canada, the European Union, and Japan. In Canada and Japan, gender is a major predictor of attitude toward government spending for scientific research, if differences in education and other factors are held constant.

In summary, this analysis of the influence that selected demographic characteristics and schemas of general attitudes have on a specific policy attitude demonstrates the complex, but comprehensible, environment in which policy attitudes are formed and maintained. The items in these four major studies form meaningful dimensions that reflect real schemas that individuals use in thinking about science and technology policy issues. Although more work needs to be done in this area, these results demonstrate both the feasibility and the value of this kind of research.

Findings and Conclusions

Several substantive findings merit a summary discussion. First, the level of public interest in new scientific and medical discoveries, new inventions and technologies, and environmental issues is relatively high in Canada, the European Union, and the United States. In the modern marketplace for time and attention, science and technology compete relatively effectively. The lower level of interest in these issues in Japan appears to reflect a combination of cultural and political factors.

Second, the level of public understanding of basic scientific concepts is relatively low in the four political systems studied. According to our index of civic scientific literacy, only one in ten adults in the United States was well informed, or civic scientifically literate. The proportions of citizens in the European Union, Japan, and Canada who qualified as civic scientifically literate was even lower. This finding raises serious questions about the ability of citizens to comprehend the arguments in serious controversies involving science and technology and makes a strong case for renewed efforts to improve the quality and effectiveness of science education in all four political systems.

Third, it is important to recognize that two schemas, the promise of science and technology and reservations about the impact of science and technology, operate simultaneously in the minds of most individuals in modern industrial societies. Some issues may activate one schema, and other issues may activate the other schema. Our analysis shows that substantial majorities of adults in Canada, the European Union, and the United States continue to hold a positive schema for science and technology, reflecting a positive assessment of the achievements and promise of science. Whereas relatively few Americans report reservations about the impact of science and technology, a substantial proportion of Canadian, EU, and Japanese respondents express some reservations about the present and future impact of science and technology. It is essential to recognize that these concerns or reservations coexist with even higher levels of expectation, except in Japan, and do not represent an antiscience sentiment as much as a wariness. Given the broad public recognition of the past and potential power of science and technology and the demonstrated low levels of understanding, some level of wariness should not be surprising.

Fourth, in the marketplace for time in complex modern societies, science and technology policy must compete with other public policy issues and with a wide array of attractive apolitical activities for the attention of individual citizens. Inevitably, a process of political and issue specialization occurs in which some individuals decide to allocate no time or attention to political issues generally, which is reflected in the relatively low rates of voter participation in many democratic political systems. Among those citizens who choose to focus some of their time and attention on public policy matters, there is an inherent need to select among the vast array of issues facing modern governments. Our analysis indicates that about one in ten Americans, Canadians, and the members of the European Union are attentive to science and technology policy issues. In all four political systems, citizens who are attentive to science and technology policy generally have significantly stronger positive schema toward science and significantly lower concerns about science and technology than other citizens.

Finally, the general attitudes of each individual toward science and technology, as expressed in these two schemas, interact with background factors such as age, gender, educational attainment, scientific literacy, and issue attentiveness to produce more specific policy attitudes. A series of structural equation models document the important intermediary role of these educational, knowledge, attentiveness, and attitude variables in individual policy preferences with regard to government spending for basic scientific research in Canada, the European Union, Japan, and the United States. It is useful to note that many of the most influential factors are educational and demographic in character and, thus, are not easily influenced in the short run by information or advertising campaigns. The implications of this line of analysis are central to our understanding of the formation, maintenance, and change of public attitudes and should be explored in future studies of the linkage between general schemas and specific policy preferences.

References

Ahmann, S. (1975). An exploration of survival levels of achievement by means of assessment techniques. In D. M. Nielsen & H. F. Hjelm (Eds.), *Reading and career education* (pp. 38–42). Newark, DE: International Reading Association.

Almond, G. A. (1950). *The American people and foreign policy.* New York: Harcourt, Brace and Company.

Almond, G. A. (1983). Introduction. In J. D. Miller, *The American people and science policy* (pp. xv–xvi). New York: Pergamon Press.

Bock, R. D., & Aitken, M. (1981). Marginal maximum likelihood estimation of item parameters: Application of an EM-algorithm. *Psychometrika, 46,* 443–459.

Bock, R. D., & Zimowski, M. F. (1977). Multiple-group IRT. In W. J. van der Linden & R. K. Hambleton (Eds.), *Handbook of modern item-response theory* (pp. 433–448). New York: Springer.

Bosch, G.; Dawkings, P., & Michon, F. (Eds.) (1993). *Times are changing: Working time in 14 industrialized countries.* Geneva: International Institute for Labor Studies.

Bosso, C. J. (1987). *Pesticides and politics: The life cycle of a public issue.* Pittsburgh: University of Pittsburgh Press.

Carson, R. (1962). *Silent spring.* New York: Houghton Mifflin.

Cevero, R. M. (1985). Is a Common Definition of Adult Literacy Possible? *Adult Education Quarterly, 36,* 50–54.

Christopherson, S. (1991). Trading time for consumption: The failure of work-hours reduction in the United States. In K. Hinrichs, W. Roche, & C. Sirianni (Eds.), *Working time in transition: The political economy of working hours in industrial nations* (pp. 171–187). Philadelphia: Temple University Press.

Converse, J. M., & Schuman, H. (1984). The manner of inquiry: An analysis of question form across organizations and over time. In C. F. Turner & E. Martin (Eds.), *Surveying subjective phenomena* (pp. 283–314). New York: Russell Sage Foundation.

Cook, W. D. (1977). *Adult literacy education in the United States.* Newark, DE: International Reading Association.

Costner, H. L. (1965). Criteria for measures of association. *American Sociological Review, 30,* 341–353.

Davis, R. C. (1958). *The Public Impact of Science in the Mass Media* (Monograph No. 25). Ann Arbor, MI: University of Michigan Survey Research Center.

Dillman, D. (1978). *Mail and telephone surveys: The total design method.* New York: Wiley.

Durant, J. R., Evans, G. A., & Thomas, G. P. (1989). The public understanding of science. *Nature, 340*, 11–14.

Durant, J. R., Evans, G. A., & Thomas, G. P. (1992). Public understanding of science in Britain: The role of medicine in the popular representation of science. *Public Understanding of Science, 1*, 161–182.

Evans, G. A., & Durant, J. R. (1995). The relationship between knowledge and attitudes in the public understanding of science in Britain. *Public Understanding of Science, 4*, 57–74.

Freudenburg, W. R., & Rosa, E. A. (1984). *Public reactions to nuclear power.* Boulder, CO: Westview Press.

Goodman, L. A., & Kruskal, W. H. (1954). Measures of association for cross-classifications. *Journal of the American Statistical Association, 49*, 732–764.

Guthrie, J. T., & Kirsch, I. S. (1984). The emergent perspective on literacy. *Phi Delta Kappan, 65*, 351–355.

Harman, D. (1970). Illiteracy: An overview. *Harvard Educational Review, 40*, 226–230.

Hayduk, L. A. (1987). *Structural equation modeling with LISREL.* Baltimore: The Johns Hopkins University Press.

Hughes, T. P. (1989). *American genesis: A century of invention and technological enthusiasm, 1970–1970.* New York: Viking Press.

Hughes, M. A., & Garrett, D. E. (1990). Intercoder reliability estimation approaches in marketing: A generalizability theory framework for quantitative data. *Journal of Marketing Research, 27*, 185–195.

Jöreskog, K., & Sörbom, D. (1993). *LISREL 8.* Chicago: Scientific Software International.

Kaase, M., & Newton, K. (1995). *Beliefs in government.* Oxford, England: Oxford University Press.

Kaestle, C. F. (1985). The history of literacy and the history of readers. In E. W. Gordon (Ed.), *Review of research in education* (Vol. 12, pp. 11–53). Washington, DC: American Educational Research Association.

Kuhn, T. S. (1962). *The structure of scientific revolutions.* Chicago: University of Chicago Press.

Labaw, P. J. (1980). *Advanced questionnaire design.* Cambridge, MA: Abt Books.

Ladd, E. C., & Bowman, K. H. (1996). *Public opinion in America and Japan: How we see each other and ourselves.* Washington, DC: American Enterprise Institute Press.

Lau, R. R., & Sears, D. O. (Eds.). (1986). *Political cognition.* Hillsdale, NJ: Lawrence Erlbaum Associates.

Milburn, M. A. (1991). *Persuasion and politics: The social psychology of public opinion.* Pacific Grove, CA: Brooks/Cole.

Miller, J. D. (1983a). *The American people and science policy.* New York: Pergamon Press.

Miller, J. D. (1983b). Scientific literacy: A conceptual and empirical review, *Daedalus, 112(2)*, 29–48.

Miller, J. D. (1987). Scientific literacy in the United States. In D. Evered & M. O'Connor (Eds.), *Communicating Science to the Public* (pp. 19–40). London: Wiley.

127

Miller, J. D. (1989, January). *Scientific literacy*. Paper presented at the annual meeting of the American Association for the Advancement of Science, San Francisco.

Miller, J. D. (1995). Scientific literacy for effective citizenship. In R. E. Yager (Ed.), *Science/technology/society as reform in science education* (pp. 185–204). New York: State University Press of New York.

Miller, J. D. (1996). *Public understanding of science and technology in 14 OECD countries: A comparative analysis*. Paper presented to a symposium of the Public Understanding of Science and Technology, OECD, Tokyo.

Miller, J. D., Pardo, R., & Niwa, F. (1997). *Public perceptions of science and technology: A comparative study of the European Union, the United States, Japan, and Canada*. Madrid: BBV Foundation Press.

Miller, J. D., & Pifer, L. K. (1995). *The public understanding of biomedical sciences in the United States*. Final Report to the National Institutes of Health. Chicago: Chicago Academy of Sciences.

Minsky, M. (1974). *A framework for representing knowledge* (MIT Artificial Intelligence Memo 252). Cambridge, MA: Massachusetts Institute of Technology (MIT) Press.

Minsky, M. (1986). *The society of mind*. New York: Simon and Schuster.

Mintz, M. (1965). *The therapeutic nightmare*. New York: Houghton Mifflin.

Montgomery, A. C., & Crittenden, K. S. (1977). Improving coding reliability for open-ended questions. *Public Opinion Quarterly, 41*, 235–243.

Morone, J. G., & Woodhouse, E. J. (1989). *The demise of nuclear energy?* New Haven: Yale University Press.

National Science Board (NSB). (1981). *Science indicators—1980*. Washington, DC: U.S. Government Printing Office.

National Science Board (NSB). (1983). *Science indicators—1982*. Washington, DC: U.S. Government Printing Office.

National Science Board (NSB). (1986). *Science indicators—1985*. Washington, DC: U.S. Government Printing Office.

National Science Board (NSB). (1988). *Science and engineering indicators—1987*. Washington, DC: U.S. Government Printing Office.

National Science Board (NSB). (1990). *Science and engineering indicators—1989*. Washington, DC: U.S. Government Printing Office.

National Science Board (NSB). (1992). *Science and engineering indicators—1991*. Washington, DC: U.S. Government Printing Office.

National Science Board (NSB). (1994). *Science and engineering indicators—1993*. Washington, DC: U.S. Government Printing Office.

National Science Board (NSB). (1996). *Science and engineering indicators—1996*. Washington, DC: U.S. Government Printing Office.

Nelkin, D. (1977). *Technological decisions and democracy: European experiments in public participation*. Beverly Hills, CA: Sage.

Northcutt, N. W. (1975). Functional literacy for adults. In D. M. Nielsen & H. F. Hjelm (Eds.), *Reading and career education* (pp. 43–49). Newark, DE: International Reading Association.

Oppenheim, A. N. (1966). *Questionnaire design and attitude measurement*. New York: Basic Books.

Perreault, W. D., & Leigh, L. E. (1989). Reliability of nominal data based on qualitative judgements. *Journal of Marketing Research, 26*, 135–148.

Pick, H. L., van den Broek, P., & Knill, D. C. (Eds.). (1992). *Cognition: Conceptual and methodological issues*. Washington, DC: American Psychological Association.

Popkin, S. L. (1994). *The reasoning voter*. Chicago: University of Chicago Press.

Popper, K. R. (1969). *The logic of scientific discovery*. New York: Basic Books.

Resnick, D. P., & Resnick, L. B. (1977). The nature of literacy: An historical exploration. *Harvard Educational Review, 47,* 370–385.

Ressmeyer, T. J. (1994). *Attentiveness and mobilization for science policy.* Unpublished doctoral dissertation, Northern Illinois University, De Kalb.

Rosenau, J. (1961). *Public opinion and foreign policy: An operational formulation.* New York: Random House.

Rosenau, J. (1963). *National leadership and foreign policy: The mobilization of public support.* Princeton, NJ: Princeton University Press.

Rosenau, J. (1974). *Citizenship between elections.* New York: Free Press.

Rutherford, F. J., & Algren, A. (1989). *Science for all Americans.* Washington, DC: American Association for the Advancement of Science.

Schank, R. (1977). *Scripts, plans, goals, and understanding.* Hillsdale, NJ: Lawrence Erlbaum Associates.

Sudman, S., & Bradburn, N. M. (1982). *Asking questions.* San Francisco: Jossey-Bass.

Verba, S., Schlozman, K. L., & Brady, H. E. (1995). *Voice and equality: Civic voluntarism in American politics.* Cambridge, MA: Harvard University Press.

Zadeh, L. A. (1987). Fuzzy sets. In R. R. Yager, S. Ovchinnikow, R. M. Tong, & H. T. Nguyen (Eds.), *Fuzzy Sets and Applications: Selected Papers by L. A. Zadeh* (pp. 29–44). New York: John Wiley & Sons.

Zimowski, M. F., Muraki, E., Mislevy, R. J., & Bock, R. D. (1996). *BILOG-MG: Multiple-group IRT analysis and test maintenance for binary items.* Chicago: Scientific Software International.

CHAPTER 5

TWO CULTURES OF PUBLIC UNDERSTANDING OF SCIENCE AND TECHNOLOGY IN EUROPE*

John Durant, Martin Bauer, George Gaskell, Cees Midden, Miltos Liakopoulos, and Liesbeth Scholten

There is much debate about the nature of scientific and technological culture in Europe. All too often, however, such debate is dominated by what is often termed arm-chair philosophy—a mixture of ill-informed speculation, ex-cathedra pronouncement, and special pleading. Only occasionally are the results of disciplined social research brought to bear in the discussion. The Eurobarometer is an important source of empirical insights into European scientific and technological culture. It is a random-sample social survey that is fielded regularly in all member states of the European Union (EU). As such, it is a rich resource for the comparative assessment of trends in public perceptions. In particular, it offers the potential for development as a reliable quantitative "indicator" of trends in European public perceptions of science and technology.

In order to fulfill this potential, it is necessary both to develop a suitable set of questions on science and technology for regular inclusion in the Euro-barometer and to undertake rigorous statistical analysis of survey results within the context of a coherent conceptual framework. Question sets on science and technology were included in Eurobarometer surveys in 1989 and 1992. These question sets were devised to be broadly comparable with earlier "indicator" survey studies in a number of other countries, including Canada and the United States. Partly for this reason, the Eurobarometer question sets gave pride of place to formal scientific knowledge (sometimes termed *scientific literacy*) in the study of public perceptions. Implicitly, they appear to have been based upon a so-called deficit model of the public understanding of science, according to which there is a direct relationship between knowledge and attitudes, with greater knowledge tending to produce higher levels of support for science and technology.

* Funded by The European Commission: Contract VETE-CT-94-0001-GB.

In this chapter, we offer an alternative framework for considering the issue of public perceptions of science and technology in Europe. This framework is drawn from macrosociological models of the industrialization process. By analyzing the Eurobarometer data within this framework, we show that the central assumption of the deficit model is supported for only a subset of the least industrialized member states of the EU. Faced with this partial disconfirmation of the deficit model, we offer a very different process model based on the transition from industrial to postindustrial society. We show that this model successfully predicts a number of important features of the 1992 Eurobarometer data and (its relative complexity notwithstanding) that it can be used to construct hypotheses about the probable trajectory of public perceptions of science and technology in different national contexts.

Thus, we use the results of an empirical study to revise the assumptions on which that study was originally conducted. We conclude by drawing out the implications of our results both for our understanding of the public understanding of science and for future research that uses the Eurobarometer in studies of this important subject.

The Eurobarometer

This study is a secondary analysis of data collected under the supervision of the Directorate General XII (Science, Research and Development) of the European Commission, in conjunction with the Eurobarometer Program of Directorate General X (Audiovisual Media, Information, Communication and Culture) of the European Commission (EC). Two surveys are considered. The first was undertaken in 1989; the second, in 1992. Both surveys were conducted with multistage, random probability samples of approximately 1,000 respondents drawn from each of the member states of the European Union; both surveys covered a rather wide range of topics, including public interest in, knowledge of, and attitudes toward science and technology; and both surveys were designed to be at least partially comparable with previous survey studies conducted in, for example, Canada, the United Kingdom, Japan, and the United States.

General Approach

We start with some remarks on our general approach to the task of analyzing public perceptions of science and technology in Europe. We are methodological pluralists. In our view, survey methodology is suited to some tasks in social science but less so to others. In the main, social surveys are useful sources of general indicator information, especially when comparative and time-series data are available. Social surveys are less well suited to fine-grained contextual analysis. To take a convenient analogy, we may say that social surveys are good at providing low-resolution portraits of the broad panorama, particularly because these portraits may compare with one another or change through time, but that they are bad at picking up subtler details of meaning, the delicate lights and shades that can be recognized within scenes only by close inspection. However, the panorama of the survey may be elaborated with more detail and meaning by the concurrent use of contextual information.

Surveys necessarily impose a rigid observational "grid," the question set, upon the social world. As a result, they detect some things and miss others. One may say that social surveys are like maps: They are representations of the public for particular audiences and purposes. Providing that this fact is clearly recognized, it need not be a serious weakness. In our work, we assume that, at best, the Eurobarometer data on public understanding of science and technology in Europe provide only a partial picture of what is inevitably the very much more complex reality of popular science. Nonetheless, the partial pictures we have in the two surveys under consideration are remarkable not least for being snapshots of what were at the time all twelve EU member states.

Survey data bear a representational relationship to the phenomena that they purport to measure. For example, responses to single attitudinal items (or even related clusters of items) are not necessarily accurate reflections of psychological "attitudes;" and psychological attitudes, in turn, are not at all the same as what is generally referred to as public opinion. Throughout this kind of research, it is vital to keep in mind that survey data do not speak for themselves. First, the data arise from question sets that embody particular theoretical assumptions and hypotheses. Change the assumptions, and both the questions and the data also change. Second, even relatively subtle changes in question ordering or wording can have substantial effects on measured results. Change the order or the wording, and once again the data may also change (Gaskell, Wright, & O'Muircheartaigh, 1993a, 1993b). Third, single-item responses are generally unreliable. Multivariate statistical analysis is usually required before any conclusions can be safely drawn from a given set of data. Lastly, survey results do not possess obvious and inherent meaning. What may appear to be a high figure to one commentator may seem a distressingly low one to another, and what may seem a solid basis for action to one policy-maker may seem like an insecure foundation to another.

Ideally, social surveys should be conceived, designed, fielded, analyzed, and reported within a single conceptual and theoretical frame of reference. Certain questions are raised; ways of addressing them are devised and tested, hopefully through an iterative process that yields increasingly robust measures; the main empirical work is conducted; and the questions are resolved. This procedure has not been followed with the Eurobarometer surveys with which we are concerned here. An implicit deficit model of the public understanding of science appears to inform the surveys, but beyond this notion no very consistent research agenda appears to be in operation. Given that quantitative indicators of public understanding of science and technology are still, relatively speaking, in their infancy, this lack of conceptual foundation is scarcely surprising. However, it is necessary to recognize that our two surveys constitute extremely large and complex data sets whose theoretical foci and practical bearings are at best unclear.

Faced with this situation, it should be apparent that we are not in the position of laboratory scientists testing carefully developed hypotheses. Rather, our position is more like that of the explorer who has come across an intriguing and little-known territory. We have chosen to investigate this territory with a view to the identification of what seem to us to be potentially useful map markers (indicators) that may help future explorers orient themselves better in what is frankly rather difficult terrain. Ideally, we will find indicator measures of European public perceptions of science that are statistically valid, sociologically meaningful, and practically useful to the scientific and the science-policy-making communities. Clearly, this ideal is somewhat lofty. Frankly, we shall be reasonably content if we manage to make some sense of the information by way of pattern-matching.

In summary, our approach to the data analysis aims to avoid being confined by whatever assumptions may have informed the development of the Eurobarometer to date. So far as possible, we attempt to be aware of and critical about our own theoretical assumptions, the assumptions that are embodied in the Eurobarometer data, the strengths and weaknesses of the research methods that we are using, and the tensions that are inherent in the project of mapping public understanding of science and technology for the primary benefit of the scientific and the science-policy-making communities.

Large data sets of the kind with which we are concerned here can be subjected to an almost indefinite variety of secondary statistical analysis. In order to make our task practicable, we have been obliged to make a number of strategic decisions. The most important of them is the decision to concentrate upon the 1992 data set and to leave aside at this stage longitudinal trend analysis. We have several reasons for taking this decision. First, the unique strength of this Eurobarometer data lies in the opportunity that it provides for systematic 12-nation comparison, which is best conducted within a single data set. Second, two data points (1989, 1992) provide very limited opportunities for significant trend analysis. Third, differences in the two questionnaires (especially in the field of attitudes, where our interest is great) set limits to the amount of trend analysis that is possible. Fourth, the primary need is to improve the survey instrument so as to provide a sound basis for comparative and trend analyses in the future. It seems desirable, in particular, that Europe should have a reliable indicator of public understanding of science and technology, as part of science indicators, analogous to that which has been provided for the United States by the National Science Foundation for many years (e.g., National Science Board, 1991). Improvement of the Eurobarometer instrument for this purpose is best done by analyzing the 1992 data set alone.

Models of Public Understanding of Science

Public understanding of science and technology is a relatively young research field. Nevertheless, a certain amount of critical attention has been given to the question of modeling the relationship between science and the public. We find particularly helpful the approach adopted by Neidhardt, who considers the public as a communication system comprising speakers, mediators, and audiences (Neidhardt, 1993). Within such a system, the reciprocal perceptions that exist may be considered "representations" at any one time between the scientific community (speakers), journalists (mediators) and the public (audiences). Leaving aside the journalists, who are not a principal object of attention in the present study, it seems helpful to ask how scientific speakers view public audiences as well as how public audiences view science and scientists. The present survey focuses on the latter perspective.

A number of commentators and critics have noted that much research in the public understanding of science, including most of the existing survey research, has tended to reflect a particular perspective on the public, according to which the chief object of interest is the public's absolute level of objective scientific knowledge (scientific literacy). Within what we term the deficit model of the public understanding of science, science is seen as a body of relatively unproblematic knowledge; and the public is seen as a body of more or less ignorant laypeople. Reducing the public's ignorance is presumed to be the key practical requirement, not least because the deficit model rests on the general presumption that a greater quantity of knowledge is associated with a higher level of support for the scientific enterprise

(Durant, Evans, & Thomas, 1992; the presumption that knowledge produces support is analyzed in Evans & Durant, 1995).

The multiple weaknesses of the deficit model have been rather widely noted. They include the failure to critically examine science itself, the failure to critically examine the relationship between professional and popular representations of science, the failure to acknowledge the role of "informal" or local knowledge, and the failure to recognize the irrelevance of scientific knowledge in many social settings (see, for example, Durant et al., 1992; Wynne, 1993; Ziman, 1991). In the present context, we note also that it is an empirical question whether and to what extent more knowledge about science and technology tends to be associated with more support for science and technology in the public domain.

The Eurobarometer surveys under consideration here contain substantial batteries of questions on both knowledge of and attitudes toward science and technology. It is of considerable interest, therefore, to explore the internal relationships between knowledge and attitudes within the data in order to test this aspect of the deficit model.

By concentrating on the issue of formal knowledge, the deficit model tends toward what may be termed an educational perspective. We note, however, that there are numbers of different "interfaces" between science and the public. Luhmann (1979) characterized "modern" societies as functionally differentiated into various spheres such as the economy, science, education, politics, art, and religion. Each of these spheres of activity may interface with others, and public opinion may be one of the linking processes. Thus science interfaces with the economy, education, and politics, among other spheres. At these different interfaces the public appears in different guises, as producers and consumers, as teachers and students, and as politicians and citizens, respectively. These different interfaces provide a heuristic device with which to judge the adequacy of a survey instrument concerned with the relations between science and the public. Acknowledging one or more of the interfaces in the design of the Eurobarometer instrument appears to be in accord with the survey's public nature.

The Eurobarometer survey instrument is constructed with the educational perspective in mind. This orientation makes some questions easier and more obvious to ask than others. For example, the educational perspective of the Eurobarometer leads it to neglect measures of citizen participation and trust in institutions. Nevertheless, for the reasons given in the preceding section, we are not obliged to remain uncritically dependent upon the implicit model contained within the data itself. By critical exploration of the data, we may hope to transcend the limitations of the deficit model and discern, at least in outline, other significant interfaces of European scientific culture.

Scientific Cultures and Levels
of Industrialization in Europe

As we have noted, it is a peculiar strength of the Eurobarometer data that they represent simultaneous snapshots of several western European countries. We know of no other coherent data sets in the field of research on the public understanding of science that provide such great opportunities for international comparative analysis. For this reason, we concentrate in this chapter on the comparative analysis of what is often referred to as European scientific culture, as it is partially revealed within the 1992 data set. In the first instance, we are faced by an obvious question: Is there a single European scientific culture, or are there several European scientific cultures?

Even the most superficial inspection of the EU member states reveals a wide variation in levels of scientific, technological, and industrial development as measured by standard international indices. For example, Denmark, Germany, and the Netherlands rank very high in the international table of industrial development, while Greece, Ireland, and Portugal rank much lower (Bairoch, 1982; Eurostats, 1992). The obvious fact of different levels of industrial development within the EU invites further exploration within the Eurobarometer data set.

The sociological literature on the dynamics of industrial development provides a useful context for such exploration. In recent years, much has been written about the cultural, economic, social, and political changes associated with the transition from industrial to what may be termed postindustrial society. In phrases such as *postindustrialism, postmaterialism,* and *the risk society,* sociologists have sought to evoke different axes of this transition. Postindustrialism emphasizes changes in science-based industries (Bell, 1973; Touraine, 1969), postmaterialism emphasizes changing cultures and value-structures (Inglehart, 1990; Rose, 1991), and risk society emphasizes globalization of risk and its distribution (Beck, 1986/1992; Boehme, 1993). In each case, the claim has been made that major structural transformations accompany the continued development of industrial society. Without theoretical preference, we note that the different theories articulate an expectation of a significant transition that is likely to happen among the most developed countries.

Macrosociological theory concerning this transition along a measure of levels of industrialization provides a useful context for interpreting our Eurobarometer data. We elaborate a tentative model of scientific culture(s) and industrialization in Europe (Bauer, 1995; Bauer, Durant, & Evans, 1994). The model projects a two-dimensional matrix within which EU member states may be located. The first dimension represents the hypothetical process of transition from industrial to postindustrial society, and the second dimension represents what may be termed ordered deviation from the idealized path of this hypothetical transition in response to local (i.e., national) circumstances. In a sense, the first dimension attempts to capture a general process, and the second dimension attempts to take account of the particularities of place and time.

For the sake of clarity, the major features of the model's first dimension, the hypothetical transition, are set out in Table 5-1 as a dichotomy of industrial and postindustrial society. In industrial society, it is expected that scientific knowledge is confined to a relatively small, well-educated elite; socioeconomic factors play a relatively large part in determining the distribution of scientific knowledge; interest in science (as a symbol of industrial and economic progress) is relatively high in all sections of the community; a relatively unified canon of popular scientific knowledge exists; and there is a positive correlation between knowledge of and support for science. The image here is of a society in which science, having achieved only a limited penetration into society, is extensively idealized as the preferred route to economic and social progress.

Table 5-1. The Hypothetical Transition from Industrial to Postindustrial Society

Industrial society	Postindustrial society
1. Scientific knowledge is confined to a small social elite.	1. Scientific knowledge is widely distributed in society.
2. Scientific knowledge is strongly socioeconomically stratified.	2. Scientific knowledge is weakly socioeconomically stratified.
3. Public interest in science is relatively high.	3. Public interest in science is relatively low.
4. Popular scientific knowledge is "unified."	4. Popular scientific knowledge is specialized.
5. The relationship between knowledge and support is positive.	5. The relationship between knowledge and support is increasingly chaotic.

By contrast, in postindustrial society it is expected that scientific knowledge is more widely distributed in society; sociodemographic factors play a relatively small part in determining the distribution of scientific knowledge; interest in science is relatively low, for science is rather generally taken for granted as part of the everyday stock of knowledge; there is a proliferation of specialist knowledges in the public domain; and the relationship between knowledge of and support for science becomes chaotic, for different sections of the community develop different points of view on particular issues. The image is that of a society in which science, having already achieved a high level of penetration into society, is not so much idealized as it is critically evaluated by a public that expects to obtain continuing benefits but is also increasingly alert to the possibility of problems or disbenefits. Thus there is an apparent paradox: As the public understanding of science expands and knowledge increases and is taken for granted, science becomes more problematic in public opinion and scientists lose confidence in the unified mission of science.

Before going any further, we wish to enter two cautions concerning the hypothetical contrast between industrial and postindustrial society, which constitutes the heart of our model. First, in using such a model, we are making a radical simplification that accentuates some significant features of what is in reality an extremely complex situation. Second, the model postulates a temporal process (the transition from industrial to postindustrial society), whereas the Eurobarometer data that it is intended to illuminate are entirely cross-sectional. The crucial assumption that we are making is that different EU member states may be regarded as being at different stages of the hypothetical transition. We recognize that this assumption is problematic and that the model's proposed second dimension can go only part of the way toward taking proper account of the diversity of cultural traditions and social circumstances across Europe. In both cases, however, the justification for making radical simplifications is that by doing so we obtain potentially illuminating and testable predictions that provide a rationale for interrogating the present and future Eurobarometer data. Thus the model is justified in terms of its heuristic value, both for analyzing the present data set and for suggesting research questions for the analysis of longitudinal data, that is to say, scenarios for future prospects in different countries.

The Indicator Scales

Using the 1992 Eurobarometer data set, we proceed with the test of the model by considering in turn each contrast in Table 5-1. For this purpose, we use a series of scalar measures of interest in science, knowledge of science, and three clusters of attitudes toward science that are derived from various Eurobarometer questions, as shown in Table 5-2.

Table 5-2. Indicators of Public Perceptions of Science and Technology

Indicator	Source questions	Statistical description
Interest in science	Two questions: "interest in new inventions and technologies" "interest in new scientific discoveries"	Range 1–6 $M = 4.28$; $SD = 1.42$
Knowledge of science	12 factual questions	Range = 0–12 $M = 6.83$; $SD = 2.66$
Attitudes toward science and technology; three attitudinal clusters:	23 attitudinal items, principal components analysis with varimax rotations: (see appendix for the items loading on each of the three factors.)	
1. Progress		Range 1–5; $M = 3.68$; $SD = .74$
2. Panacea		Range 1–5; $M = 2.32$; $SD = .82$
3. Future shock		Range 1–5; $M = 3.38$; $SD = .88$

In order to test our dynamic model, we use a socioeconomic indicator of industrialization developed by Bairoch. Bairoch (1982) offers a comparable index of the per capita industrial output for 1980 (Great Britain 1900 = 100). This indicator correlates highly with other indicators such as GDP 1991 ($r = .90$) (Eurostats, 1992), R&D expenditure as percentage of GDP in 1991 ($r = .84$), and national aggregates of Inglehart's postmaterialism index ($r = .82$) (Inglehart, 1990). In the following explorations of the model, we fit linear and curvilinear plots onto data based on national aggregates.

Testing the Industrialization Hypothesis

We may now test our model of the transition from industrial to postindustrial society. First, we consider the distribution of scientific knowledge in the general public. Common sense suggests that as science and technology come to have a more central place in industrial society, scientific knowledge will tend to spread among the general public. Our expectation, then, is that through the hypothetical transition the average level of the general public's scientific knowledge will increase. This shift is indeed what we find across the EU (see Figure 5-1).

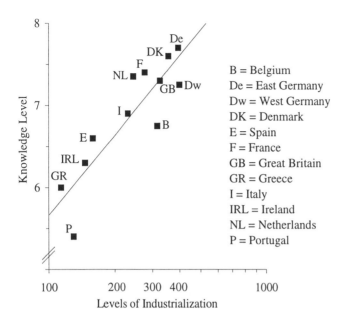

B = Belgium
De = East Germany
Dw = West Germany
DK = Denmark
E = Spain
F = France
GB = Great Britain
GR = Greece
I = Italy
IRL = Ireland
NL = Netherlands
P = Portugal

Figure 5-1. Relation between the level of public scientific knowledge and the level of industrialization in the European Community (data base: Euro-barometer, 1992).

Similarly, our model predicts that with increasing levels of industrialization scientific knowledge will come to be more widely dispersed among the public. We call this the *normalization hypothesis* because it leads us to expect a bimodal distribution of knowledge in less industrial societies compared with a more normal distribution of knowledge in postindustrial societies. This hypothesis is confirmed in two separate analyses showing that (a) sociodemographic variables are better predictors of knowledge in the less industrialized countries than in advanced industrialized countries, and (b) the variance of knowledge increases across the hypothetical industrialization dimension (see Table 5-3).

A further implication of normalization concerns the public's interest in science and technology. When scientific knowledge becomes generally distributed, it might be expected to become more taken for granted. This change of public attitude, in turn, suggests that levels of interest in science and technology might decline in the course of the hypothetical transition. In fact, the relationship between interest and knowledge declines through the hypothetical transition. However, the direct relationship between interest and industrialization is more complex: from low to medium levels of industrialization,

Table 5-3. Testing the Industrialization Hypothesis: The Relation between Knowledge and Attitudes

Measure	GR	P	IRL	S	I	NL	F	B	GB	DK	D-W
Industrialization: industrial output per capita	114	130	147	159	231	245	277	316	325	356	395
Prediction knowledge from sociodemographics (R^2 of regression)	.46	.35	.29	.39	.37	.35	.25	.23	.35	.27	.24
Variance of knowledge	2.74	2.74	2.87	2.85	2.48	2.56	2.46	2.40	2.62	2.37	2.41
Percentage explained by first knowledge factor	29.9	39.4	29.9	25.8	21.1	17.6	17.7	17.3	19.8	15.3	21.4
Knowledge and "progress": Partial correlation	.05	.12	.07	.08	.08	.01	.01	.04	−.02	−.05	.01
Knowledge and "panacea": Partial correlation	−.08	−.15	−.22	−.21	−.23	−.27	−.23	−.17	−.29	−.18	−.23
Knowledge and "future shock": Partial correlation	.03	.13	−.03	−.01	.03	−.01	−.01	−.03	−.04	−.08	−.02
General scientific knowledge: Mean	6.00	5.41	6.16	6.60	6.91	7.32	7.34	6.74	7.28	7.61	7.23
Environmental knowledge: Mean	2.60	2.64	3.26	2.88	3.05	4.06	3.19	3.41	3.63	3.88	3.85
Biotechnological knowledge: Mean	4.55	4.60	5.36	5.18	5.62	6.93	6.71	5.96	6.89	7.34	6.06

Note. B = Belgium; DK = Denmark; D-W = West Germany; F = France; GB = Great Britain; GR = Greece; I = Italy; IRL = Ireland; NL = the Netherlands; P = Portugal; S = Spain. Data base: *Eurobarometer 1992* and *Eurobarometer 39.1—Biotechnology and genetic engineering: What Europeans think about it,* by Institut National de Recherche Agricole (INRA) & E. Marlier, 1993, Report DG XII/E/1, Brussels: Commission of the European Communities.

interest levels rise; but interest levels again fall away as the postindustrial society approached (see Figure 5-2). This pattern suggests that we should distinguish between two qualitatively different forms of low interest in science and technology: "disinterest out of ignorance" in industrial societies and "disinterest out of familiarity" in postindustrial societies.

It is well known that the expansion of expert knowledge systems is associated with increasing levels of specialization. As the expert knowledge base becomes increasingly differentiated, individual experts come to know more and more about less and less, creating a paradoxical situation in which ignorance increases alongside knowledge. Bauer and Durant have argued that this specialization of knowledge is not confined to experts but is increasingly found at the level of popular knowledge (Bauer et al., 1994). To see whether this specialization tends to happen through the hypothetical transition, we can investigate the structure of the knowledge scale in relation to levels of industrialization. Our expectation is that as the public begins to develop particular interests and knowledges, a general or composite knowledge scale will become less adequate. In fact, this change is the case. The percentage of explained variance of a single factor on a number of knowledge items decreases as we move through the hypothetical transition toward postindustrial society. In Portugal a single-factor solution explains 40% of the knowledge variance as compared to 15% in Denmark.

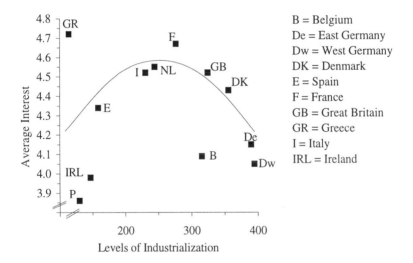

Figure 5-2. Relation between level of public interest in science and level of industrialization in the European Community (data base: Eurobarometer, 1992).

We turn now to the crucial question of the relationship between knowledge and attitudes. As we have indicated, a naive application of the deficit model leads us to expect those who know more science to be more supportive of it. In short, this model leads us to predict that knowledge and attitudes are positively correlated. A number of studies have demonstrated such a positive correlation on the national level (Evans & Durant, 1995). This having been said, simple measurement of national averages masks a great deal of interesting information at both the national and the international levels. Evans & Durant (1995) have argued that with increasing knowledge,

143

individuals develop clearer and more strongly held attitudes (both positive and negative) on scientific issues.

If this relationship is a general phenomenon, then we might expect (a) attitudinal variance to increase with increasing levels of knowledge and (b) the relationship between knowledge and attitudes toward science and technology to become less predictable as levels of knowledge increase. Both of these expectations are confirmed by our analyses. As levels of knowledge increase across the countries, so does the variance of the attitudes toward "progress" and "future shock." On the relationships between knowledge and attitudes, we analyzed the partial correlations, controlling for all the socio-demographic variables (see Table 5-3). For knowledge and progress, panacea, and future shock, the results are very similar; the relationship is small and statistically insignificant across the hypothetical transition. These results lend weight to the idea that the relationship has become chaotic. In a post-industrial context, it seems that the common-sense deficit model, according to which more knowledge brings more support, is not confirmed in any very obvious or straightforward fashion.

The presence within the 1992 Eurobarometer of a battery of questions on knowledge of the environment allows us to compare knowledge in general and in specific scientific domains. Furthermore, the comparison may be extended by including the results of the 1993 Eurobarometer Biotechnology Survey (Institut National de Recherche Agricole [INRA] & Marlier, 1993). Table 5-3 shows the mean scores for each country on general scientific knowledge, knowledge of the environment, and knowledge of biotechnology, ordered with respect to the industrialization index. Although there are clearly some outliers, the general tendency is for knowledge on all three indicators to increase with increasing levels of industrialization. The regression of general knowledge on environmental knowledge shows $r^2 = .60$; the regression of environmental knowledge on biotechnological knowledge shows an $r^2 = .73$. These positive correlations between general and specific knowledge are not in themselves surprising. However, the systematic links between scientific knowledge and levels of industrialization reinforce the significance of industrialization as a key variable in understanding scientific culture in Europe.

Science and Society: Assimilation and Accommodation

As reported earlier from the attitudinal items in Eurobarometer, three clusters of items were found that were labeled *progress, panacea,* and *future shock*. Further analysis of the second and third clusters showed a strikingly similar curvilinear fit to the index of industrialization. This finding suggested that the statistically derived clusters labeled panacea and future shock might be indicative of a single attitudinal orientation toward science. This interpretation was confirmed by multivariate analyses where a single cluster with acceptable scaling characteristics was found. It captures the public's sense of pessimism versus optimism about science. The optimists agree that science can play a role in industrial development and in protection of the environment and do not believe that science is making modern life change too fast. The pessimists take the contrary view. For them science has a limited role in industrial development, will not solve environmental problems, and threatens modern life. For convenience this indicator is labeled pessimism.

More generally, the two belief clusters, progress and pessimism, may be regarded as different beliefs about the relations between science and society. Progress and optimism are the views of those who see science as working for humanity. To them, science as an enterprise is under control and is contributing to the economy, the environment, and improved life styles. In Piaget's terms, they believe that human beings can assimilate science within aspirations for development. By contrast, there are people who do not think that science is contributing to progress and who are pessimistic because they believe that science is making modern life change too rapidly. Such a view represents the Piagetian pressure of *accommodation,* the conviction that society is at the mercy of science, because *in extremis* science is working against humankind (Piaget, 1967/1971, pp. 172–182).

The Piagetian analysis holds that assimilation and accommodation are inevitable components in the process of individual development. However, Piaget's genetic epistemology is equally relevant to the process by which societies adapt to science and technology. Whereas a child has little option but to accept the frustrations of accommodation, a society, or groups within a society, may view the prospect of accommodating to new science and technology as a social choice, a moment of decision, and possibly the occasion for resistance. Such moments of choice, which are likely to appear at different times in different countries as developments in science and technology, are seen to challenge deeply held values and pose "unacceptable risks." Faced with the pressures of accommodation, public reactions may vary from acceptance to fatalistic yielding or even active resistance (Touraine, 1995).

To explore this line of investigation further, Table 5-4 shows, for groups of countries, the index of industrialization and relative scores on the progress and pessimism dimensions.

Table 5-4. Relation between Industrialization and Public Views of Science

Countries	Level of industrialization	Progress	Pessimism
GR, P	Low	High	High
B, F, GB, NL	Medium/high	Low	Low
DK	High	Medium/high	Low
D-W	High	Medium/high	High

Note. Only those countries that scored either high or low on industrialization, progress, and pessimism are shown. Classification of high, medium, and low is based on a ranking of the countries on the particular scale. B = Belgium; DK = Denmark; D-W = West Germany; F = France; GB = Great Britain; GR = Greece; NL = the Netherlands; P = Portugal. Data base: Eurobarometer 1992.

In the context of assimilation and accommodation to science and technology, Table 5-4 is illustrated in Figure 5-3, to show what may be called the *industrialization paradox*.

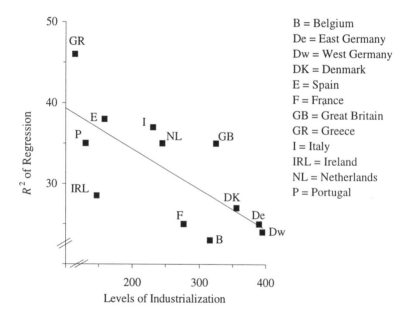

Figure 5-3. The industrialization paradox in the European Community: The relation between industrialization and the accommodation to science and technology (data base: Eurobarometer, 1992).

We find the members of the public in industrially developing EU countries to have ambivalent attitudes toward science. On the one hand, they see science and technology as the route to progress, but they are pessimistic about the extent to which science can solve broader societal issues and are concerned about the influence of science on lifestyles. The public in established industrial countries is less inclined to think that science is the way forward, is more optimistic about the contribution of science to societal problems, and is less concerned about science as a threat to lifestyles. In these countries the view is that science is being adapted to the needs of humankind. For the advanced industrial countries an interesting bifurcation is observed. Both Denmark and West Germany share a belief in the contribution of science to progress; but whereas the Danish are optimistic assimilators, the West Germans are pessimistic about the costs of accommodating to science.

It is beyond the scope of the survey data to explain these differences. However, the results raise some interesting questions. Will Greece and Portugal move toward the position of the established industrial nations as industrialization gathers pace? What is the trajectory of the established industrial nations, toward the German model or toward the Danish? Why have the two leading industrial nations such different views on the relations between science and society?

The first two questions may be best answered by comparative time-series data. Many factors contribute to changes in public attitude, so predicting the direction of change from cross-sectional data is fraught with uncertainty. One plausible scenario is based on the notion that the views of the elite, who are sometimes described as opinion formers or opinion leaders, are in advance of the views of the general public. All things being equal, the opinions of the elite will be adopted by the public at some later time. Clearly, this statement is a gross simplification of the complex dynamic of public opinion, but it is worth exploring. We define the elite on the basis of two sets of questions in the survey measuring social class and self-ascribed opinion leadership. About 5% of the total sample fall into the category of the elite, that is, social class A/B and high self-ascribed opinion leadership. Across Europe the elite are predominantly male, 35 to 55 years of age, live in urban or middle-sized towns, and have political orientations reflecting the national government.

How do the elite of the different countries compare to the general public on attitudes to science? Across most European countries the elite have the highest scores on the knowledge scale, higher expectations about the contribution of science to progress, and lower scores on pessimism. However, there are exceptions to this general pattern. In West Germany, Denmark, Ireland, and Greece, the elite group shows lower expectations about scientific "progress" than the average person. With respect to the pessimism orientation, the elite in Greece, Luxembourg, and West Germany shares with the general public relatively high levels of pessimism about the wider impacts of science. In contrast, the elite in Britain, Denmark, East Germany, France, Belgium, and Italy are considerably more optimistic than the public. If elite opinion is the nascent opinion of the wider public, we might expect relative stability in the pessimistic orientation in Greece, West Germany, and Luxembourg and a shift toward greater optimism about the wider impacts of science in the other countries.

A partial answer to the last question concerning Denmark and West Germany may be found in some analyses we conducted on the Eurobarometer 1993 Biotechnology Survey (Bauer & Gaskell, 1994). In Denmark 18% of the respondents said they had confidence in the public authorities to tell the truth about biotechnology, and a further 21% had confidence in schools and universities. In West Germany the percentages were 4 and 10, respectively. In the same survey it was found that German and Danish respondents reported the EU's highest levels of the perceived risk associated with seven applications of biotechnology and genetic engineering. However, in terms of global support for these seven applications, Denmark was above the EU country average, whereas Germany showed the lowest level of support registered in any country.

Perhaps the issue is one of trust in the established institutions of science, science education, and science policy and in their ability to manage the risks involved in some new technologies. Those who trust these institutions are more likely to believe that science can be assimilated to the needs of society, to be optimistic about the future, and to support research in what are seen to be high-risk areas. Those who distrust the institutions of science think that society will have to accommodate to science, and because this change involves costs they are pessimistic and show little support for research in areas of perceived risk.

If this admittedly speculative analysis has some truth, it shows the importance of science policy in fostering public attitudes that are supportive of science and its role in social progress. Two cultures of science communication may be in operation in the two countries. Public participation and involvement has been encouraged in Denmark through, for example, regular consensus conferences on controversial issues (Joss & Durant, 1995). Scientific research and political decisions are seen to accommodate the concerns of the public. In Germany one might characterize the process as one of low institutionalized public involvement and sometimes open conflict. The scientists and decision-takers know best, and the public is expected to accommodate to expert opinion. Trust is a complex social process. Laypeople may be willing to defer to experts on account of experts' superior knowledge. But when the experts are shown to be partisan or fallible, trust can quickly evaporate. If trust is at the center of public attitudes toward science, then the role of communication strategies that treat members of the public as participants in decision-taking acquires a particular, long-term significance.

In a concluding statement about a review of technological risks and benefit that was undertaken at a 1979 conference on technology and risk organized under the auspices of the European Commission, Dierkes (1980) observed that "people are demanding greater public accountability of decision-makers whom they perceive to be operating in ways that are neither open nor responsive to public interests" (p. 28). The slow progress toward establishing what Nelkin and Pollack (1980) term instruments for the "balanced and open dialogue that will enhance the constructive sense of collective responsibility necessary for legitimate and acceptable decisions" (p. 75) may, in part, explain our findings.

Conclusions

Our conclusions are listed under the following headings: substantive, strategic, and methodological.

Substantive Conclusions

— The Eurobarometer provides a unique and potentially important resource for comparative study of scientific and technological culture in Europe. In particular, it offers the potential for development into a robust "indicator" of public perceptions of science.

— The Eurobarometer survey instrument appears to be based on a "deficit model" and an educationalist agenda, according to which knowledge is the key construct because it is presumed to be positively correlated with support for science and technology. This "common-sense" view is supported by our analysis only for a subset of the least industrially developed European member states; in the more highly industrially developed states, the relationship between knowledge and attitudes is "chaotic."

— We propose a process model of the transition from industrial to post-industrial society as a basis for interpreting the changing pattern of relationships between interest, knowledge, and attitudes across the European Union. The two-dimensional model combines a hypothetical transition in one dimension and deviations due to local cultures in the other. The model successfully predicts a number of structural features within the Eurobarometer 1992 data set, and it allows us to construct hypotheses concerning the likely trajectory of the public understanding of science in different national contexts. Robust longitudinal data will be required to test these hypotheses.

— Closer analysis of attitudes suggests three different attitude complexes typical of low, medium, and high levels of industrialization. Interestingly, in two of the most highly industrialized EU member states—Germany and Denmark—science appears to interface with public opinion in rather different ways. These contrasting ways of organizing the interface between science and public opinion point to the importance of the development of trust relationships between scientific institutions and the public.

— Our analysis suggests that the dominant view of public understanding of science within the scientific communities of the most highly industrialized EU member states is actually more appropriate to the scientific and social circumstances of the least highly industrialized EU member states. Paradoxically, we may be faced with an older common-sense model of the public understanding of science that, though it displays considerable "inertia," is no longer well suited to the circumstances of postindustrial societies. The implications of the resulting mismatch between a common-sense theory and reality deserve to be taken seriously.

Strategic Conclusions

— So far as the governance of science is concerned, survey research of this kind is a potentially useful source of guidance to policy-makers. In a sense, it provides a functional equivalent of the familiar political opinion poll. It can alert science policy-makers to the differentiated state of public understanding and attitudes toward science across the European Union, it highlights significant differences between member states to fine-tune public understanding of science activities, and it provides early warning signs of impending disquiet or even active opposition to particular scientific and technological developments.

— The trajectory of public attitudes toward science and technology across the hypothetical industrial transition suggests, at least so far as the majority of more highly industrialized European member states are concerned, that the time has long gone when science and technology policy could be safely conducted by professional policy analysts and politicians alone, in isolation from and disregard of public opinion. Ambivalence and qualified assent are the hallmarks of postindustrial public attitudes toward science and technology, and in this context the success of particular policies will depend as much upon public opinion as upon professional or commercial endorsement.

— Public attitudes toward science and technology operate on two levels: the general-science cultural indicator and more specific issue-related indicators such as biotechnology and information technology. There is a case for survey research that combines the assessment of both levels within the same sample. This link would provide the opportunity to explore the relations between scientific culture, the longer term "climate," and contemporary issues, the "weather" of the particular "season." Combining the two levels of analysis would also bring small economies of scale. Small because each level must retain independent viability.

— We have provided indicative evidence that public participation is important in the building of trust between science and the public. We suggest that this association is of the greatest potential significance to the policy-making community. There is growing interest in the development of new, more participatory technology assessment procedures. We suggest that future social surveys pay more attention to the political interface between science and the public and in particular to the evaluation of the role of participation in the development of public perceptions and attitudes. Of course, participation has implications for the ways in which policy-making communities conduct their business.

— We note that survey data in and of themselves seldom provide a picture detailed enough to permit adequate interpretation of either national or international comparative public perceptions of science and technology. For this reason, we strongly recommend that future survey research of this kind be carried out alongside complementary studies of scientific culture at both the national and the European levels. By scientific culture in this context, we have in mind such things as significant historical events, significant features of the political culture with respect to science and technology, significant features of the regulatory system, and significant patterns of media reportage (see Bauer in this volume). These studies should include in-depth analysis of the public's representations of science, using qualitative methodologies such as interviews and focus groups as well as analysis of documents and media reportage. In terms of constructing indicators of science cultures, all of these elements are vital ingredients in the adequate interpretation of survey data on public understanding of science.

Methodological Conclusions

— The Eurobarometer science indicator needs more explicit and more carefully considered conceptual and theoretical foundations than it presently has.

- The questionnaire should be redesigned to take account of multiple inter-faces between science and the European public, and not only the educa-tional interface, which carries the status of common sense among scien-tists and policy-makers.
- The 23 attitude items contained in Eurobarometer 1992 are a very hetero-geneous selection, and they would benefit from careful redesign with a view to the creation of more robust scales. In the course of redesign, attention should be paid to developing both general and specific attitude items, and the four general attitudinal dimensions *progress, panacea, future shock,* and *pessimism* should be measured more reliably.
- Some of the topics that need to be addressed in future questionnaires are the knowledge of scientific institutions; images of science and scientists; levels of confidence and trust in scientific, social, and political institu-tions; relevant consumer behavior; and levels of political participation with respect to science and technology.
- The Eurobarometer would benefit from better background measures in a number of areas, including media consumption, religious orientation and practice, and general value orientation.
- Although we have not dealt with the issue here, we urge the reinstatement of the open-ended item that was dropped in the 1992 survey: "Please tell me in your own words, what does it mean to study something scientifi-cally?" (Bauer et al., 1992, Bauer & Schoon, 1993), because it yields potentially interesting information about symbolic appreciation of sci-ence.

References

Bairoch, P. (1982). International industrialization levels from 1750 to 1980. *Journal of European Economic History, 11,* 299–331.

Bauer, M. (1995, October). *Industrial and post-industrial public understanding of science.* Paper presented at the Scientific Literacy Conference of the Chinese Association for Science and Technology (CAST), Beijing.

Bauer, M., Durant, J., Allansdottir, A., Fluegel, P., Gervais, M. C., Jovchelovitch, S., Mello, A. M., Porter, A., Schoon, I., & Stathopoulou, A. (1992). *Mapping Euro-pean public understanding of science: A reanalysis of the open question included in the Eurobarometer survey no. 31 on science and technology from 1989* (Tech-nical report). Brussels: European Communities.

Bauer, M., Durant, J., & Evans, G. (1994). European public perceptions of science: An exploratory study. *International Journal of Public Opinion Research, 6*(2), 163–186.

Bauer, M., & Gaskell, G. (1994, November). *European public opinion on biotech-nology.* Paper presented at the International Conference on Biotechnology in European Society at the European Federation of Biotechnology, Den Haag, the Netherlands.

Bauer, M., & Schoon, I. (1993). Mapping variety in public understanding of science, *Public Understanding of Science, 2,* 141–155.

Beck, U. (1992). *The risk society: Towards a new modernity* (M. Ritter, Trans.). London: Sage. (Original work published 1986)

Bell, D. (1973). *The coming of post-industrial society.* New York: Basic Books.

Boehme, G. (1993). *Am Ende des Baconschen Zeitalters.* Frankfurt on the Main: Suhrkamp.

Dierkes, M. (1980). Assessing technological risks and benefits. In M. Dierkes, S. Edwards, & R. Coppock (Eds.), *Technological risk: Its perception and handling in the European Community* (pp. 21–30). Cambridge, MA: Oelgeschlager, Gunn & Hain.

Durant, J. R., Evans, G. A., & Thomas, G. P. (1992). Public Understanding of science in Britain: The role of medicine in the popular representation of science. *Public Understanding of Science, 1,* 161–182.

Eurostats (1992). *Basic statistics of the community.* Brussels.

Evans, G. A., & Durant, J. R. (1995). The relationship between knowledge and attitudes in the public understanding of science in Britain. *Public Understanding of Science, 4,* 57–74.

Gaskell, G., Wright, D., & O'Muircheartaigh, C. (1993a). Measuring scientific interest: The effect of knowledge questions on interest ratings. *Public Understanding of Science, 2,* 39–57.

Gaskell, G., Wright, D., & O'Muircheartaigh, C. (1993b). Reliability of surveys. *The Psychologist, 6,* 500–503.

Inglehart, A. (1990). *Culture shift in advanced industrial society.* Princeton, NJ: Princeton University Press.

Institut National de Recherche Agricole (INRA), & Marlier, R. (1993). *Eurobarometer 39.1—Biotechnology and genetic engineering: What Europeans think about it* (Report DG XII/E/1). Brussels: Commission of the European Communities.

Joss, S., & Durant, J. (Eds.). (1995). *Public participation in science: The role of consensus conferences in Europe.* London: Science Museum.

Luhmann, N. (1979). *Trust and power.* Chicester: Wiley.

National Science Board. (1991). *Science and engineering indicators—1991.* Washington, DC: U.S. Government Printing Office.

Neidhardt, F. (1993). The public as a communication system. *Public Understanding of Science, 2,* 339–350.

Nelkin, D., & Pollack, M. (1980). Consensus and conflict resolution: The politics of assessing risk. In M. Dierkes, S. Edwards, & R. Coppock (Eds.), *Technological risk: Its perception and handling in the European Community* (pp. 65–75). Cambridge, MA: Oelgeschlager, Gunn & Hain.

Piaget, J. (1971). *Biology and knowledge: An essay on the relation between organic regulation and cognitive processes* (B. Walsh, Trans.). Chicago: University of Chicago Press. (Original work published 1967)

Rose, A. R. (1991). *The post-modern and the post-industrial.* Cambridge, England: Cambridge University Press.

Touraine, A. (1969). *La societe post-industrielle.* Paris: Denoel.

Touraine, A. (1995). The crisis of 'progress'. In M. Bauer (Ed.), *Resistance to new technology—Nuclear power, information technology, biotechnology* (pp. 45–55). Cambridge, England: Cambridge University Press.

Wynne, B. (1993). Public uptake of science: A case for institutional reflexivity. *Public Understanding of Science, 2,* 321–337.

Ziman, J. (1991). Public understanding of science. *Science, Technology, and Human Values, 16,* 99–105.

Appendix

Wording of Questions Forming the Scales for Interest, Knowledge, and Attitudes

Interest in science

"Please tell me if you are very interested, moderately interested, interested or not at all interested in . . .

1. new inventions and technologies?"
2. new scientific discoveries?"

The Factual Knowledge Scale

1. The center of earth is very hot.
2. The oxygen we breath comes from plants.
3. Radioactive milk can be made safe by boiling it.
4. Electrons are smaller than atoms.
5. The continents on which we live have been moving their location for millions of years and will continue to move in the future.
6. It is the father's gene which decides whether the baby is a boy or a girl.
7. The earliest humans lived at the same time as the dinosaurs.
8. Antibiotics kill viruses as well as bacteria.
9. Lasers work by focusing sound waves.
10. All radioactivity is man-made.
11. Does the earth go around the sun or does the sun go around the earth?
12. How long does it take for the earth to go around the sun?

Progress

1. Thanks to science and technology, there will be more opportunities for future generations.
2. The application of science and new technology will make work more interesting.
3. The benefits of science are greater than any harmful effects.
4. Science and technology are making our lives healthier, easier and more comfortable.

Panacea

1. Scientific and technological research do not play an important role in industrial development.
2. New technology does not depend on basic scientific research.
3. Scientific and technological research cannot play an important role in protecting the environment and repairing it.

4. Thanks to scientific and technological advances, the earth's natural resources will be inexhaustible.

Future Shock

1. Science makes our way of life change too fast.
2. Because of their knowledge, scientific researchers have a power that makes them dangerous.
3. We depend too much on science and not enough on faith.

Pessimism

1. Scientific and technological research do not play an important role in industrial development.
2. Scientific and technological research cannot play an important role in protecting the environment and repairing it.
3. Science makes our way of life change too fast.

CHAPTER 6

"SCIENCE IN THE MEDIA" AS A CULTURAL INDICATOR: CONTEXTUALIZING SURVEYS WITH MEDIA ANALYSIS

Martin Bauer

In this chapter I argue the case for the establishment of a cultural indicator of popular science. To use an analogy, a cultural indicator is a measure of the scientific "climate." The description of the climate rather than the daily weather requires a long-term commitment to measurement. Furthermore, such an indicator needs to be based (a) on surveys of public understanding of science, to gauge reception, and (b) on the systematic analysis of intensity and the contents of the media coverage of science, henceforth called *media science,* to gauge the circulation of images of science, within the same contexts.

In Britain my colleagues and I* have established the Science Museum/LSE (London School of Economics) media monitor for documenting media coverage of science since 1946. Our aims are (a) to explore the changing patterns of science coverage in British media over the last 50 years, (b) to pursue systematical collection of news material as a research archive, (c) to develop a standard instrument with which to measure the quantity and quality of media coverage of science and technology, (d) to explore the changing relationship between science and the public, and (e) to test specific hypotheses about trends in media reportage of science and technology. The windows of science open and close over time, and the architecture of the windows— their frames and the contents visible through them—changes with it.

* I wish to acknowledge the combined efforts of those in the Media Monitor Project—Asdis Ragnarsdottir, Annadis Rudolfsdottir, Agnes Allansdottir (the Icelandic connection)—and of others over the last 4 years. I thank John Durant for his continuous institutional backing. The project was jointly funded by the Welcome Trust for the History of Medicine and the Science Museum between 1993 and 1995. In writing this chapter I am indebted to George Gaskell and several anonymous reviewers for their helpful comments.

My argument is fourfold. First, recent surveys of public understanding of science across many countries will gain from an interpretation contextualized within *cultural indicators*. Second, the idea of cultural indicators commits us to a research program that combines survey and media data over an extended period. Third, results from our analysis of the previous 50 years of British press science coverage show that such commitment is relevant. Lastly, cultural indicators of science should be created, a call that I support by discussing practicalities of establishing an international media-science monitor.

To start with some terminological notes, for the present purpose I drop the distinction between science and technology. From the point of view of "pure science" or "engineering" such a distinction may be desirable. Science and technology historically developed from different concerns and with a different ethos. In modern society, however, the development of science and the development of technology are closely intertwined: Scientific research relies on new technologies, and the development of the latter is almost entirely based on scientific developments. Both in public policy, under the heading of R&D, and in popular imagination the difference is fading. The term *technoscience* seems appropriate. Hence, I use the terms *science* or *technology* interchangeably.

My colleagues and I define popularization of science as all activities that distribute scientific ideas, facts, or methods among people who are not closely involved in their production and for whom scientific research is not the main preoccupation in life. Popularization is driven by various and historically changing motives and agendas (Burnham, 1987). *Popular science* is the traceable product of this distribution in various channels and genres of communication such as public lectures, books, newspapers, magazines, radio, television, museum exhibitions, or, recently, the internet (Lewenstein, 1995). This process of distribution is constructive. Guided by production rules, such as "news value," the mass media, their authors, editors, and producers select, accentuate, highlight, dramatize, and sensationalize events, including those with a scientific or technological reference (Hansen, 1994; Neidhardt, 1993). This process *represents* and *transforms* the source activity, scientific research, into retrievable texts of various natures. Therefore, media coverage cannot be a faithful mirror image of scientific activity and its achievements and failures; it mediates between science and other spheres of life. The representation of science commands relative autonomy from the scientific activity and serves various, and at times contradictory, functions in society: the legitimation of science in public (Hilgartner, 1990; Wynne, 1995), the resistance of social groups to ideologies such as scientism or technocracy (Bauer, 1994), and the epistemological self-reassurance of the scientific community (Jurdant, 1993).

Media Indicators of Scientific Culture

Most people do not directly participate in scientific research. With the exception of some hands-on experiments, engineering courses, or field excursions during primary or secondary education, science reaches most people in mediated forms: at the user-interfaces of cars, household machines, computers, labels of food ingredients; at the hospital; or in the form of stories and exhibitions about science, its achievements, or problematic consequences. These situations are expository windows of science (Roqueplo, 1974; Whitley, 1985). People are invited to stare in awe, and consume with pride. The stained glass of mediation separates most viewers from where the action lies. This symbolic construction of science and technology is an essential part of culture. The celebration or fearful apprehension of scientific achievements are core elements of a modern culture, both as a resource and a constraint for future developments. A cocktail of symbols and imaginings constitutes the basis of people's attitudes and interests in new developments, influences people's consumer choices, and in some cases guides the ballot box should new developments become a political issue (Buchmann, 1995). Nuclear power and visions of the "atomic society" of the 1950s and 1960s (Gamson & Modigliani, 1989; Weart, 1988), microprocessors and the "information society" of the 1980s (Nelkin, 1995), and recent genetic engineering and its visions of a "biosociety" mobilize images of science in society in the public arenas. To study images of science as part of culture is legitimate and necessary. Scientific culture is a constraint of the productive system and a resource of modern democracy.

Since the early 1980s many countries have conducted surveys into "science literacy" (Beveridge & Rudell, 1988; Miller, 1983) or the "public understanding of science" (Durant et al. in this volume). For years the U.S. National Science Foundation (NSF) has included a section on public attitudes toward science and technology in its science indicator reports. This endeavor has widened and proliferated international data—some entirely comparable—from the United States, Canada, France, Britain, Spain, Sweden, Bulgaria, India, Brazil, Korea, Japan, China, and the European-wide surveys of 1979, 1989, and 1992 (see Durant et al. in this volume). Researchers are increasingly in a position to undertake systematic cross-national comparisons of survey data on knowledge of, attitudes toward, and interests and trust in science, as well as on patterns of attention to media science. Interesting patterns of cross-national similarities and differences emerge that require interpretation, not least in the context of the differential cultivation of media science in the national contexts.

The publication in many countries of science indicator reports is an attempt to make science publicly accountable for its performance as a key factor of national productivity. In parallel, science is being rediscovered as a cultural resource (Schiele, Amyot, & Benoit, 1994). The dissemination of popular science through newspapers, magazines, radio programs, television, science centers, festivals, celebration weeks, and on the World Wide Web and the revival of science and technology museums constitute a growing segment of the culture industry.

It is unsatisfactory to rely solely on survey data for the construction of indicators of science culture. Survey data are not self-explanatory; one needs contextual information in order to interpret them realistically. Researchers in this field need simultaneous and systematic analysis of the mass media that circulate those modern myths of origins (the big bang; the gene), great challenges (war on cancer), heroes (Einstein), and axial transformations (the genetic revolution) that fuel people's imaginations, opinions, and attitudes toward science (Caro, 1994; Moscovici, 1992). A measure of media science is necessary (a) to interpret the differences and similarities between groups of people as uncovered by survey data and (b) to study the long-term dynamics of these images as they are cultivated in the mass media. In order to achieve these aims, cultural indicators of science need to be based in parallel on surveys of people's opinions and attitudes and in systematic content analysis of media-science stories.

The Cultivation of Science: A Triangle of Mediation

The debate on cultural indicators is extended and controversial. Namenwirth (1984) distinguished action systems from symbol systems. Social indicators refer to action systems. They measure the performance of various activities and indicate how to improve future performance. By contrast, cultural indicators do not measure performances; they refer to conditions of performance. Culture, as a symbolic system, does not act; it merely is. It is the constraining background for activities. Compared to the cycle of change characteristic of action systems, culture's cycle of changing is long. The task is one of observing and recording culture, and less one of deliberately changing it. An action system such as the free-market economy rests on a cultural basis. Minimal cultural standards need to apply. For example, the notion that property and contracts ought to be respected is a familiar precondition of a modern economy.

Cultural indicators of science and technology characterize constraints for research and development activities that vary across countries; they measure the prevailing Weltanschauung and the place of science and technology therein. Culture is relatively autonomous of other spheres of society and is not easily changed by design. Culture does not reflect reality; images— efficiently combining fact, fiction, and values—fuel actions with prophetic, sometimes utopian, aspirations with which a dull mirror image of some historical reality cannot compete. It is impossible to identify the functions of images by their contents; their consequences have to be observed in context. Iconic or symbolic images of science serve at least two functions: (a) potentially ideological, they frame people's imagination in the service of domination; and (b) potentially empowering, they enable people to resist uninvited influences or to liberate themselves from "natural" or "superstitious" constraints. Culture is not a consistent belief system, though such a system may be a refined part of it. Its contradictions and their tolerance mark the differences that make identities (Jovchelovitch, 1996).

Scientists need to distinguish values from the evaluation of concrete issues. Attitudes link abstract values to concrete objects and achieve an evaluation of the specific, an evaluation that is relevant for action. Attitudes are deemed to be closer to action than values are, although notoriously low correlations between attitudes and related behavior show that the modality of this closeness is controversial. Attitudes are social indicators and are the target of change management. Cultural indicators frame the variation of attitudes. They characterize *la longue duree* of change, the climate rather than the weather of social life (Klingemann, Mohler, & Weber, 1982; Rosengren, 1984). Daily or seasonal variability of the weather are characteristic of a particular climate. By analogy, the structure and variance of attitudes rather than their level within a community is indicative of that community's scientific culture (see Durant et al., in this volume). Levels may change in the short run, structures only in the long run. To understand a culture, a long-term perspective is a sine qua non. With the quest for cultural indicators, scientists widen the research horizons into larger spaces and longer time spans.

The systematic study of media science has gained momentum in recent years. One can broadly classify these efforts in the "triangle model" of public understanding of science (Bauer, 1994):

1. *Production studies* How much popular science is produced? How do stories break? Under what conditions? With what motives do scientists, journalists, and other activists engage in popularizing? How are stories selected and framed? How do journalists and scientists see each other?

What is the operative image of the public, or audience, among media people?

2. *Contents of mediation* This type of study involves the analysis of news material, popular books, exhibits, images, stories, and reportage across different media and across time; the comparison of forms and contents; and the analysis of dominant frames and themes, metaphors and associations that objectify and anchor scientific ideas in popular imagination.

3. *Reception studies* How are the mass media segmented? What do people do with particular stories? How do audiences relate to scientists and mediators? This area includes comparing audiences, their enthusiastic receptions, their divergent interpretations, or their resistance to new developments in science and technology.

The model puts each type of study at one edge of a triangle of mediation between production, message or image, and reception. The schema of the triangle helps one avoid thinking in linear causation and points toward circular influences: Production brings images to audiences, images link production and audiences successfully together, and audiences render images relevant for production. For a particular social segment users of the triangle model assume a process of coevolution over time. It is difficult to decide what has priority—the production, the images, or the audience. Mediation is understood as constraining and enabling: Images frame and therefore motivate; production needs to be selective in order to produce anything at all; audiences have limited capacity for information and meaning, thus forcing a choice and the freedom of interpretation.

The approach that my colleagues and I take has affinities with the influential cultivation analysis described in Morgan and Signorelli (1990), the continuation of the earlier cultural indicators program (Gerbner, 1969) in communications studies. We separate the media content from its production and reception, thereby highlighting the relative autonomy of images and rejecting simple causality from production to images and from images to reception. Second, we are skeptical about the short-term manageability of cultural images. In recent years, the notion of organizational culture (Smircich, 1983) has moved culture within the grasp of design probably because of the managerial promise: the charismatic leader showing the way into a more efficient and profitable future. It is recognized that influencing the symbolic conditions of corporate activity is as important as controlling productive activity itself. However, taking into account the notorious intransigence of the cultural sphere, and setting unrealistic or exaggerated expectations aside, researchers should measure culture not for the purpose of design but for the purpose of sensitizing themselves and political actors to the prevailing climate. The climate is something people take into account, not try to change, when they plan activities. Society may, however, wake up to the fact that its activities had an impact on the climate, but not as intended. For cultural monitoring, something similar to continuous weather observation is required.

The approach taken at the LSE media monitor differs from cultivation analysis in two important respects. First, cultivation analysis is exclusively concerned with television and its impact in modern society. It is "a set of theoretical and methodological assumptions and procedures designed to assess the contribution of television viewing to people's conception of social reality" (Morgan & Signorelli, 1990, p. 15). By contrast, my colleague and I continue to focus on newsprint and therefore prefer the older term cultural indicators to mark the difference. For practical reasons we consider newsprint, in particular the daily or weekly general press, to be a valid proxy for mass media trends. There is also empirical support for this view, at least in the European context. I return to this issue later.

Second, in cultivation analysis television is considered a unified single source of dominant images confronting modern society. Audience effects may differ because of different exposure and processing variables. By contrast, we start from a world where media production, reception, and messages still covary, segmented onto self-referential social groups, old and new. Images, ideas, values, attitudes, and beliefs about science circulate within social groups, who champion different themes at different times. These differences may be observed by comparing production activities, audience surveys, and the media contents *circulated out of, for, and within specific groups.* Newsprint traditionally had this group specificity, as in the difference between quality and popular newspapers, or as in the distribution of pamphlets, newsletters, and magazines. More recently, local TV and radio stations or WWW networks are produced by and for specific social groups. The coordinated observation and content analysis of segmented media processes shows the expressions of particular societal groups and constitutes a paradigm for the study of "social representations" of science (Bauer & Gaskell, 1997; Cranach, 1995; Farr & Moscovici, 1984; Jodelet, 1989; Moscovici, 1976). Science continues to pose a challenge for groups in society who receive scientific ideas, elaborate them, and reconstruct them in particular ways according to the traditions of a multicultural society. The image of the stone thrown into the pond comes to mind: The resulting ripples are unpredictable, but they reveal the tectonics underneath the surface.

Within this conceptual context media science is a social fact. To theorize adequately on the production of images, their form and content and their reception need to be studied and compared across time and groups. This constitutes an ambitious research program that may simultaneously enhance the prospects of the public-understanding-of-science movement and contribute to the study of modern mass communication.

Science in the British Press, 1946–1990

Much of the research on public understanding of science lacks a critical historical perspective and seems preoccupied with activism and efficiency of communication to the neglect of a wider perspective. Only recently is this omission being rectified (see Burnham, 1987, for a history of popular science in the United States; see LaFollette, 1990, for an analysis of *Science* from 1910 through 1955 in the United States; see Raichvarg & Jacques, 1991, for popular science in France). In London, my colleagues and I have collected and archived a representative sample of science reportage in the British national press from 1946 through 1990,[1] and have conducted a systematic content analysis of the science coverage over this period. We are now in a position to explore the whats and hows of science in the press over the last half century and to characterize the openings, the closings, and the architecture of the mass-media window of science.

For a variety of reasons we take science in the press as a proxy for media science in general. Mass media comprises outlets other than the press. However, comparisons of science coverage across various outlets have shown that, aggregated over an extended period, the distribution of thematic content is similar (Hansen & Dickinson, 1992). The newspapers are still the source where TV and radio journalists look for news that is not to be missed. We can expect that the trends in the key newspapers faithfully represent the trends of science coverage in other media over a lengthy period.

Our media-science archive currently contains over 6,000 hard copies of press articles from a stratified cluster sample representing the years from 1946 through 1990. Newspapers are stratified by quality and popular press, political orientation, opinion-leadership function, and circulation figures. We select 10 random days for a set of newspapers, our sampling frame. For each random day we identify all relevant material by using an open definition of media science (see Bauer, Durant, Ragnarsdottir, & Rudolfsdottir, 1995, vol. 2).[2]

---

[1] The data sample is currently being updated for 1992 to 1996. Regular updates are envisaged.

[2] The Science Media Monitor Archive (Bauer & Ragnarsdottir, 1996) is publicly available to researchers of popular science. The Science Museum Library in London hosts the archive. Researchers are invited to make use of these materials for their

For our analysis we developed a modular coding frame based on the idea of "news narratives." Each science article is coded for up to 100 variables covering (a) formal elements such as article size, page number, page location, and size of headlines and (b) narrative elements such as authors, the actors, the news event, a rating of the valuation tone, the world location of the event, the scientific field involved, the handling of controversies, reported consequences in terms of risks and benefits, and the moral of the story (Bauer et al., 1995). I briefly present results on the postwar cycles of science in the British press, the decline in front-page science news, the changing evaluation of science, the changing structure of academic fields, and the emerging discussion of risk since World War II.

The Opening, Closing, and Architecture of the Window of Science

To judge from much that is lamented about science in the present British press, people might suppose that there has been a steady decline in the amount of science coverage in the postwar period. In fact, the amount of science reportage depends on both the precise periods and the type of newspaper that are considered. The broadsheet newspapers, or quality press, and the tabloid newspapers, or popular press, go through cycles of coverage, which may be summarized as shown in Table 6-1.

Table 6-1. Cycles of Science Coverage in the British Press, 1946–1990

Characterization	Period and volume
Quality press	
Postwar take-off	1946–1960 (400% increase)
1960s decline	1960–1974 (50% decrease)
Recovery	1974–1990 (50% recovery)
Popular press	
Postwar take-off	1946–1962 (300% increase)
1960 stability	1962–1978 (rough stability)
1980s decline	1978–1990 (60% decline)

If there has been a decline in science coverage in the British press, it has been gradual, beginning in the late 1950s and early 1960s. However, the

own purposes and to collect similar materials in their own country to allow comparisons. Contact E-mail: Bauer@lse.ac.uk pattern is more complex than the lament about declining science reportage suggests (see Figure 6-1). The total space dedicated to science news, taking into account the changing news space, was around 4% on average, ranging between 7% of total newsprint in "good" times and as low as 2% of total newsprint in "bad" times.

Figure 6-1. Overall trend in science news in the postwar British press (*n* = 6,080 news articles). The frequencies are weighted to a standard of "two newspapers" (20 random days per year) and are to be read as index values. The curve is smoothed to show the broad trends. Adapted from *Science and technology in the British press, 1946–1990* (vol. 1, p. 42), by M. Bauer, J. Durant, A. Ragnarsdottir, & A. Rudolfsdottir, 1995 (Tech. Rep., vols. 1–4), Science Museum: London.

Overall levels of science coverage in the quality press peaked in the early 1960s, declined throughout the 1960s, and increased again as of the mid-1970s. According to similar studies from Australia and Germany (Australian Department of Industry, Technology and Commerce [ADITC], 1991; Kepplinger, 1989), the pattern of growth in science coverage after the mid-1970s was an internationally consistent phenomenon. For all of the postwar period, the quality press carried more science news than the popular press did.

Front-page Science News

A key editorial decision concerns the positioning of a story within the newspaper. The space of newsprint is limited; decisions have to be taken about which story is front-page material. The positioning of a story reflects editorial judgments of news value, not least as a potential selling proposition ("Newspapers are in the business of selling newsprint," as stated by Lord

Beverbrook, the famous British press baron). This position of a story includes a judgment of a news item's public importance. The volume of front-page science news in the popular press is much smaller than in the quality press. Quality-press front-page science news increased from 1946 through 1952 and has since decreased to about a quarter of its peak in 1952. Front-page science news may indicate the decreasing public excitement over scientific and technological developments and (but only for the front-page-news) echoes the laments about declining media science within living memory.

Valuating Scientific Achievements

The evaluation of scientific achievements in the British press underwent a different cycle of change over the postwar period (see Figure 6-2).

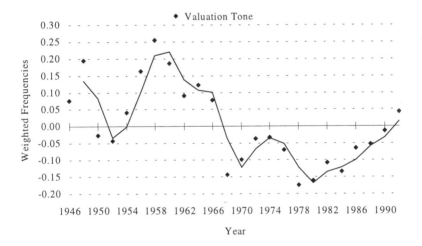

Figure 6-2. Standardized ratings of the "valuation tone" of science news. Ratings are standardized for each coder. The overall mean is 3.51 on a scale ranging from 1 (*discourse of promise and triumph*) to 6 (*discourse of concern*). Aggregates are based on between 80 and 400 articles per year. Overall differences are statistically significant (ONEWAY: $F = 3.45$, $df = 23$, $n = 5,985$; $p = .000$). Adapted from *Science and technology in the British press, 1946–1990* (vol. 1, p. 49), by M. Bauer, J. Durant, A. Ragnarsdottir, & A. Rudolfsdottir, 1995 (Tech. Rep., vols. 1–4), Science Museum: London.

Broadly speaking there are two phases: a first phase (1946–1965) during which the overall tone of science coverage leaned toward the "positive and celebratory," with a short exceptional period in the early 1950s; and a second phase (1965–1990) during which the overall tone leaned toward the negative and toward "public concern," with a short exceptional period in the early 1970s. Turning points in the trend of the evaluative tone occur in the early 1950s (from negative to positive), in the late 1950s (from positive to negative), and in the late 1970s (from negative to positive). At the cut-off point of the present study (1990), the trend toward the positive side is unbroken. Should this trend continue, our up-dated analysis will show that a celebratory tone of science reportage has been recovered in Britain in recent years.

Fields of Scientific Inquiry

Classifying fields of scientific inquiry over a long period of time is a tricky business. First, there is no generally acceptable scheme of classification at any moment in time. Second, even if there were such a scheme, fields of inquiry come and go in a constant flux over extended periods. It is generally agreed that only a very small selection of scientific activity enters the public window at any given moment. For these reasons, it is revealing to look at the long-term trends in this selection using a simplified clarification. The architecture of the public window changes significantly, reflecting a complex mix of changes in scientific research, media activities, and public concerns and preoccupations (see Figure 6-3).

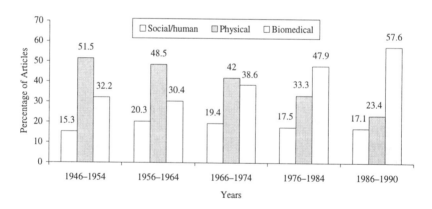

Figure 6-3. Coverage of academic fields of investigation in the postwar popular British press. The classification covers social sciences (social sciences, history, and philosophy), physical science (technology and engineering, physics, chemistry, astronomy, and earth sciences), and biomedical science (biological sciences and medical sciences) and is based on 1,326 articles. Adapted from *Science and technology in the British press, 1946–1990* (vol. 1, p. 54), by M. Bauer, J. Durant, A. Ragnarsdottir, & A. Rudolfsdottir, 1995 (Tech. Rep., vols. 1–4), Science Museum: London.

Into the 1960s, scientific coverage in the popular press was dominated by the physical and engineering sciences. Over 50% of news was related to physical sciences. Thereafter, there was a steady shift away from the physical and toward both the biomedical sciences and to the social sciences. In the popular press biomedical science and health issues came to dominate during the 1970s. Our data, which ends in 1990, does not yet show the crossover between physical and biomedical science for the quality press. However, if the trend continued unchanged, the crossover is likely to have occurred by now. Social science coverage increased in the quality press after 1945, but not in the popular press, where it remained at a constant and generally low level.

Four Base Technologies in the Media Window

Nuclear power, space exploration, information technology, and genetic engineering were four strategic developments of the postwar period. Each of them fueled popular imagination of an axial transition to a new type of society: From "Atomic" to "Space" to "Information" to "Bio-society" or "Gen-society." Analysis of the press coverage devoted to each of these visions of society shows that they succeeded each other in their media prominence: 1954 to 1962 was the media time of nuclear power; 1962 to 1970, of space exploration; 1978 to the late 1980s, of information technology and computing; and after 1986, of new genetics, and the latter's end is not yet in sight. This succession and periodization of new technologies in the public imagination and of their structure and changing detail warrants further comparative analysis.

Defining the British Entry into the "Risk Society"

The last set of data represents the emerging discussion of risk. Each article is coded for consequences in terms of benefits or risks attributed to scientific activity. A discourse about benefits dominated science coverage in the press until the end of the 1960s. During the 1960s the discourse about risk increased rather sharply. Risk and benefit stories achieved near parity after 1970, with around 45% of all stories containing either a risk argument or a benefit argument or both. Figure 6-4 shows the linear trend throughout the period: a decline in benefit-only stories, a marked increase in risk-only stories, and an increase in risk-and-benefit stories. We take the trends of benefit-only and risk-only stories to indicate what has come to be termed the *risk society* (Beck, 1986/1992). This expression is taken to mean a society in which the successes of science lead to ever increasing uncertainties and awareness of manufactured as opposed to natural risks in everyday life, thus undermining the foundations of this success story and resulting in new lines of social conflict over the distribution of these risks in society. The crossover point of benefit-only and of risk-only stories could be taken as an operationalization of the British entry into the risk society during the early 1990s, a period that, curiously enough, coincides with the reception of the very concept in the Anglo-Saxon social science literature.

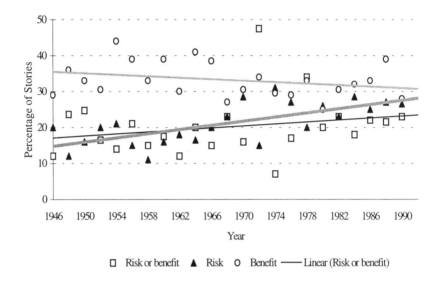

Figure 6-4. Annual percentages of stories containing a risk-and-benefit, a risk-only, or a benefit-only argument in British science news (*n* = 6,080). The lines show the linear trends over the whole period. Adapted from *Science and technology in the British press, 1946–1990* (vol. 1, p. 41), by M. Bauer, J. Durant, A. Ragnarsdottir, & A. Rudolfsdottir, 1995 (Tech. Rep., vols. 1–4), Science Museum: London.

Empirical results of this kind—the cycles of intensity, the cycle of valuation, the changes in reported research activities, and the emergence of risk stories—raise more questions than they answer. None of these "facts" is self-explanatory but rather calls for investigations into what lies behind them. However, describing these patterns and identifying periods with some precision facilitates international comparison and sharpens the ability to distinguish between relevant and irrelevant explanations. This chapter is not the place for explanations, not least because my colleagues and I continue to be puzzled by our results. The search for underlying structures and processes continues. All we can be sure of is that candidate explanations need to take into account the complexities of the changes in scientific activities, in the workings of the media, and in the understanding of science in public.

Monitoring "Science in the Media"

I am aware of four studies that represent an attempt to characterize global science reportage and that have cultural-indicator quality. Lafollette's study (1990) is based on weekly magazines in the United States from 1910 through 1955; Kepplinger (1989) and his colleagues based their investigation on the German press covering the years from 1965 through 1985; the study by the ADITC (1991) is based on newspapers, magazines, television, and radio from 1980 through 1990; and our own British efforts encompassing the whole period from 1946 through 1990. These studies have their strength and weaknesses, which have been discussed elsewhere (Bauer et al., 1995, vol. 2). Nevertheless, the work is a sound foundation on which to proceed in establishing comparable longitudinal data on media science. Several issues deserve further attention in this exercise. To most people, constructing cultural indicators on the basis of media material is not cost-effective. The application of sampling theory—the idea of making reasoned trade-offs between the costs of collecting information and the costs of being incomplete—is crucial to the argument.[3]

The selection of media science deserves a broad definition. There is a wide range of radio, television, and press genres that cultivate science and technology, and the main focus of the reportage may not be scientific in a technical sense, or in the sense the scientists prefer to see it. Crime news, celebrity news, sports, leisure, and lifestyle are major anchors of scientific information for the wide public. Elements of science news are found in unexpected quarters and by no means only in specially dedicated sections such as the science pages of a newspaper or in science programs on radio or televi-

---

[3] To judge the feasibility of comparative mass-media studies, one needs to look at the costs of survey data. A reasonable survey indicator of the public understanding of science requires 50 or more question items every 2 to 3 years. In comparison, media analysis by newspaper proxy can be conducted with considerably lower running costs. A national media sample can probably be conducted at 20% to 25% of a person-year equivalent, which is a realistic basis for the running costs of a reliable yearly sample of 300 to 400 articles. The initial investment for creating a sampling frame, establishing reliable procedures for collection, archiving, coding, analyzing data, and reporting results is a one-off capital investment. An overall cost advantage could come from extending the interval between surveys on the public understanding of science to 5, 6, or 7 years and establishing a media indicator during the interval years.

sion. The cultivation of science is not restricted to the documentary genre. The sample of relevant materials cannot be restricted to those specially demarcated outlets. This reality has several implications for the sampling of science news.

A first sampling issue concerns the choice of medium. Television is the most widespread mass medium in modern societies. Hence, its science content seems to deserve priority for the construction of indicators of scientific culture. However, television analysis is time-consuming, and the archival access is notoriously complicated and expensive, particularly for historical material. I suggest, for various reasons, that newspapers are valid proxis for the main mass media as a whole. Although television, radio, and general newspapers clearly differ in format, there is good evidence that their distribution of science content is similar when aggregated over a period of time. Overall, the bias toward the visual in TV news does not seem to result in thematic distribution different from that in radio or newspaper coverage (Hansen & Dickinson, 1992). These three media differ from lifestyle magazines, which devote an overwhelming amount of space to biomedical science (ADITC, 1991). Current trends toward visualization across media and toward cross-media ownership may lead to even more convergence in the content of science coverage. Furthermore, newspapers still serve as agenda-setters for other media. In our current studies of news coverage on genetic engineering in the British press, we receive inquiries from radio and TV journalists who read the daily newspapers and wonder about the "news value" of newspaper stories, such as the import of Monsanto soy beans into Europe in late 1997. Science journalists are looking around for stories they have not yet covered. They desperately want to avoid missing the potentially big story. Over an extended time span it is therefore not surprising that news content is similar across media outlets. Newspapers continue to be well placed to indicate the trends in news (Tunstall, 1996).

Second, newsprint is easy accessible in archives on hardcopy or on microfiche, in recent years also on-line or via CD-ROMs. Furthermore, content analysis of text material is a mature social research method and is much less time-consuming than the analysis of television programs which requires the simultaneous analysis of text, image, and sound.

Third, the sampling of press material can take various routes. On the converging evidence from the studies mentioned above, all of which take different sampling approaches, it is known that a much reduced number of random days per year provides a reliable estimate of science news over time. The sample requires stratification according to the segments of the newspaper market in any one country. In view of the observations by Ruhrmann (1991) and Haller (1991), I recommend taking newspapers as the sampling clusters and the single article as the unit of analysis; sampling newspapers and collecting all relevant articles within any one issue leads to a *stratified cluster sample*. Smaller units of analysis, such as paragraphs or sentences, produce intricate difficulties for the aggregation of indices across different sizes of articles and editorial styles. In the British study, we sampled 10 different random days for 4 to 6 newspapers per year, totaling 40 to 60 news days a year. A sample containing anywhere from 100 to 400 articles per year is sufficient to characterize the cultivation of science across extended periods.

Fourth, a word of caution about convenient on-line data is in order. Newsprint is now available on-line, as with FT-Profile or REUTERS for Britain or NEXIS for the United States, or via CD-ROMs in many countries on a regular cycle of updates. These sources make newspaper data very easily accessible. However, several serious problems arise for the construction of cultural indicators as proposed in this chapter. (a) On-line data goes back at most only to the mid-1980s. The capacity for retrospective construction of cultural indicators drawn from these sources is therefore limited. (b) On-line news services such as FT-Profile serve mainly business purposes. This focus often means that popular newspapers are excluded or incompletely covered. However, science stories in popular newspapers are an essential part of cultural indicators. To exclude the popular press from the analysis is to bias any indicator toward the cultural elite. (c) The cultural analysis of science needs the news context. Current on-line services provide only limited contextual information. For example, the newspaper section within which the story appears is not identified, and pictures or illustrations are not included. This information is culturally highly significant, however. (d) On-line data is retrieved on the basis of keywords and keyword combinations. This procedure, the keyword search, is not very useful for the construction of cultural indicators as proposed in this chapter. The researcher attempting to characterize media science comprehensively will find it impossible to define a practical set of keywords to cover an open definition of *science material*. A keyword search is effective only for case studies on such topics as BSE/CJD or AIDS, for which keywords are easy to specify. A keyword search is not reliable for a comprehensive study of science news. It remains possible to scan sampled hard copies of newsprint and identify the relevant materials visually, practices for which high reliability can be achieved (Bauer et al., 1995). It may be possible to identify articles manually and retrieve them later from the on-line source by using unique keywords such as the title of the newspaper, the date, and the name(s) of the author(s). For constructing cultural indicators of science, the manual selection of material will remain the only valid method of sample gathering until on-line scanning of actual newsprint, not only text, becomes widespread. Hence, for both substantive and practical reasons, newsprint is a valid proxi for the construction of indicators of media science.

Summary and Conclusions

In this chapter the case was made for complementing survey data on the public understanding of science with systematic analysis of science reportage in the mass media. This approach is the way forward to construct indicators of scientific culture. In recent years the collection of survey data has gained international momentum, and interesting comparisons of the structure and dynamics of popular science are within reach. However, it is increasingly clear that survey data are not self-explanatory; they lead to further questions. One needs contextual information in order to interpret complex data patterns not least in the context of the media cultivation of science.

Cultural indicators of science measure the stock of images, beliefs, and values in a population with reference to science and technology. These social facts are a promiscuous mix of past popularization, mythical and religious images, and other traditions and constitute the background for the future development of science. Cultural facts have a long cycle of change, and they command recognition as constraints on productive activities. Images, beliefs, and values relate to individual opinions and attitudes as the climate relates to the daily weather. The climate defines the variability and the overall weather pattern aggregated over time. Cultural indicators define the frame, not the consensus, within which short-term perturbations of opinions and attitudes are contained and within which people may disagree. By implication, cultural indicators specify the questions that can sensibly be asked at any moment in history. To establish the climate, a commitment to long-term monitoring is required. Such a monitor is most effectively established on the basis of newsprint. Empirical evidence and practical reasons suggest that newsprint is a valid proxi for the important mass media as a whole. Media indicators have two advantages over surveys of public understanding of science. First, time-series data are hard to come by. To my knowledge there are only French, British, and U.S. survey data that are suitable for time-series analysis over the last 15 years at most. Second, surveys that have not been conducted cannot be constructed after the passing of the moment. By contrast, media data on those moments are simply awaiting retrieval in archives.

To illustrate this point, I summarized the British trends in media science over the postwar period. It can be seen that the media window of science has opened and closed in two cycles over the last 50 years. The evaluation of scientific achievements has been equally cyclical, albeit not contemporaneous. Science in the public window changed its face significantly over the period. For example, overall risk-benefit arguments were on the increase; risk-only arguments markedly increased, whereas benefit-only arguments decreased. This change marked a defining criterion for Britain's entry into the risk society in the early 1990s, a moment that coincided with the time when the very idea of a risk society entered the wider sociological imagination.

Experience shows that complementing survey data with media data appears to be more than feasible and cost effective. My colleagues and I argue in favor of (a) widening the intervals between national surveys on the public understanding of science, (b) establishing a media indicator time-series at least back to the 1960s, and (c) continuously monitoring the current years. Media analysis offers the opportunity to reconstruct the past and establish the contextual trends within which to interpret present survey data. This double strategy may even be cheaper overall than relying solely on surveys and will certainly achieve a more realistic picture of current scientific culture in various national contexts. Several practical observations were made. First, a media indicator needs an open definition of media science. Relevant material appears in news coverage of crime, gossip, culture, sports, and business. Second, sampling theory allows one to reduce the material to a manageable corpus without losing the basis for valid conclusions. Third, on-line newspaper services are presently not effective for the purpose of cultural indicators. A keyword search does not support the open definition of media science that is required for the construction of cultural indicators. Manual selection of hardcopy newsprint continues to be the only reliable method.

Enthusiastic research teams need to take the necessary steps to construct cultural indicators based on media science. The methodology of content analysis is as mature as that of survey research. However, procedures of media analysis are less industrialized and hence seem to have a somewhat lower public profile. Under the call of "Let a thousand flowers grow," national research efforts supported by local funding agencies may emerge. Teams need to coordinate their analysis toward at least partially comparable data, the lack of which is historically the weak spot of content analysis. In the European context the role of the Commission may be similar to the one it played in the establishment of an indicator in the specific area of biotechnology: the role of encouraging concerted actions by which surveys studies are complemented with media analysis and other context data across the EU member states (Gaskell, Bauer, & Durant, 1997). The coming years will be a unique opportunity for Europe and other world regions to establish indicators within a multicultural context and to measure the climates of science and technology as an essential orientation for all those who are active in research and development and in the public-understanding-of-science movement.

References

Australian Department of Industry, Technology and Commerce (ADITC). (1991). Science and technology news in the media. *Australian Science and Innovation Impact Brief,* section 5, 35–43.

Bauer, M. (1994). A popularizacao da ciencia como imunizacao cultural: a funcao de resistencia das representacoes sociais [Popular science as cultural immunisation: The resistance function of social representations]. In S. Jovchelovitch & P. Guareschi (Eds.), *Textos em representacoes sociais* (pp. 229–260), Petropolis, Brazil: Vozes.

Bauer, M., Durant, J., Ragnarsdottir, A., & Rudolfsdottir, A. (1995). *Science and technology in the British Press, 1946–1990* (Tech. Rep., vols. 1–4). London: Science Museum.

Bauer, M., & Gaskell, G. (1997). *Towards a paradigm for research on social representations.* Manuscript submitted for publication.

Bauer, M., & Ragnarsdottir, A. (1996). A new resource for science communication studies. *Public Understanding of Science, 5,* 55–57.

Beck, U. (1992). *The risk society: Towards a new modernity* (M. Ritter, Trans.). London: Sage. (Original published 1986)

Beveridge, A. A., & Rudell, F. (1988). An evaluation of 'public attitudes toward science and technology' in Science Indicators. *Public Opinion Quarterly, 52,* 374–385.

Buchmann, M. (1995). The impact of resistance to biotechnology in Switzerland: A sociological view of the recent referendum. In M. Bauer (Ed.), *Resistance to new technology—Nuclear power, information technology, biotechnology* (pp. 207–226). Cambridge, England: Cambridge University Press.

Burnham, J. C. (1987). *How superstition won and science lost: Popularizing science and health in the U.S..* New Brunswick, NJ: Rutgers University Press.

Caro, P. (1994). *Le mandala de la vulgarisation des sciences* [The channels of the popularization of the sciences]. Paris: Albin Michel.

Cranach, M. v. (1995). Über das Wissen von sozialen Systemen [On the knowledge of social systems]. In U. Flick, *Psychologie des Sozialen. Repräsentationen in Wissen und Sprache* (pp. 22–53). Hamburg: Rowohlt.

Farr, R., & Moscovici, S. (Eds.). (1984). *Social representations.* Cambridge, England: Cambridge University Press.

Gamson, W. A., & Modigliani, A. (1989). Media discourse and public opinion on nuclear power: A constructivist approach. *American Journal of Sociology, 95,* 1–37.

Gaskell, G., Bauer, M., & Durant, J. (1997, June 26). Europe ambivalent on biotechnology, *Nature, 387,* 845–847.

Gerbner, G. (1969). Toward 'cultural indicators': The analysis of mass mediated public message systems. In G. Gerbner, O. Holsti, K. Krippendorf, W. I. Paisley, & P. I. Stone (Eds.), *The analysis of communication content* (pp. 123–132). New York: John Wiley.

Haller, M (1991). Über Böcke und über Gärtner. Kommentar zu Kepplinger [About foxes and geese: Comments on Kepplinger]. In J. Krueger & S. Russ-Mohl (Eds.), *Risikokommunikation* (pp. 175–196). Berlin: sigma.

Hansen, A. (1994). Journalistic practices and science reporting in the British press. *Public Understanding of Science, 3,* 111–134.

Hansen, A., & Dickinson, R. (1992). Science coverage in the British mass media: Media output and source input. *Communication, 17,* 365–377.

Hilgartner, S. (1990). The dominant view of popularization: Conceptual problems, political uses. *Social Studies of Science, 20,* 519–539.

Jodelet, D. (1989). *Folies et représentations sociales* [Madness and social representations]. Paris: Presses Universitaire de France (PUF).

Jovchelovitch, S. (1996). In defense of representations. *Journal for the Theory of Social Behaviour, 26,* 121–136.

Jurdant, B. (1993). Popularisation of science as autobiography of science. *Public Understanding of Science, 2,* 365–373.

Kepplinger, M. (1989). *Künstliche Horizonte, Folgen, Darstellung und Akzeptanz von Technik in der Bundesrepublik* [Artificial horizons, impacts, representation, and acceptance of technology in the Federal Republic of Germany]. Frankfurt on the Main: Campus.

Klingemann, H.-D., Mohler, P. P., & Weber, R. P. (1982). Cultural indicators based on content analysis: A secondary analysis of Sorokin's data on fluctuations of systems of truth. *Quality and Quantity, 16,* 1–18.

LaFollette, M. C. (1990). *Making science our own: Public images of science, 1910–1955.* Chicago: University of Chicago Press.

Lewenstein, B. (1995). Science and the media. In S. Jasanoff, G. E. Markle, J. C. Peterson, & T. Pinch (Eds.), *The handbook of science and technology studies* (pp. 343–360). Thousand Oaks: Sage Publications.

Miller, J. D. (1983). Scientific literacy: A conceptual and empirical review. *Daedalus, 112*(2), 29–48.

Morgan, M., & Signorelli, N. (1990). Cultivation analysis: Conceptual issues and methodology. In N. Signorelli & M. Morgan (Eds.), *Cultivation analysis: New direction in media effects research* (pp. 8–32). Newbury Park: Sage.

Moscovici, S. (1976). *La psychanalyse—son image et son public* [Psychoanalysis—Its image and its public]. Paris: Presses Universitaire de France (PUF).

Moscovici, S. (1992). The psychology of scientific myths. In M. v. Cranach, W. Doise, & G. Mugny (Eds.), *Social representations and the social bases of knowledge* (pp. 3–9). Bern: Hogrefe.

Namenwirth, J. Z. (1984). Why cultural indicators? A critical agenda. In G. Melischek, K. E. Rosengren, & J. Stoppers (Eds.), *Cultural indicators: An international symposium* (pp. 85–96). Vienna: Verlag der Österreichischen Akademie der Wissenschaften.

Neidhardt, F. (1993). The public as a communication system. *Public Understanding of Science, 2,* 339–350.

Nelkin, D. (1995). Forms of intrusion: Comparing resistance to information technology and biotechnology in the USA. In M. Bauer (Ed.), *Resistance to new technology—Nuclear power, information technology, biotechnology* (pp. 379–391). Cambridge, England: Cambridge University Press.

Raichvarg, D., & Jacques, J. (1991). *Savant et ignorants. Une histoire de la vulgarisation des sciences* [Knowledge and ignorance: A history of the popularization of science]. Paris: Le Seuil.

Roqueplo, P. (1974). *Le partage du savoir. Science, culture, vulgarization* [Sharing knowledge: Science, culture, popularization]. Paris: Edition du Seuil.

Rosengren, K. E. (1984). Cultural indicators for comparative study of culture. In G. Melischek, K. E. Rosengren, & J. Stoppers (Eds.), *Cultural indicators: An international symposium* (pp. 177–194). Vienna: Verlag der Österreichischen Akademie der Wissenschaften.

Ruhrmann, G. (1991). Analyse der Technik- und Risiko-Berichterstattung—Defizite und Forschungsperspektiven, Kommentar zu Kepplinger [Analysis of technology reporting and risk disclosure—Shortcomings and perspectives. Comments on Kepplinger]. In J. Krueger & S. Russ-Mohl (Eds.), *Risikokommunikation* (pp. 145–174). Berlin: sigma.

Schiele, B., Amyot, M., & Benoit, C. (Eds.). (1994). *When science becomes culture.* Ottawa: University of Ottawa Press.

Smircich, L. (1983). Concepts of culture and organizational analysis. *Administrative Science Quarterly, 28,* 339–58.

Tunstall, J. (1996). *Newspaper power: The new national press in Britain.* Oxford, England: Oxford University Press.

Weart, S. R. (1988). *Nuclear fear: A history of images.* Cambridge, MA: Harvard University Press.

Whitley, R. (1985). Knowledge producers and knowledge acquirers. In R. Whitley (Managing Ed.) & T. Shinn & R. Whitley (Vol. Eds.), *Sociology of sciences: Vol. 9. Expository science: forms and functions of popularization* (pp. 3–28). Dordrecht: Reidel.

Wynne, B. (1995). Public understanding of science. In S. Jasanoff, G. E. Markle, J. C. Peterson, & T. Pinch (Eds.), *The handbook of science and technology studies* (pp. 361–388). Thousand Oaks: Sage Publications.

CHAPTER 7

STUDYING PUBLIC PERCEPTION OF BIOTECHNOLOGY: HELICOPTER OR MICROSCOPE?

Anneke M. Hamstra

There are various ways to study public perceptions and attitudes toward a new technology. In this article some of these variations are explored, and reflections are made about their backgrounds and uses. The chapter focuses upon the field of public perceptions and attitudes toward biotechnology. This field is very topical. Since the discovery of the structure of the double helix by J. D. Watson and F. Crick in 1953, modern biotechnology has progressed from a fundamental science into a considerable number of application areas. New biotechnological methods of producing medicine have already become common practice but have not resulted in much social discussion. Medical applications such as gene therapy, on the other hand, are raising great concern. And with the products of modern food biotechnology now actually coming onto the market, a similar stage has begun in the development of biotechnology.

The perspective taken in this chapter is that of the applied researcher looking for studies and results that could inform decision-makers of the main actor groups: governments, science and technology actors, industry, and nongovernmental organizations. In the first part of the chapter, some of the background factors in the variations that can be found in the study of public perceptions of biotechnology are explored in light of the Eurobarometer on biotechnology and other studies on public perceptions and attitudes toward biotechnology, with particular attention to the author's own empirical work on consumer acceptance of the application of biotechnology in food production. This examination then leads to some evaluative thoughts about the various approaches and some recommendations for the future.

Members of the public can come into contact with the developments of a new technology in their roles as citizens, consumers, or employees. For this chapter, the roles of citizen and consumer are the most relevant. As citizens, people react—actively or passively, via political actions such as voting and protesting—to the political decisions that are being made about technology policy. As consumers, people are directly or indirectly confronted with the results of technological developments in the form of goods and services. They may react in various ways: with passive acceptance (by buying, but not feeling positive about it), with positive appreciation (by showing clear consumer preference for the new product), with reluctance (by leaving the new product mostly on the shelves), or with clear protest behavior (by boycotting the new product). Although the roles can be distinguished clearly, they are, of course, combined within each person. A transformation of citizen arguments or opinions to consumer behavior or the other way around may take place, but internal conflicts between citizen responsibility and consumer needs are also quite possible (see, for example, Loor, Midden, & Hisschemöller, 1992).

Public perceptions of and attitudes toward a new technology have become increasingly relevant to the various involved parties in society, especially to governments, industry, science and technology actors, and nongovernmental organizations. Each group of actors has specific reasons for its interest in public perceptions and attitudes.

In democratic societies, governments are to a large extent dependent on public support for their policies. Whereas technological developments used to be seen as more or less autonomous, they have now become the focus of policy—not only to stimulate the development of technology for economic reasons but also to take negative effects of technological developments into account (Smits & Leyten, 1991). The public, although generally advocating the use of technology, has become more aware of the adverse effects that the application of technologies can entail. Well-known cases are the public reactions toward nuclear energy and the unrest over the environmental and health effects of pesticides. It has therefore become important for governments to get a feel for public perceptions and opinions about a new technology.

The actor groups in the private sector, especially the companies applying a new technology, have other reasons to be concerned about public attitudes toward a new technology. Eventually, the market success of products made with the help of a new technology may to a large extent depend on public attitudes toward that technology or product. This relation affects not only companies making and selling consumer products but, indirectly, also the producers supplying other companies. For example, the yeast-producing industry may have to face the concerns of bakers, who may fear that consumers will not buy bread baked with the help of genetically modified yeast. Apart from direct sales results, companies may have to reckon with impacts on their corporate image if they use a controversial new technology. Hence, industry is obviously interested in information about public perceptions of and attitudes toward new technology and its application. This interest may range from general indications of public opinion about a given technology to very specific reactions to products in which the technology is applied.

Public perceptions and attitudes may also be of interest to the science and technology actors working on the development of a new technology. In Denmark, for example, a 1987 consensus conference on gene technology in industry and agriculture led directly to a decision by the Danish Parliament to eliminate public funding of research into transgenic animals from 1987 to 1990 (Klüver, 1995). However, the influence that public opinion has upon decisions about the public funding of research is usually more indirect.

Nongovernmental organizations, such as environmental or consumer organizations, have an important intermediary role between governmental or industrial actors and the general public. They articulate public concerns about the negative effects of new technologies and their applications and make the public more aware of potential or real problems. However, their success in influencing decision-makers depends on public support as well. This means that nongovernmental organizations also use, and even initiate, studies into public perceptions of and attitudes toward a new technology.

Given the various interests that different actor groups have in public perception and attitude studies, the existence of a "market" for a range of various types of studies is not surprising. In the case of biotechnology, there are studies initiated by governmental bodies (including the European Commission), industrial parties, retailers, and consumer organizations. Although the researchers conducting a study may in general have ample freedom with which to do so in a reasonably independent way, it is likely that the scope and perspective of the studies vary with the questions and interests of the party financing the work. For example, some studies primarily focus on members of the public in their roles as citizens (e.g., Borre, 1989, 1990; Heijs & Midden, 1994, 1995; Heijs, Midden, & Drabbe, 1993; Institut National de Recherche Agricole [INRA] & Marlier, 1993; Marlier, 1992); others focus mainly on their consumer roles (e.g., Folkers, 1992; Hamstra, 1991a, 1993, 1995). Still others combine these approaches (e.g., Hoban & Kendall, 1992; Kelley, 1995). Some studies provide a "helicopter" view to give an impression of public perceptions of biotechnology in general (e.g., INRA & Marlier, 1993; Marlier, 1992; OTA, 1987); others zoom in on specific application fields (e.g., food) or even the reactions to concrete products (Hamstra, 1991a, 1993, 1995).

But there is a further difference between the studies that is also worth mentioning. The various actors who initiate the studies and the researchers conducting the work may have dissimilar perceptions of the public's role in technological development. These differences are expressed in the choice of independent variables explaining public perceptions and attitudes. On the one hand, there are attempts to explain variation in public attitudes toward biotechnology by looking at characteristics of the public: demographic features (age, gender, education); personal orientations (religion, political views); interest in, knowledge of, and attitudes toward technology in general; and media use. Finding out whether and how these characteristics are related to the public's perceptions of and attitudes toward technological development gives users of the study an insight into the need and potential for educating and informing the public. A traditional expectation is that a more educated and informed public will be more positive about the new technology. This point will be elaborated upon later.

On the other hand, the public attitude may be related to characteristics of the technology or its applications. In that case, public perceptions of and attitudes toward expected effects or aspects of the technology (e.g., economic and environmental effects, ethical aspects) can be used to explain differences in attitudes. In addition, the perceived characteristics of the technology or its applications (e.g., natural, healthy, expensive) can be chosen as the independent variable to explain variation in attitudes. This approach may contrast attitudes toward biotechnology with attitudes toward other technologies, or it may contrast attitudes toward one application of biotechnology with attitudes toward another application. Explanations of this kind inform the actors about the acceptability of applications, enabling them to adapt decisions about the direction of further development rather than to attempt to influence the public.

Of course, both approaches may be used simultaneously. In many cases, however, the emphasis will be more on one or the other.

Looking for Explanations of Public Perceptions and Attitudes

A general issue in studies on the public perception of and attitude toward biotechnology may be whether the public will accept it. This question is directly followed by another: What influences this level of acceptance? As already indicated these questions can be approached in different ways. In this section I first examine the approach of using the public's characteristics as independent variables. The focus is on attitudes toward science and technology in general and on knowledge of biotechnology because of their potential for influence through information and education. Personal predispositions, broad attitudes (e.g., the attitude toward the environment), and, to a lesser extent, demographic features can also be of interest when explaining attitudes. (Demographic characteristics are easy to measure but generally explain little about public attitudes toward new technology; see Loor et al., 1992.) However, these last three groups of variables will not be discussed in this section, for they are less easily influenced by policy instruments.

Attitudes toward Science and Technology, and Knowledge of Biotechnology

From one traditional viewpoint, the attitude toward science and technology in general is seen as an explanatory variable for levels of acceptance of specific technologies (i.e., public understanding of science; for comments on this approach, see Wynne, 1995, among others). However, a "general attitude" toward technology does not necessarily predict the attitude toward a specific technology. In a study by Daamen, Lans, and Midden (1990), for example, the correlation between the general attitude toward technology and the specific attitude toward genetic engineering was .13 if the general attitude was measured first, and .29 if the general attitude was measured after genetic engineering was presented. For other technologies, these figures ranged between .20 and .39.

In several cases, studies on public acceptance of new biotechnology have had questions about technology in general. One example is the study by the Office of Technology Assessment (OTA) in the United States (OTA, 1987). Macer (1992) used the OTA questions in his public opinion survey on biotechnology in Japan and New Zealand. The two Eurobarometers on biotechnology began with a question that presented seven technologies: solar energy, computers and information technology, biotechnology, genetic engineering, telecommunications, new materials and substances, and space exploration. The next question was: "Do you expect this technology to make your life better or worse in the next 20 years?"

Although it has generally been found that a positive attitude toward science and technology is positively related to the acceptance of a new technology (e.g., Alvensleben, 1989; Hamstra, 1991a; Kelley, 1995; OTA, 1987), this result does not explain why some technologies are viewed quite differently from others by the same respondents. Findings of this type (positive attitudes toward science and technology correlate positively with attitudes toward specific technologies) may lead to general conclusions about the need to educate the public in matters of science and technology, but they give no indication of the reasons for the relative acceptability of the technology in question. Additionally, the general attitude toward technology does not necessarily allow any prediction of attitude toward an application or product of a specific technology (Dongelmans, 1994).

Rather than being a unidimensional predictor, the attitude toward science and technology seems to be a multidimensional space in which various technologies, or even specific applications of those technologies, have different positions. In other words, approximating the midpoint of the space does not predict the position of a specific technology or application very well.

A second factor in this research tradition is the knowledge of the technology in question. In this approach it is often presupposed that increased knowledge of a given technology is connected with an increased level of its acceptance. In order to study this relation, some researchers of public perceptions and attitudes toward biotechnology have used a set of questions to measure public knowledge of biotechnology (e.g., Hamstra, 1995; Heijs & Midden, 1995; INRA & Marlier, 1993; Marlier, 1992).

In the Eurobarometer study, the biotechnology knowledge scale showed a clearly positive correlation between knowledge of biotechnology and the attitudes toward a set of seven applications of biotechnology (INRA & Marlier, 1993). It should be noted, though, that this correlation was found by taking all respondents of the Eurobarometer in all EU member states together. Analysis has shown, however, that the relation between knowledge of science and technology on the one hand and attitudes toward science and technology on the other may very well be positive in one country but not in another (see the analysis of the Eurobarometer on the Public Understanding of Science and Technology in the chapter by Durant et al. in this volume). Almås and Nygård (1995), analyzing the Eurobarometer data, found that in the northern countries of Europe higher knowledge of biotechnology coincided with higher awareness of risks to humans and the environment.

Furthermore, the relation between knowledge and attitude is not necessarily equal in strength and direction for each of the seven application areas that were presented. Zechendorf (1994) pointed out that support for the genetic engineering of animals was lower in the countries with higher levels of knowledge about biotechnology. Heijs and Midden (1995), who originally developed the biotechnology knowledge scale that was later used in the Eurobarometer, applied the scale in their Dutch Public Monitor study. They found that the relation between the rating on the knowledge scale and the attitude toward 10 different clusters of applications was much less clear than in the Eurobarometer analysis. In two clusters of applications (both related to stock-breeding), there was even a negative correlation between knowledge of biotechnology and the attitude toward the application.

Measuring knowledge of biotechnology in the Netherlands, I used a similar set of questions to evaluate a set of 20 products (Hamstra, 1995). Of the 20 products involved, 8 had been genetically engineered. For only three products was a slight positive relation found to exist between the knowledge of biotechnology and the level of acceptance. Two of those products (sauer-kraut and farm cheese) had been made with traditional biotechnology, and one (ecologically grown potatoes) had been made with a traditional physical method. For the other products, there was no relation at all. Interestingly, if the arguments about the acceptance of the products were categorized according to benefits and risks (positive and negative expectations about product characteristics), it turned out that a higher rating on the knowledge questions was linked to a lower perception of risks but not to a higher per-ception of the benefits of the products (Hamstra, 1995).

In other studies, the knowledge level of biotechnology has been measured subjectively. In a study carried out for the Food and Drink Federation (FDF) in the United Kingdom, the subjective knowledge of biotechnology was positively related to both the number of benefits that respondents saw in the use of biotechnology in food production and the number of disadvantages mentioned (FDF, 1995). In the Australian study by Kelley (1995), subjective knowledge of genetic engineering was not related to support for applications of biotechnology, but respondents with higher subjective knowledge were more positive about using genetically engineered products. In Denmark, Borre (1990) found that the group with more supportive attitudes toward genetic engineering and the group with more opposing attitudes toward that field both tended to be more informed about genetic engineering than respondents with more neutral opinions.

What is the relevance of the question about knowledge of a particular technology? Research results thus far seem to indicate that the level of knowledge—objective or subjective—may be a rather marginal factor in explanations of variance in the levels of acceptance of specific applications or products. The original hypothesis that increased knowledge leads to an increased level of acceptance does not seem to hold. The expectation of a positive link between knowledge and attitude is linked to a rather techno-cratic view of society, a view in which scientists and technological experts are seen as the parties who can give the correct answers for the benefit of society. However, apart from the empirical results on macro- or mesolevels, it has been observed on the microlevel that individuals may not like what they learn as they increase their knowledge[1] (see also Hoban & Kendall,

---

[1] At the Dutch consensus conference on transgenic animals, the members of the lay panel were asked to indicate their attitudes toward transgenic animals before, during, and after the process of knowledge acquisition and discussion. Although all 15 panel members had received the same information and had taken part in the same discus-sions, the development of the attitudes differed greatly: Roughly equal groups in the panel became more positive or more negative—independent of their original attitudi-nal position (Hamstra & Feenstra, 1993).

1992). Certain developments or applications of technology will be perceived by the public as going against basic values (Tait, 1988). Increased knowledge may also cause the public to become more uncertain or concerned (Loor et al., 1992). In such cases, education and information may lead to more resistance than to more acceptance. The fact that public resistance should not necessarily be seen as a negative situation, that it could be regarded as a constructive element in the development of technology, is another matter (e.g., Bauer, 1995; Rip, 1987).

However, education and information will remain very important for reasons other than boosting the acceptance of a technology. People have a right to know about new developments and potential risks and benefits related to them and have a right to know about what is being done with public funding for research. Citizens need information if they are to be involved in democratic opinion-forming and decision-making. Moreover, in an increasingly technological society—in everyday life at work and at home—people need at least some access to technical knowledge in order to cope and participate in society (e.g., Loor et al., 1992).

But are attitudes toward and knowledge about general science and technology relevant, important explaining factors for the variance in attitudes toward biotechnology? And what kind of knowledge may be relevant for opinion and attitude formation? Is it the "scientific" or schoolbook-type information such as that used in the knowledge scales mentioned earlier (e.g., "A virus can be infected by a bacterium"—is this true or false?)? Is it knowledge about the scientific method and systems of regulation and control (e.g., "If a company wants to do a field test with genetically modified plants, it should have a permit"— true or false?)? Or is it knowledge about the newest developments (e.g., "Do you think that genetically engineered foodstuffs already exist in the shops?")?

Other factors perhaps more directly connected to aspects and characteristics of technology and its applications may be more promising. Indications of this potential are found in research into the adoption of innovations. Although demographic and attitudinal factors indicate which groups will adopt innovations sooner than others, the perceived characteristics of the innovation, such as the relative advantage of the innovation and the extent to which the innovation fits into existing values, experiences, and needs, are at least as important for adoption (Loor et al., 1992).

Characteristics of Biotechnology, Applications, and Products

If the explanations for variations in attitudes are sought within the objects of the attitude instead of solely in characteristics of the public, it becomes very important to ask how to define these objects, or rather, how to categorize them. It has already been suggested that the public is likely to distinguish between biotechnology and other technologies, and not use only its attitude toward science and technology in general (if such an attitude exists) to form an opinion. But will the public also differentiate between application areas, various biotechnological methods, subjects of biotechnology, effects, and goals of biotechnology? To explore this question, researchers need methods that go into greater depth than is the case with precategorized survey questionnaires in which distinctions among categories are based on the frame of reference and perceptions of the researchers or initiators of the studies.

In 1988, Hamstra and Feenstra started with a series of studies into consumer acceptance of biotechnology in the Netherlands. The first approach to the problem used focus-group discussions and semistructured individual interviews about biotechnology in general in order to find out how members of the general public talk about biotechnology, which questions they have, and which arguments they start using when forming an initial opinion. The results showed that biotechnology was not a concept that respondents could give only one opinion about. Although the level of knowledge was very low, after a brief introduction they immediately said: "Well, it depends what you use it for!" Clear differentiation was made between applications of biotechnology in the areas of food, medicine, and the environment. An additional distinction was made as to whether genetic modification would be applied to animals (or humans) or to plants or microorganisms. In the case of animals or humans, ethical aspects prevailed in the forming of opinion. In the case of plants or microorganisms, the main focus was on the usefulness of the applications (Hamstra & Feenstra, 1989).

Because of the difficulty with talking about public attitudes toward biotechnology in general, the next study was focused on genetic engineering in food production. This delimitation also enabled me to present a series of examples of potential products that would result from the technology. It is very difficult for anybody—experts and laypeople alike—to judge a technology according to its own benefits or risks. The benefits and risks express themselves as effects of applications in relation to needs, wishes, and concerns. Although there may be reasons for people to expect that a broad use of biotechnology might have certain cumulative desired or undesired effects on the economy or environment, for example, these expectations will always relate to the expected applications. The example products were chosen in such a way that they reasonably represented the field of food-related biotechnology. The required information was collected in a separate study (Hamstra, 1991b; updated by Smink & Hamstra, 1994). Additionally, the set of product examples used in the subsequent studies was evaluated for representativity by experts from different actor groups (government, industry, and consumer organizations) active in the biotechnology field (Hamstra, 1993; 1995).

In a preliminary study in which respondents were asked to react to a large series of examples of potential applications (medicine, diagnostics, food industry, environment, and agriculture), Heijs et al. (1993) resolved the problem of differentiated attitudes toward different applications of biotechnology. Factor and cluster analysis showed that these 10 areas would be evaluated differently by the public, according to certain characteristics: life span, techniques used, target objects, origin, sector, and end product and purpose. The results were elaborated with cluster analysis, which resulted in 10 clusters within the total field of biotechnology. In the actual biotechnology monitor, each of these clusters was represented by the application example with which they were most associated (Heijs et al., 1993).

The Eurobarometer on biotechnology introduced seven different application areas referring to types of biotechnology or genetic engineering and mixed in questions testing for variations in attitudes toward different organisms (plants, microorganisms, animals, and humans), toward application areas (food, medicine, and the environment), and toward goals (e.g., quicker, more precise, more useful, better quality, higher in protein). For illustration, the questions are quoted in the appendix. The question arises as to whether these items could not be improved if use were made of insights into the natural tendencies for categorization that are observed in respondents in more open-ended (qualitative) studies.

In the group discussions and individual interviews used in my studies, reactions to various groups (or categories) of applications were observed to belong to one of five attitudinal groupings. One group of people may decide that the application of, for example, genetic engineering to food production is fundamentally wrong; they will reject all applications in that field. The second group may decide that they do not feel comfortable with the principles of the technology (putting genes from one species into another) and thus generally will reject the technology but may keep the door open for applications that are genuinely important to them and that represent effects unachievable through methods other than gene technology. The third group may have no reason to accept or reject any category of application but will evaluate each application on a case-by-case basis; they will balance risks and benefits in a pragmatic way. The fourth group may react positively to applications of genetic engineering to food production in general but will make exceptions if, for example, the production implies environmental or health risks. The fifth group may be fundamentally enthusiastic about the idea of using genetic engineering in food production and may see no immediate reason to limit its application.

An individual respondent may feel comfortable with one attitudinal grouping when the application of genetic engineering in food is concerned but take another position on environmental or medical applications. Such a difference is clearly seen when it comes to the genetic engineering of different organisms (see Table 7-1).

Table 7-1. Attitudes toward the Genetic Engineering of Different Organisms for Food Production: Percentage of Respondents in Each Answering Category

Attitude	Plants	Micro-organisms	Pigs and cows	Fish
Acceptable for all applications	16	18	4	5
Acceptable, with exceptions	39	42	18	13
Assessable only on a case-by-case basis	31	26	22	23
Unacceptable, with exceptions	10	10	28	28
Unacceptable for all applications	4	4	28	30
N	421	422	422	422

Note. From *Consumer Acceptance Model for Food Biotechnology* (Final Report No. Z0011, p. 36), by A. M. Hamstra, 1995, Leiden: SWOKA Institute for Strategic Consumer Research. Reprinted with permission.

"Zooming in" even further and looking at various food products made with the help of genetic engineering, one also finds a clear distinction between the attitudes toward these products (see Table 7-2). Although there is, on average, a tendency for products made with the help of genetic engineering to be somewhat less accepted than products made with traditional biotechnology or other methods,[2] the implication is not that any single product made with genetic engineering is equally acceptable or unaccept-

---

[2] Of the 20 products in the sample (Hamstra, 1995), 12 had been made with modern biotechnology, and 8 of those had been made with genetic engineering. The other products had been made with traditional biotechnology or with physical or chemical methods. The average acceptance for the 12 products of modern biotechnology was significantly lower than that for the other products (t test, $p = .0001$), but this difference is mainly accounted for by the two transgenic animal products. The most accepted products were made with traditional biotechnology (e.g., sauerkraut, farm cheese) and those made with physical methods (like ecologically grown potatoes).

able. The general attitude toward the application of genetic engineering to food production is only a very limited predictor of the eventual acceptance level of an individual product. Only when transgenic animals are involved does the general attitude become a clear indicator of a low acceptance level.

Table 7-2. Mean *(M)* Levels and Standard Deviations *(SD)* of Social Acceptance (SA) and Willingness to Try (WT): Eight Food Products Made with the Help of Genetic Engineering

Product and advantage	SA M 1–20	SA SD	WT M 1–20	WT SD
Yoghurt made with genetically engineered organisms (involves more stable production process)	13.5	5.3	13.7	5.5
Pork from a transgenic pig (reduces level of pork fat)	6.7	5.6	8.4	6.4
Cheese made with rennet from genetically engineered organisms (replaces calf rennet)	14.7	4.8	14.0	5.2
Transgenic tomato, insect resistant (needs less spraying)	13.7	5.4	13.3	5.8
Soft cheese, made with genetically engineered bacteria (makes nisine, against pathogens)	14.2	4.9	13.1	6.0
Transgenic carp (grows much faster than normal carp)	6.3	5.1	5.9	5.4
Transgenic tomato, long shelf life (becomes soft much later)	13.6	5.4	13.6	5.7
Transgenic potato, herbicide-resistant (uses less toxic herbicides)	11.8	5.8	11.9	6.0

Note. The products are presented here in a summarized, technical manner. In the study itself, the respondents received several sentences explaining in understandable terms how the products were made. The mean level of social acceptance for sauerkraut was 16.1; for ecological potatoes, 17.1. On the other hand, a nonbiotech product like instant mashed potatoes had a mean social acceptance of 11.2. Adapted from *Consumer Acceptance Model for Food Biotechnology* (Final Report No. Z0011, p. 133), by A. M. Hamstra, 1995, Leiden: SWOKA Institute for Strategic Consumer Research.

Thus, relevant domains or categories seem to be "genetic engineering of plants and microorganisms for food production" or "genetic engineering of animals for food production" and not "biotechnology" or even "biotechnology for food production." Even within the differentiated domain, there are multiple dimensions that should be taken into account in attempts to predict acceptance of specific applications or products.

Applying this microscope approach even further, one finds which perceived characteristics of the products caused the differentiation between the products. The evaluation of each product was based on a set of expected characteristics such as taste and safety. The decision about which characteristics to use in the study was based on a qualitative investigation in which these characteristics had been identified by consumers. In brief, the procedure implied talking with respondents about food and food production in general and identifying the aspects relevant to consumers as they evaluate food and its production, followed by the introduction of biotechnology, and relate this method to the general scheme. In total, 13 aspects were identified as relevant for the evaluation of foodstuffs made with biotechnology. These 13 aspects were taken up in statements and used in a preliminary phase with 40 respondents in order to evaluate the set of 20 product examples, which included 8 examples of foodstuffs made with the help of genetic engineering (Hamstra, 1993). Two overall product evaluation statements were used to measure the attitudes toward the product: "I think it is acceptable when this product is made this way" (social acceptance, addressing the citizen role) and "If this product is available in the shop, I would like to try it" (willingness to try, addressing the consumer role).

It appears that eight of the aspects related significantly to the overall evaluations (see Table 7-3). Five were positive: taste, the possibility of indulging oneself with the product, safety, health, and sustainable production. Three were negative: expectations about physical complaints resulting from consumption of the product, expectations about disturbance of the natural balance (e.g., of the environment), and the perception that production would only benefit the producer. In a multiple regression model, the eight aspects explained 70% of the variance of the total evaluation (social acceptance and willingness to try). Other explanatory variables, such as demographic variables, general attitudes (e.g., toward food production or nature), and awareness and knowledge of biotechnology and genetic engineering, improved the model only marginally (2%). Taken separately, these variables accounted for 9% of the variance (Hamstra, 1995, p. 87).

The level of detail in this study also makes it possible to look at the profile of individual products in order to see in which of the aspects a product is doing well and in which of the aspects the product or its image can be improved. Two example product profiles are presented in Figures 7-1 and 7-2.

Table 7-3. Eight Aspects (and the Statements as Used in the Survey) that are Significantly Related to Total Evaluation (Social Acceptance and Willingness to Try) of Products

Aspect	Statement
Taste	I think this product will have a nice taste.
Indulge	With this product I can indulge myself.
Safe	This product is safe for me.
Healthy	This product is good for my health.
Sustainable production	If this product is made in this way, my children and grand-children will also be able to have a reasonably good life.
Natural balance	The production of this product disturbs the natural balance.
Physical complaints	When I eat this product, I get physical complaints.
Benefit producer	[By] making this product, only the manufacturer profits.

Note. From Consumer Acceptance Model for Food Biotechnology (Final Report No. Z0011, p. 28), by A. M. Hamstra, 1995, Leiden: SWOKA Institute for Strategic Consumer Research. Reprinted with permission.

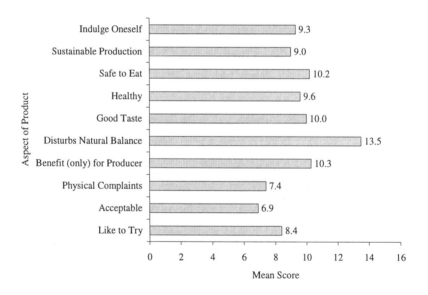

Figure 7-1. Product profile: Public evaluation of pork genetically engineered to reduce the level of pork fat. From *Consumer Acceptance Model for Food Bio-technology* (Final Report No. Z0011, p. 76), by A. M. Hamstra, 1995, Leiden: SWOKA Institute for Strategic Consumer Research. Reprinted with permission.

The description used in the Dutch version of the survey read:

"LESS FAT PORK

These days, many people like to eat lean meat. Therefore, pig-breeders are looking hard for new races of pigs that have less pork fat. A new way to reduce pork fat is to take the hereditary characteristics for the production of growth hormones from a cow and transfer them to a pig. The pig itself subsequently makes bovine growth hormones, which makes it grow faster and produce leaner pork.

The product:

Pork with less fat, achieved by means of bovine growth hormones." (My translation)

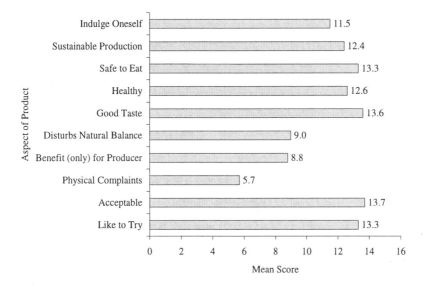

Figure 7-2. Product profile: Public evaluation of an insecticide-producing tomato. From *Consumer Acceptance Model for Food Biotechnology* (Final Report No. Z0011, p. 64), by A. M. Hamstra, 1995, Leiden: SWOKA Institute for Strategic Consumer Research. Reprinted with permission.

The description used in the Dutch version of the survey reads:

"TOMATOES MAKING THEIR OWN INSECTICIDE

Tomato producers often use chemical pesticides to exterminate insects such as fruitworms. Recently, another method has been developed. A certain bacterium makes a toxine that is harmless to humans but lethal to fruitworms. The hereditary characteristic of this special bacterium can be built into the tomato plant. The tomato plant consequently makes a toxin very much like the toxin from the bacterium. This way, the tomato is resistant against the fruitworms.

The product:

Intact tomatoes, grown with less reliance on chemical pesticides."

Another approach to addressing the evaluation criteria for products made with the help of genetic engineering was used by Frewer and her colleagues (Frewer, Howard, & Shepherd, 1997). They found that open-ended interviews did not lead to results sufficient for articulating evaluative aspects. Instead, they used Kelley's (1995) repertory grid method in combination with Procrustes Analysis followed by a quantitative survey to validate the results. With Kelley's repertory grid, each respondent was presented three applications of genetic engineering and was asked to rank them according to a certain criterion (e.g., most concern to least concern). The respondent was then asked to explain why he or she had made this ranking. The results of this exercise were then transformed into a personalized questionnaire in which each of the application examples was rated on the aspects that resulted from the grid method. After that, Procrustes Analysis was applied, a step that yielded key determinants of the attitude.

Detailed studies (e.g., Frewer et al., 1997; Hamstra, 1991a, 1993, 1995; Heijs & Midden, 1994, 1995; Heijs et al., 1993) show that generalized opinions about biotechnology should be regarded with caution. For politicians, the general picture may show tendencies, especially when it can be compared over time, for example, in order to see how the public attitude toward biotechnology is changing. An instructive example is the Danish studies by Borre, who measured general opinions toward gene technology several times, trying to find out whether the opinions tended to become polarized or whether an adoption process was taking place. He found both effects: a general tendency in a more accepting direction as well as an increase in articulated opinions by both supporters and opponents (Borre, 1989, 1990).

Generalized opinions are also interesting for comparisons between countries. This is clearly a very useful asset of Eurobarometers. For example, it can be seen that the public in southern European countries tends to be less concerned about biotechnology and its potential risks than the public in northern European countries. For individual companies that want to decide about the way forward for specific applications of biotechnology, however, these general findings on the public attitude toward biotechnology can be very misleading, for one field or set of applications may be much better accepted than the other. Basing a decision about a specific application on

Eurobarometer findings that "the public" is negative about "biotechnology" in general might mean unnecessary loss of business opportunities.

Conclusions and Discussion

In this article, different approaches to the study of public perceptions of and attitudes toward a new technology such as biotechnology have been discussed. A broad, helicopter view of the public first in its role as citizens, an approach employed in both Eurobarometer surveys on attitudes toward biotechnology, will give a general picture of public opinion. It will be helpful especially in a comparative sense (public attitudes compared over time or between countries). Results of such comparisons are most likely to be useful to politicians and policy at a high level of abstraction.

For more specific decision-making about the directions of developments, including the chances for the market success of applications, the broad approach will not be very helpful. An approach that looks for differentiation of public attitudes toward application fields and products is then more functional.

However, the value of detailed, microscope-type studies in which researchers look at specific products will have more value if the detailed findings can be placed against a broad background such as that provided by Eurobarometers. For example, the Eurobarometer shows a clearly lower general level of acceptance of biotechnology applications in Germany than in other European countries. If taken literally, this finding could very well discourage a company from using biotechnology in Germany. Detailed studies such as the ones carried out in the Netherlands indicate which applications are more acceptable than others and, to some extent, why they are more acceptable. The combined results of the two studies suggest that, although the range of products that might be acceptable may well be smaller in Germany than in the Netherlands, there might be certain products that would also be acceptable to the German public.

A helicopter approach to research can inform research of the microscope type, but the reverse is also true. It has already been suggested that the structure of the questionnaire and the choice of response categories in helicopter studies like the Eurobarometer seem to be based more on the perceptions of representatives of the initiators of the study (in this case, the European Commission) than on qualitative information from the public. To improve the validity and correct interpretation of Eurobarometer results, I recommended that the questionnaire be based on microscope-type insights into the public's frame of reference. Moreover, the interpretation of helicopter results can be informed by more in-depth, microscope-type insights. For example, studies that delve into the details can show whether seemingly neutral attitude ratings are based on obviously ambivalent feelings (respondents who see clear benefits but also clear risks), lack of interest, or the inability to give an articulate answer for lack of knowledge on the subject. Detailed research has already shown that supporting and opposing groups may account for only a relatively small percentage of the total population but that they may have a strong influence on decision-making and the further development of public perceptions and opinions (Martin & Tait, 1992).

The second line of investigation presented here is the exploration of suitable independent variables to explain the variation in public attitudes toward biotechnology. It seems that the general variables *attitude toward science and technology* and *knowledge of biotechnology* are not very successful candidates for this explanatory or predictive role. Although the relation seems to be positive on a very general or macrolevel, both meso- and microstudies and both quantitative and qualitative studies have suggested that the relation is more complex, at least in the northern European countries.

Educating and informing the public about a technology and its applications remains an important activity—important for supporting democracy and the right to be informed and to make balanced choices rather than for boosting an acceptance level. Informing and educating the public may lead to more differentiation in views or more articulation of opinions. For the government, ensuring reasonably open and balanced information for the public is an important task that cannot, or at least should not, be left to market forces. For companies, the choice to be open and inform the public about technology and products is risky precisely because of its partly uncertain outcome. From a company's viewpoint making the "right" (i.e., acceptable) products will make it much easier to mount successful information campaigns than would otherwise be the case.

For the explanation of attitudes toward biotechnology, specific, object-linked perceptions about the effects, benefits, and risks of the applications appear to be much more promising than variables linked directly to personal characteristics. This observation does not explain the general level of acceptance of biotechnology. (Biotechnology seems to be a rather meaningless concept to the general public anyway.) But it does indicate which applications are more acceptable than others. These types of results show the applications to which the public seems to be open and indicate the applications for which additional information will only serve to strengthen public opposition. Such results may thus be used in the preparation of public information campaigns. But information on public perceptions and attitudes may also be taken into account in decision-making about new directions for technological developments and applications. This assertion may imply that discussants will focus on addressing the question of which products and production methods are desirable from a societal point of view (including market demands as well as broader social aspects) rather than on trying to solve acceptance problems by assuming that they stem from public ignorance or scientific illiteracy.

References

Almås, R., & Nygård, B. (1995). *New biotechnologies: Attitudes, social movements and regulation* (Paper No. 2/95). University of Trondheim, Norway: Centre for Rural Research. ISSN 0801-7263.

Alvensleben, R. von (1989). *Die Beurteilung moderner Agrartechnologien durch den Verbraucher* [Consumer judgment of modern agricultural technologies]. University Conference Lectures, Vol. 71. University of Kiel, Faculty of agricultural sciences.

Bauer, M. (Ed.). (1995). *Resistance to new technology—Nuclear power, information technology, biotechnology.* Cambridge, England: Cambridge University Press.

Borre, O. (1989). *Befolkningens holdning til gentechnologi. Diffusion eller mobilisering?* [Public attitudes toward gene technology: Diffusion or mobilization?] (Teknologi Nævnets Rapporter 1989/3). Copenhagen: Danish Technology Board.

Borre, O. (1990). *Befolkningens holdning til genteknologi II, Kommunikation og tillid* [Public attitudes toward gene technology II: Communication and trust]. (Teknologi Nævnets Rapporter 1990/4).Copenhagen: Danish Technology Board.

Daamen, D. L., Lans, I. A. van der, & Midden, C. J. H. (1990). Cognitive structures in the perception of modern technologies. *Science, Technology, and Human Values, 15,* 202–225.

Dongelmans, P. C. A. (1994). *Substraatteelt, genetisch modificatie en voedseldoorstraling. Acceptabele technologieën voor de produktie van consumptiegewassen?* [Hydroculture, genetic engineering, and food irradiation: Acceptable technologies for the production of food crops?] (SWOKA Research Report No. 145). The Hague: SWOKA Institute for Consumer Research.

Folkers, D. (1992). Verbraucherbefragung zur Gen- und Biotechnologie im Ernährungsbereich [Consumer survey on gene engineering and biotechnology in the food industry]. In K. D. Jany & B. Tauscher (Eds.), *Biotechnologie im Ernährungsbereich. Berichte der Bundesforschungsanstalt für Ernährung* (No. BFE-R-92-03, pp. 37–45). Karlsruhe: Bundesforschungsanstalt für Ernährung.

Food and Drink Federation (FDF) (1995). *Consumers and Biotechnology.* Report on research conducted for The Food and Drink Federation, U.K. London: Food and Drink Federation.

Frewer, L. C., Howard, C., & Shepherd, R. (1997). Public concerns in the United Kingdom about general and specific applications of genetic engineering: Risk, benefit, and ethics. *Science, Technology, and Human Values, 22,* 98–124.

Hamstra, A. M. (1991a). *Biotechnology in foodstuffs: Towards a model of consumer acceptance* (SWOKA Research Report No. 105). The Hague: SWOKA Institute for Consumer Research.

Hamstra, A. M. (1991b, June). *Impact of the developments of the new biotechnology on consumers in the field of food products* (SWOKA Final Report). The Hague: SWOKA Institute for Consumer Research.

Hamstra, A. M. (1993). *Consumer acceptance of food biotechnology: The relation between product evaluation and acceptance* (SWOKA Research Report No. 137). The Hague: SWOKA Institute for Consumer Research.

Hamstra, A. M. (1995). *Consumer Acceptance Model for Food Biotechnology* (SWOKA Final Report No. Z0011). Leiden: SWOKA Institute for Strategic Consumer Research.

Hamstra, A. M., & Feenstra, M. H. (1989). *Consument en biotechnologie. Kennis en meningsvorming van consumenten over biotechnologie* [Consumer and biotechnology: Consumer knowledge and opinion about biotechnology] (SWOKA Research Report No. 85). The Hague: SWOKA Institute for Consumer Research.

Hamstra, A. M., & Feenstra, M. H. (1993). *Publiek Debat Genetische modificatie van dieren, mag dat? Projectverslag en evaluatie* [Public debate about "The genetic modification of animals: Should it be allowed?" Report and evaluation] (SWOKA onderzoeksrapport No. 153). The Hague: SWOKA Institute for Consumer Research.

Heijs, W. J. M., & Midden, C. J. H. (1994). *Biotechnologie, houdingen en achtergronden. Tweede meting* [Biotechnology: Attitudes and influencing factors, second survey]. The Netherlands, Eindhoven University of Technology, Department of Philosophy and Social Science.

Heijs, W. J. M., & Midden, C. J. H. (1995). *Biotechnology: Attitudes and influencing factors, third survey.* The Netherlands, Eindhoven University of Technology, Department of Philosophy and Social Sciences.

Heijs, W. J. M., Midden, C. J. H., & Drabbe, R. A. J. (1993). *Biotechnology: Attitudes and influencing factors.* The Netherlands, Eindhoven University of Technology. (Report also available from the Ministry of Economic Affairs, The Hague)

Hoban, T. J., IV, & Kendall, P. A. (1992). *Consumer attitudes about the use of biotechnology in agriculture and food production* (Draft of final report). Raleigh, NC: North Carolina State University.

Institut National de Recherche Agricole (INRA), & Marlier, E. (1993). *Eurobarometer 39.1—Biotechnology and genetic engineering: What Europeans think about it in 1993* (Report DG XII/E/1). Brussels: Commission of the European Communities.

Kelley, J. (1995). *Public Perceptions of Genetic Engineering: Australia, 1994* (Final report to the Department of Industry, Science and Technology). n. p.: Institute of Advanced Studies, Australian National University.

Klüver, L. (1995). Consensus conferences at the Danish Board of Technology. In S. Joss & J. Durant (Eds.), *Public participation in science: The role of consensus conferences Europe* (pp. 41–51). London: Science Museum.

Loor, H. M. de, Midden, C. J. H., & Hisschemöller, M. (1992). *Publieksoordelen over nieuwe technologie. De bruikbaarheid van publieksonderzoek voor technologiebeleid* [Public opinions on new technology: The usability of public opinion research for technology policy]. The Hague: NOTA [Dutch Office of Technology and Assessment (The Rathenau Institute)].

Macer, D. R. J. (1992). *Attitudes to genetic engineering: Japanese and international comparisons*. Christchurch, New Zealand: Eubios Ethics Institute.

Marlier, E. (1992). Eurobarometer 35.1. Opinions of Europeans on biotechnology in 1991. In J. Durant (Ed.), *Biotechnology in public: A review of recent research* (pp. 52–109). London: Science Museum for the European Federation of Biotechnology.

Martin, S., & Tait, J. (1992). Attitudes of selected public groups in the U.K. to biotechnology. In J. Durant (Ed.), *Biotechnology in public: A review of recent research* (pp. 28–41). London: Science Museum for the European Federation of Biotechnology.

OTA (Office of Technology Assessment). (1987). *New developments in Biotechnology* (Background paper 2, U.S. Congress Office of Technology Assessment, OTA-BP-BA-45). Washington, DC: U.S. Government Printing Office.

Rip, A. (1987). Controversies as informal technology assessment. *Knowledge, 8,* 349–371.

Smink, G. C. J., & Hamstra, A. M. (1994). *Impacts of new biotechnology in food production on consumers* (SWOKA Research Report No. 170). The Hague: SWOKA Institute for Consumer Research.

Smits, R., & Leyten, J. (1991). *Technology Assessment: waakhond of speurhond? Naar een integraal technologiebeleid* [Technology assessment: Watchdog or bloodhound? Toward an integrated policy] (TNO Study Center for Technology and Policy). Zeist, the Netherlands: Kerckebosch, BV.

Tait, J. (1988). Public perception of biotechnology hazards. *Journal of Chemical Technology and Biotechnology, 43,* 363–372.

Wynne, B. (1995). Public understanding of science. In S. Jasanoff, G. E. Markle, J. C. Petersen, & T. Pinch (Eds.), *The handbook of science and technology studies* (pp. 361–388). London: Sage Publications.

Zechendorf, B. (1994, September 12). What the public thinks about biotechnology. Better than synthetic food but worse than organ transplantation: A survey of opinion polls. *Bio/Technology, 12,* 870–875.

Appendix
Application Areas as Described in the
Eurobarometers 1991 and 1993

"I would like to ask your opinion about some examples of biotechnology/ genetic engineering research.

[1. PLANTS] Scientists are trying to use biotechnology/genetic engineering to change plants in ways that may be quicker or more precise than traditional breeding programs in order to make the plants more useful. For example, to make them resistant to diseases or pests, make them ripen faster, or give them the ability to grow in dry or salty soils.

[... 2. 'A' MICRO-ORGANISMS] Let us now talk about micro-organisms, such as the yeast we use to make bread, or beer, or yoghurt; or the micro-fungi we use to make medicines such as penicillin.
Scientists know how to change these micro-organisms through biotechnology/genetic engineering in order to improve their performance— that means, getting them to work faster or even to produce new products.

[... 3. 'B' MICRO-ORGANISMS] Some of these micro-organisms are used to break down sewage and other waste products and to turn them into materials harmless to the soil. Here again, scientists are trying, through biotechnology/genetic engineering, to improve these micro-organisms. They are trying to make them work faster or to make them clean up oilslicks or other contaminants in the environment.

[... 4. FARM ANIMALS] Another development is the application of biotechnology/genetic engineering to farm animals, to change them in quicker or more precise ways than traditional breeding programs in order to make them more useful. For example, to make them resistant to diseases, or grow faster, or produce more or better quality meat or milk.

[... 5. FOOD] These new methods of biotechnology/genetic engineering are also being applied to the production and processing of foods. Scientists say that they can improve the quality of food and drink. For example, by making it higher in protein, or lower in fat, or making it keep longer or taste better.

[... 6. MEDICINES/VACCINES] Yet another application of biotechnology/ genetic engineering is the development of new medicines and vaccines to improve human health. For example, the production of human insulin for the treatment of diabetics.

[... 7. HUMAN BEINGS] Science is also trying to apply some of the new methods of biotechnology/genetic engineering to human beings, or to their cells and their tissues, for various purposes such as detecting or curing diseases and characteristics we might have inherited from our parents."

For each of the areas, three issues were addressed:
– Is this research worthwhile and should it be encouraged;
– Does this research involve risks to people or to the environment;
– Should this research be controlled by the government.
(INRA & Marlier, 1993, Appendix 4)

PART THREE

"SCIENCE" AND "THE PUBLIC"— REVISED CONCEPTS

INTRODUCTION

Meinolf Dierkes and Claudia von Grote

The three chapters in this part have the same point of departure—a diffusion model that underlies the Eurobarometer survey questionnaire. The assumption behind this model, also known as the cognitive-deficit approach, is that a certain level of scientific literacy is required in modern societies and that an increase in the public understanding of science and technology will come mainly from transferring relevant knowledge to the general public or specific subgroups. The authors challenge this assumption to different extents. By reinterpreting this approach, E. Einsiedel tries to reconcile it with its critics, whereas S. Yearley and K. Sørensen, M. Aune, and M. Hatling's objection to the diffusion model extends beyond Einsiedel's arguments. The discussion in all three chapters centers on what the diffusion model means for the use of the two central concepts, *science* and *public,* in the investigations of the public understanding of science and technology. Who is defined as the public? Which image of science is involved when one asks about public understanding or lack thereof?

In chapter 10, Einsiedel takes on the difficult task of utilizing constructivist, or interactive, critique to discuss the neglected dimensions and heterogeneity of the concept of public without accepting this extension of the concept as a refutation of the cognitive-deficit approach. The second part of her essay is an attempt to examine the expanded concept of the public for the theoretical and practical implications of its application and contribution to the cognitive-deficit approach as compared to the constructivist approach. She asks which public is captured better by the one or the other and which methodological consequences must be kept in mind.

Whereas Einsiedel's article specifies the contribution that the cognitive-deficit model makes to the analysis of the public understanding of science and technology, Yearley and Sørensen et al. formulate a basic theoretical critique of what, in this model, are taken to be relevant aspects. Starting out from the restricted model of science as conceived of in the diffusion approach, Yearley takes up Wynne's point that the public never experiences science without some sort of context. There are two ways in which this initial assumption figures in Yearley's analysis of the understanding of science. First, he studies two structural elements underlying the processes of scientific work and exchange—the trust that scholars must place in the operations of scientific work and in the discursive process of exchange within the scientific community, and the judgment resulting from the assessment of data and analyzed facts. Second, Yearley shows the relation that exists between these elements of scientific practice and their impact on social and institutional contexts. Because of the increasing importance of science in modern societies, the public is experiencing more and more science in such institutional contexts as courts of law, which systematically and effectively allow for contradictions and divergent opinions.

This analysis of the relation between scientific practice and society provides Yearley with a framework for studying attitudes toward and knowledge of science and technology. That framework raises relevant questions other than those about the population's level of education, namely, those about the mechanisms of building trust as a formative process in the way the population deals with scientific and technological issues. Both points addressed by Yearley, the building of trust and the active intercourse with science and technology, are key elements of analysis in later chapters of this reader as well (see part 4).

This concept of an active public, which stems from Yearley's criticism of the traditional concept of science, serves as the starting point for Sørensen et al. Their rejection of the linear model of providing society with scientific facts and technological artifacts centers on the concept of actively appropriating knowledge and technical innovations. In this approach, the public is conceived of as a user who must construct the meaning of scientific results and technical products in daily life. Borrowing the concept of domestication from anthropology, Sørensen et al. study the cultural appropriation of facts and artifacts on the basis of three empirical cases—not as integration in a cultural setting but as a process of negotiating meaning and practical use.

Sørensen et al. thereby take the traditional, diffusion approach to studying the public understanding of science and technology and turn it upside down. As in Yearley's essay, the question of how much formally codified knowledge the public has is judged by Sørensen et al. to be of secondary importance in the overall process of understanding science and technology. The traditional approach's educational bias toward the question of the public understanding of science and technology prevents insights into the process by which knowledge is transformed and passed on for use in daily life. From the public's viewpoint, that issue is precisely what the domestication approach centers on. Its very core is the analysis of understanding as a multidimensional undertaking—a cognitive, emotional, and practical process of appropriation.

CHAPTER 8

UNDERSTANDING "PUBLICS" IN THE PUBLIC UNDERSTANDING OF SCIENCE

Edna F. Einsiedel

Research on the public understanding of science initially presumed three neatly circumscribed arenas: science, the public, and understanding. Early research began with much concern about "understanding," that is, what the public knows and does not know about science. Underlying much of this concern was the modernist, or enlightenment, assumption that a certain level of scientific literacy or understanding was requisite in order for modern democratic societies to flourish. This approach was called the "cognitive deficit" model of public understanding (Layton, Jenkins, McGill, & Davey, 1993; Ziman, 1991). Practitioners of this approach considered science to be a fixed body of knowledge; saw scientists as having the primary, if not the sole, claim to expertise; and viewed the public as passive recipients of "scientific wisdom." In the 1970s and 1980s, studies that were focused on scientific literacy and grounded in large-scale surveys of national populations were representative of this genre.

Following this flurry of research, the postmodern approach was adopted for a critical examination of the key elements in the public understanding of science: the notions of *public understanding* and *science* (see, e.g., Wynne, 1995). This approach, variously called the constructivist approach and the "interactive science" perspective, has called attention to the uncertainty of scientific knowledge, the inseparability of science from its social and institutional contexts, the lack of demarcation between scientific knowledge and other kinds of knowledge that are needed by decision-makers who are not scientific experts, and the functionality and defensibility of public "ignorance." The contrasting images of the public evoked by these two perspectives point to the need to elaborate on the understanding of *publics* in order to further this discussion of science and the public. I also address the seemingly intractable gap between these two models of the public.

Many scholars have argued over the term public. As Habermas (1962) maintained, "the use of 'public' and of 'the public' betrays a multiplicity of competing meanings" (p. 13). One of the word's meanings alludes to "openness" (Noelle-Neumann, 1984, p. 63), as understood in the expressions *public place* and *public trial,* which denote the opposite of a private space or sphere. This meaning is relevant here insofar as one might argue that many of the issues involving science are indeed public issues in several respects. Issues that often involve a scientific dimension are often discussed in the public domain because they involve public interests, such as what to do with CFCs, the addictive qualities of nicotine, and the ozone problem. The public nature of science is also due to the fact that scientists themselves and their activities are matters of public interest. The public funding of scientific research, the ethical parameters within which science must operate, and scientific standards that have impacts on particular political judgments (e.g., risk assessments) are a few examples.

Another attribute that needs to be accounted for is the heterogeneity of publics. John Dewey (1927) maintained that there are many publics, each consisting of individuals who, together, are affected by a particular action or idea. Thus, each issue creates its own public, and each public will not normally consist of the same individuals who make up any other particular public, although every individual will, at any given time, be a member of many other publics. For example, a person may be a church member, a rider of mass transit, a reader of printed mass media, and a member of an arts group. When an issue concerning arts groups arises, the individual may join with other theater members in coalition with other artistic communities; the opinion of these people becomes, for this issue, one public opinion. The resolution of this issue may then dissolve the public that coalesced in response to it.

Technological developments have also drawn the attention of specific groups at various points in the process. Bijker (1995) labeled them "relevant social groups," indicating that they can be influential in shaping the directions that technology takes. His case studies of mundane technologies such as the bicycle have demonstrated how particular groups of users, emphasizing values ranging from safety to masculinity or high fashion, can affect technological design and evolutionary paths at particular times. This potential has been further demonstrated in the area of technological resistance. For instance, deaf people succeeded in impeding the further deployment of a cochlear implant (or "bionic ear," as its producers and the media labeled the product), partly because the product's designers had failed to recognize that deaf persons form an established culture and community that has its own norms and forms of communication (see Blume, 1997).

In summary, this view of the public as issue-centered, transient, occasional, or value-directed is another dimension of publics that needs to be considered.

In addition, one might see the notion of public in terms that reflect or coincide with identity (national, ethnic, and political are only a few of the possible subcategories). References to "the European public" and "the British public" fall within this domain, for they illustrate national or regional groupings. Ethnicity has spawned groupings that sometimes regroup under new names (African Americans) or combine with religious subidentities (Bosnian Muslims). By the same token, identity politics create publics. The feminist and the gay and lesbian communities are just two examples. The state may create labels that impose identities, such as "guest workers" in Germany and "visible minorities" in Canada.

Noelle-Neumann (1984, p. 61) also speaks of what she metaphorically referred to as "a social skin" to publicness. That is, "an individual does not live only in that inner space where he thinks and feels. His life is also turned outside, not just to other persons, but also the collectivity as a whole" (p. 61). To be a member of a given public is to bring into that membership the web of social relationships that surrounds one and to imbue that membership with these broader social frameworks.

These facets of the word public suggest that there is not necessarily one homogeneous public but many and heterogeneous publics that act in social contexts and shift their attention and levels of knowledge with the rise and fall of a variety of issues. That is not to say that there is no such thing as a public for science. Indeed, as Miller (1983) pointed out in an early depiction of publics for science, one may talk about a public attentive to science in the same way that one may talk about a public attentive to politics or sports. The success of science sections in many elite newspapers bears Miller out. But again, these publics are among the specialized ones that exist for a variety of issues, times, and places. For the majority of people, whom Miller (1983, p. 8) labeled the "inattentives," I argue that this inattention may vary with time, place, and circumstance.

One of the contextual factors researchers need to know of in order to understand this public's relationship with science is the junctions where the public and science meet. There are many such junctions. The political, the market-centered, the occupational, and the leisure-oriented junctions are only a few examples. At the political junction, citizens may be called on to make judgments in referendums on issues such as the siting of a nuclear facility, the introduction of water fluoridation, or the building of a toxic-waste disposal plant. In these instances, a background in science may be one of several factors that come into play in the decision-making process. The junction represented by the market, or consumption, may be another factor. With genetically engineered products on the supermarket shelf, the shopper may have to decide whether he or she wants a "natural" product or whether a genetically engineered one offers a better value. The person must take such things as price, physical quality, and safety into account. At the occupational junction, the geologist working for an oil and gas company may know just as much about geology as a scientist, if not more; the fisherman may have situational understandings of marine ecology that may be more precise than the marine ecologist's. At the leisure junction, people may take their families to a science museum to "learn about science," but the primary reason for the visit may be family entertainment.

And what about the social context of understanding science, the social skin to which Noelle-Neumann so poetically referred? It exerts some control over what one may choose to learn, the opinions one dares to express, the contexts in which one finds oneself inadvertently learning, and the knowledges one is prohibited from having. Social taboos or group norms may exert pressures on what young people may be able or willing to learn about sex, their bodies, or reproduction.

The boundaries between scientific experts and the lay public also may not be fixed, and levels of expertise may vary among various publics. Shapin (1990) noted in his examination of the historical make-up of the divide between laypeople and experts that the question of who possesses cultural competence "is one of the most obvious means by which we, and people in the past, discriminate between 'science' and the 'public'" (p. 993). Although experts may think of this boundary as clear, there is considerable evidence to the contrary. Indeed, the lay publics for various topics have well-known experts. The acquisition of expertise may be fueled by a risk situation, defined by medical diagnosis or an external event (e.g., the discovery of a hazard in one's neighborhood), or determined simply by interest (e.g., the astronomical discoveries made by amateur astronomers or the expertise of amateur bird-watchers or amateur paleontologists).

A number of case studies have documented the development of expertise among lay publics (see, e.g., Epstein, 1995). In an examination of development of expertise among AIDS activists, Epstein (1995) found that the learning process involved the acquisition of the language and culture of the medical community—its vocabulary, its working procedures and practices, and even a working knowledge of its scientific foundations (e.g., drug mechanisms and the viral replication cycle of HIV). Once these activists learned the basics, including the language of biomedicine, they were able to make the AIDS research community address some of their concerns. More important, by using the credibility tactics just described, these activists succeeded in establishing themselves as *partners* in the research process and then in winning changes in medical protocols, such as the broadening of the representation of the recipient population for drug tests (as opposed to the principle of a more homogeneous and hence more numerically limited pool of subjects). Thus, the activists achieved what was agreed to be a more scientifically credible policy for gathering generalizable data and what they perceived to be a fairer research approach. In the end, they effectively challenged traditional notions about knowledge-making, about the public as a passive audience, and about science as a closed or autonomous arena, and thereby shifted definitions about who has cultural authority and currency in science and medicine.

The acquisition of expertise among lay publics has also been spurred by public concern about scientific disagreements. Nelkin (1977) has maintained that the very existence, rather than the substance, of technical debate can sometimes galvanize a public into learning and action. In controversies involving the proposed locations for a nuclear power plant and an airport, "the fact that there was disagreement among experts confirmed the fears of the community and directed attention to what they felt was an arbitrary decision-making procedure in which expertise was used to mask questions of political priorities" (p. 199).

At the Public Agenda Foundation in Washington, DC, researchers have conducted interesting experiments comparing the ways in which scientists and laypeople deal with complex policy questions such as global warming and solid-waste disposal. The results show that laypeople, when given an opportunity and sufficient information, were perfectly capable of coming to considered judgments about complex technological issues and that their judgments sometimes dovetailed with those of the scientists. Divergence between the judgments of the two groups stemmed not from different levels of technical expertise but from different values (Doble, 1995).

I have made a case for understanding the complexities inherent in the concept of "the public" in the public understanding of science. Underlying this argument are the beliefs that there are many publics and that their compositions and natures vary with the circumstances under which publics and science meet. What are the implications of this understanding of publics? In the following pages I discuss one theoretical and one practical implication.

In asserting that publics are heterogeneous, I may have inadvertently made a case against the cognitive deficit model and in favor of the interactive science model. This is not my intention, however, for both models have things to contribute to the ongoing discussions about the public and science. In the cognitive deficit model, as mentioned above, science is viewed as a set body of knowledge that one ought to know for reasons of civic mindedness, cultural savvy, and so forth. Contrasting this model with the interactive science model may have analytical value, but one thereby tends to over-emphasize the stark differences between the two and to overlook the possibility that these frameworks may be complementary rather than mutually exclusive.

Applying the cognitive deficit model does not necessarily mean that one accepts the notion of science as a fixed and unchanging body of knowledge. It implies instead a recognition that different degrees of consensus on particular problems or areas of knowledge exist. There is generally great consensus on the connections between cigarette smoking and lung cancer (despite the tobacco industry's proclamations to the contrary) and on the links between sunlight and skin cancer. There is considerably less consensus on the links between alcohol consumption and breast cancer. There is considerable consensus on many of the "factual" items employed in such surveys (e.g., that lasers function on the basis of light rather than soundwaves, that hot air rises, and that the oxygen humans breathe comes from plants). What do the responses as a whole indicate? No more and no less than the distribution of knowledge as seen through particular indicators. Is it possible to acquire a sense of the context for differences in understanding? At a general level, the answer is yes. A survey of young people in Surrey, England, showed that men's interest in a job in the field of science was accounted for by such factors as their liking school, watching science programs on television, and having higher educational aspirations than their peers. For the young women in the survey, attitudes were more predictive. That is, those who did not want a job in the field of science were more likely to be cynical about the scientific establishment and to see practices of sexual discrimination in the scientific enterprise (see Ziman, 1991, p. 104).

Comparisons of measures of scientific literacy in Europe show interesting cross- national differences that have been explained in terms of cultural and structural dissimilarities (see Bauer, 1996). In Canada, I found significant gender differences showing that women were consistently more likely than men to score poorly on measures of knowledge (Einsiedel, 1991, 1994), a pattern documented by other surveys as well (see Durant, Evans, & Thomas, 1989; Miller, 1991). At first, one might simply point to the need for greater efforts to educate women about science. Further analysis, however, reveals that women also take fewer science courses than men do (a fact well documented elsewhere; Etzkowitz, Kemelgor, Neuschatz, Uzzi, & Alonzo, 1994; Industry, Science and Technology Canada, 1991) and that women are more distrustful of science than men are. There are many studies documenting women's negative experiences with science, documentation suggesting that sociostructural factors, not individual capacities, are to be held accountable for women's distrust of science. One policy response has been to critically examine the way science is taught at the primary and secondary school level and to understand how girls deal with or respond to different subjects. The way science is often taught—as an abstract theoretical entity unmoored from social contexts and relationships—has succeeded in frustrating and alienating girls (Edwards, 1995). This impact has been exacerbated by the highly competitive behavior of boys in science classes, which increases the marginalization of girls in science classes despite the fact that boys and girls perform at similar levels in earlier years. Some school boards have now begun to encourage experimentation with all-girl science courses in junior high school or middle school and with alternative teaching styles.

What is the value of such surveys? They are certainly useful as one indicator of what people say, think, or know about science as an enterprise or about some concepts with which some publics may or may not be familiar. As an indicator of what people know, one could argue (see, e.g., Wynne, 1991, 1995; Ziman, 1991) that these types of surveys are not very reliable and informative measures of what the public knows about science. But as indicators, they do tap one dimension of knowledge without necessarily negating the fact that there are many other ways to tap understanding. The rationale for knowing the answers to factual questions is no more or no less important than the rationale one might give for having citizens know who the head of state is, what he or she does, or what the capital of China is.

Large-scale surveys can also be useful as a tool for understanding attitudes on specific policy issues or for identifying the extent of specific behaviors. One needs to be careful, however, to discriminate between general and specific attitudes toward science (or technology). The former do not necessarily explain or predict the latter.

The cognitive deficit model to public understanding has been widely criticized for treating the public like empty vessels waiting to be filled. That view of the public implies that a top-down communication approach should be preferred when correcting for collective ignorance. There are many arguments against this authoritarian approach. Publics are diverse, capable of expertise, equipped with information-seeking skills, and attentive and motivated in many instances. But this recognition makes another understanding of publics just as compelling: Publics can also be inattentive, unmotivated, and, yes, ignorant. There are situations when the imperative for public education becomes acute, a shift that makes a top-down flow of information to the public at large more pragmatic than other approaches.

Many of the examples or case studies used to support the interactive science model deal with situations in which members of the public experience an issue directly. Wynne (1989), for example, looked at farmers dealing with a radiation problem and at workers in a nuclear plant; Layton et al. (1993) looked at families coping with a form of Downs' syndrome and at elderly persons dealing with a domestic energy issue. In these instances, there is no question that uses of scientific knowledge intersect with other stocks of public knowledge or that knowledge-seeking depends on motivation, social networks, and so forth.

On the other hand, what if the public largely lacks knowledge about a topic and this ignorance is detrimental to some acknowledged social good? The cognitive deficit model has proved quite useful in such cases, particularly for specific environmental or health issues with which the public has little or no experience. How does a concept enter public consciousness unless the concept's root issue has a champion? And how can these champions succeed without some gauge of public understanding?

Public understanding of AIDS is a useful example of my point. Health educators began with the finding of public uncertainty about the causes of AIDS, the manner in which it is transmitted, and the people who contract it. Because the consequences of public ignorance about the causes of AIDS, the kinds of people it can afflict, and the types of behaviors that are considered risky may be devastating for society as a whole, the identification of the levels and nature of public knowledge became a critical first step in public education. These efforts were based not on a final version of knowledge or truth but on "best knowledge," that is, a version of knowledge that is subject to change as new findings replace the old.

As noted earlier, however, publics are diverse and complex, and there is a need for localized, high-resolution snapshots of the ways in which they experience and use science. The constructivist tradition has made a significant contribution by providing a picture of science and scientific activity that is flexible and uncertain rather than fixed, one based on interests and values. It is an image that, in turn, shapes the public's vision and use (if not rejection) of science.

What about the practical implications of this broader understanding of "publics" in the public understanding of science? The recognition that publics are heterogeneous has a corollary. In the social and political arenas, public participation is also necessarily heterogeneous (Arnstein, 1969). And opportunities to become involved should ideally be available to those for whom understanding needs to be coupled with action. However, in democratic societies, where civic participation is a value, participation remains discretionary and opportunity, for it is still unevenly provided by the different levels of government. Such participation remains largely the purview of political interest groups, the educated, and the affluent.

Experience has demonstrated that individual efforts to create opportunities for action must be combined with institutionally created and supported efforts. Denmark, for example, has pioneered the use of consensus conferences and "citizen juries" for technology assessment (Joss & Durant, 1995; Klüver, 1995). Sclove (1995) detailed other ways by which technological development, implementation, and assessment may be made more democratic and participatory.

I began this chapter by laying out the complexities of publics and emphasizing the need for further insight into the topography of the image and understanding that the public has of science. I briefly described the variety of contexts in which the public and science meet and argued that various publics have different interests and motivations at different times. The premise of the interactive science model is that certain publics are indeed quite capable of acquiring expert knowledge and often do so. The studies conducted by Wynne (1989, 1991, 1995) go a long way toward an improved understanding of these experiences. These works are the more obvious sources with which to start, for their authors look at instances where the public may have a need to interact with or think about science directly.

What about those instances when direct experience is precluded or made difficult by certain issues of the day (e.g., the ozone problem and biodiversity)? In these cases the need to understand the public's responses and the influences of mediators (e.g., the mass media and interest groups) becomes critical. The recognition that different publics may entail different approaches to the public understanding enterprise strongly suggests that more inclusive theoretical frameworks can and ought to be used in research in this area. These diverse frameworks make it necessary to widen the range of methodologies currently used to examine the varieties of public understandings and the different contexts for their expression.

Lastly, I discussed the need to couple public understandings of science with the examination of opportunities to increase active participation in science and technology issues (see Durant, 1995). Only then might this line of research enhance its contribution to the thinking, acting, and transacting citizen in a democratic, technological society.

References

Arnstein, S. (1969). A ladder of citizen participation. *American Institute of Planning, 35*, 216–224.

Bauer, M. (1996). Socio-demographic correlates of the 'don't know' responses in knowledge surveys. *Social Science Information, 35*, 1.

Bijker, W. E. (1995). *Of bicycles, bakelites, and bulbs: Toward a theory of socio-technical change.* Cambridge, MA: Massachusetts Institute of Technology (MIT) Press.

Blume, S. (1997). The rhetoric and counter-rhetoric of a "bionic" technology. *Science, Technology, and Human Values, 22,* 31–56.

Dewey, J. (1927). *The public and its problems.* Chicago: Swallow Press.

Doble, J. (1995, April). Public opinion about issues characterized by technological complexity and scientific uncertainty. *Public Understanding of Science, 4,* 95–118.

Durant, J. R. (1995, November 28). *A new agenda for the public understanding of science.* Inaugural lecture, Imperial College, London.

Durant, J. R., Evans, G. A., & Thomas G. P. (1989). The public understanding of science. *Nature, 340,* 11–14.

Edwards, P. (1995). From impact to social process: Computers in society and culture. In S. Jasanoff, G. E. Markle, J. C. Petersen, & T. Pinch (Eds.), *The handbook of science and technology studies* (pp. 257–285). Thousand Oaks: Sage Publications.

Einsiedel, E. F. (1991). *Science literacy among Canadian adults* (Report to the Social Sciences and Humanities Research Council). Calgary, Alberta: University of Calgary.

Einsiedel, E. F. (1994). Mental maps of science. *International Journal of Public Opinion Research, 6,* 35–44.

Epstein, S. (1995). The construction of lay expertise: AIDS activism and the forging of credibility in the reform of clinical trials. *Science, Technology, and Human Values, 20,* 408–437.

Etzkowitz, H., Kemelgor, C., Neuschatz, M., Uzzi, B., & Alonzo, J. (1994). The paradox of critical mass for women in science. *Science, 266,* 51–54.

Habermas, J. (1962). *Strukturwandel der Öffentlichkeit* [Structural change of the public]. Neuwied on the Rhine: Luchterhand.

Industry, Science and Technology Canada. (1991). *Women in science and engineering.* Ottawa: Supplies and Services.

Joss, S., & Durant, J. (Eds.). (1995). *Public participation in science: The role of consensus conferences in Europe.* London: Science Museum.

Klüver, L. (1995). Consensus conferences at the Danish Board of Technology. In S. Joss & J. Durant (Eds.), *Public participation in science: The role of consensus conferences in Europe* (pp. 41–51). London: Science Museum.

Layton, D., Jenkins, E., McGill, S., & Davey, A. (1993). *Inarticulate science? Perspectives on the public understanding of science and some implications for science education.* East Yorkshire, England: Studies in Education Ltd.

Miller, J. (1983). *The American people and science policy: The role of public attitudes in the policy process.* New York: Pergamon Press.

Miller, J. (1991). *The public understanding of science and technology in the United States.* Washington, DC: National Science Foundation.

Nelkin, D. (1977). The political impact of technical expertise. In G. Boyle, D. Elliott, & R. Roy (Eds.), *The politics of technology* (pp. 189–205). New York: Longman.

Noelle-Neumann, E. (1984). *The spiral of silence.* Chicago: University of Chicago Press.

Sclove, R. (1995). *Democracy and technology.* New York: Guilford Press.

Shapin, S. (1990). Science and the public. In R. C. Olby, G. N. Cantor, J. R. R. Christie, & M. J. S. Hodge (Eds.), *Companion to the history of modern science* (pp. 991–1007). London: Routledge.

Wynne, B. (1989). Sheep farming after Chernobyl: A case study in communicating scientific information. *Environment Magazine, 31,* 10–15, 33–39.

Wynne, B. (1991). Knowledges in context. *Science, Technology, and Human Values, 16,* 111–121.

Wynne, B. (1995). The public understanding of science. In S. Jasanoff, G. E. Markle, J. C. Petersen, & T. Pinch (Eds.), *The handbook of science and technology studies* (pp. 361–388). Thousand Oaks, CA: Sage Publications.

Ziman, J. (1991). Public understanding of science. *Science, Technology, and Human Values, 16,* 99–105.

CHAPTER 9

WHAT DOES SCIENCE MEAN IN THE "PUBLIC UNDERSTANDING OF SCIENCE"?

Steven Yearley

In the last ten years the public understanding of science has stealthily emerged to become a focus for a wide variety of activities. It has come into use as a term that crystalizes worries in the scientific community about public attitudes toward and support for science. It has been taken up as a useful slogan by high-technology sectors of industry concerned about the public acceptance of their products. It has simultaneously become a rallying point for the activities of science journalists and other science communicators, with new training programs and degrees bearing the public-understanding-of-science label. It is also, inevitably, a new arena for academic, specifically social scientific and humanistic, research on science. When a new portmanteau term such as PUS (public understanding of science) appears and achieves wide circulation, social scientists commonly talk of a new discourse having been introduced (Yearley, 1995).

My aims in this chapter are threefold: (a) to review the rise of this discourse about the public understanding of science and thus to examine what model of science is assumed in most policy-level discussion of issues relating to the public understanding of science; (b) to suggest weaknesses in this interpretation of what is meant by *science;* and (c) to suggest that an alternative interpretation of scientific knowledge-making, one based on recent work in the sociology and history of science, offers a more promising framework for analyzing public understandings of, and responses to, science and the scientific community. To achieve these aims I use evidence and case-study material from recent qualitative research on the public understanding of science and of environmental issues to suggest that the dominant model is limited because it overlooks key aspects of knowledge-making in the scientific community.

The U.K. Context for the Discourse of Public Understanding

Specific features of the U.K. context made the appeal of the discourse on the public understanding of science in the mid- and late-1980s easy to understand. Sustained and systematic interest in this issue can reasonably be traced to the "Bodmer Report" on the public understanding of science, issued in 1985 by the Royal Society of London. The report was apparently motivated by a feeling within the scientific community of a loss of its rightful influence and standing in official circles, a loss compounded by a perceived decline in the public's support for science.

The Conservative administrations of the 1980s, well kitted-out with a comprehensive political ideology, had little need of scientific advisers on the issues that mattered to them most. The 1980s were politically polarized years in which governments were disinclined to refer controversial policy questions to bodies of outside experts; the great and the good enjoyed something of a rest from public service. In all, scientists were not at the heart of policy or public spending. Contrary to the technocratic fears of the preceding decades, it was not scientists but accountants who seemed poised to take over practical political control, whereas accountancy and neoclassical economics appeared set to become the "queens of the sciences." Worse still, there was minimal official interest in any science that was of no demonstrable short-term commercial potential. Though Conservative administrations had no ideological antipathy to natural science, they tended to take the view that industries know best what they need in the way of R&D and should accordingly pay for it themselves. Politicians saw no compelling reason to increase public generosity toward the sciences.

Alarmed by dwindling political support, the scientific community also found the public uninterested and, if anything, rather inclined to question scientific advice on matters of public policy; on occasions the public even seemed to treat scientists as an unsavory element in civil society. As Bodmer subsequently reflected:

> a large, and apparently growing, section of the community seems to believe that scientists take pleasure in carrying out sadistic and pointless animal experiments in the name of medical research: if such statements are made frequently enough over a long enough period, many will come to believe them as the truth. It follows that misrepresentations, both of the science and of the motives of those carrying it out, must be refuted whenever and wherever they appear. (Bodmer & Wilkins, 1992, p. 8)

Worst of all from the viewpoint of the scientific community, the trends in public and governmental attitudes threatened to be mutually reinforcing.

Because the public would not speak up for the scientific community it was harder for scientists to get politicians to accord them appropriate respect or to increase their research funds. And if even the government did not seem to treat science as a priority, what lesson was the public supposed to derive from that attitude?

This difficult situation was relieved to some extent after the general election in 1992. Not only was there a less doctrinal look to the U.K. government but a White Paper on science was in preparation. When it finally appeared in 1993, this document—though it met with resistance from some sectors of the scientific community—was generally welcomed because it assigned science a definite and central role in the business of the modern British state (U.K. Government, 1993). With a new mission statement and, for a short time, a prominent minister with responsibility for science outside the overweening departments of industry and education, British science was back in from the cold.

At the same time, however, the scientific community in Britain was facing an educational challenge. Market-inspired reform of higher education had switched the emphasis in the Conservative party from the encouragement of subjects the government approved of (the natural sciences and the traditional humanities, and definitely not the social sciences) to the idea that students—*consumers* of education—should be allowed to pursue whatever subjects they wished. In the 15 years from 1980 to 1995, there was a significant switch in students' preferences away from the physical sciences and engineering and toward the social sciences and humanities.

There were related changes in the teaching of science in schools, changes that also created a context for anxieties about the public's understanding of science. First, since the spread of comprehensive education in the 1970s, there has been a great increase in the number of students, girls as well as boys, doing "balanced science" courses, a mix of the sciences that can be pursued up to the age at which compulsory attendance at school ends. By 1990, more GCSE students (16-year-olds) were taking this subject than any other single-science qualification.

In the twin lights of this subject choice and of the fact that relatively few students go on to study science subjects in higher education (and of those who do only about a quarter go on to work in science), commentators are increasingly justifying school science education as training in scientific literacy. In other words, for the majority of students, science education is not a preliminary to becoming expert scientists; accordingly, courses can beneficially be directed to shaping citizens who are competent to deal with the role of science in modern societies. Eventually, this approach might mean that the scientific community will find a better informed public, but in the short term it meant that the percentage of teaching time devoted to science shrank and that many post-GCSE students were turning away from scientific options.

The final contextual element relates to high-technology firms that have also perceived problems that they associate with the need for greater public understanding and appreciation of science. Despite governments' commitments to deregulation and successive bonfires of red tape, such firms still believe they face limitations that competitors in other countries have avoided, limitations that in many cases they believe to be ill founded. Particularly in the area of biotechnology, many scientists have felt that public opinions are out of line with their own (objective) estimations of safety and risk. These scientists feel frustration that the public often seems as happy to believe Greenpeace as to accept the word of leading figures in the scientific community (see Martin & Tait, 1992). For such firms, a commitment to cooperating in efforts to increase public understanding of science fit their commercial agenda because they saw their way blocked by irrational limitations that the public was reluctant to relax.

This whole complex of concerns came together in the White Paper's endorsement of the importance of initiatives to foster the public understanding of science. The White Paper quoted approvingly the Royal Institution's claim that "any national policy for science and technology must contain, as a necessary foundation, the diffusion among the public at large of an appreciation of what science is" (U.K. Government, 1993, p. 65). This solution seemed to suit all the concerned parties because it simply demanded that knowledge be diffused, albeit with new vigor and innovation, down to a receptive public.

The Workings of Science, and Problems with the Diffusion Model

Although the context and the attractions of the discourse on the public understanding of science are readily understood, the assumptions of the diffusion approach have lately come in for severe questioning (see Irwin & Wynne, 1996; Wynne, 1991; Yearley, 1994). The main point of criticism relates to the question of trust and the associated notion of credibility. As Wynne has made clear from his analyses of the public understandings of science in a variety of contexts, the issue of overriding importance is that people do not experience scientific expertise in a pure context, freed from imputed interests and other background expectations. It is people's typical experience that they receive scientific information for a purpose, such as that of persuading them that one washing powder is better than another, that meat is (or is not) a core ingredient of a rounded diet, or that nuclear power is a safe and dependable component of a national energy strategy. Since expertise is so commonly related to the practical agenda of the experts (or of their bosses), people evaluate the information in the light of their regard for the organization disseminating it and of any ulterior purpose that they believe they can spot. To arrive at a fuller appreciation of the significance of this point, one should consider two factors, one within the workings of science itself and the other in the nature of the "contract" that science has with society. I begin with the first factor.

I am far from alone in suggesting that one focus on the workings of science. It is quite common even for spokespersons for science in the public understanding of science movement to find a reason for public misunderstanding in the nature of science itself. However, the key variable they tend to focus on is what one might call the *provisionality* of science. Spokespersons such as Richard Dawkins insist that one of the characteristics of science is that scientific knowledge is never certain.[1] In essence this uncertainty is the point, famously associated with Popper, about the inductive, rather than deductive, nature of empirical science. Researchers can never know for sure that the next instance they see will not behave in some unexpected way, undermining their grand generalizations. Hence, one can never be certain that scientific beliefs are correct. This point lay at the core of Popper's recommendation to scientists to make bold conjectures and then to try to refute them. The argument that can be developed at this stage is that politicians or policy-makers pretend to greater certainty about scientific evidence than is justified and thereby over-egg the pudding of public expectation. New, countermanding evidence will, accordingly, be a big disappointment to a public fed on talk of certainties and may breed disillusion with scientific expertise, although refutations are exactly what scientists do (from time to time) expect.

Without wanting to dismiss entirely the overall issue of science's provisional nature and the uncertainty arising from it, I find it less telling than is commonly thought, in part because the "uncertainty" that is admitted is of a special sort. Scientists do not straight away respond to apparent refutations by giving up their cherished theories. They weigh the apparently disconfirming evidence against the supporting evidence, assess the source of the

---

[1] Dawkins holds a chair in the Public Understanding of Science at Oxford University. His views were, for example, represented in a television program entitled *Breaking the science barrier with Richard Dawkins,* which was broadcast on the U.K.'s Channel 4 on Sunday, 1 September 1996.

evidence, and so on. Hence, Dawkins can express himself with a great deal of certainty against Creationists and not be troubled by the in-principle pro-visional nature of evolutionary theory. My second reason for downplaying this point is that two other aspects of scientific practice seem to me much more important: trust and judgment.

These two characteristics might not be thought of as the typical features of science, and it will be necessary to take a little time to explain what I mean. My point is that the scientific community, even within the natural sciences, depends on trust, trust in people, in machines, and in assumptions. Despite the business of peer review and the formal methods learned by scientists during their long training, scientific life turns on trust. In particular, scientists cannot independently check every detail of every claim made by every other relevant scientist, even during a controversy. Published findings in high-quality journals are generally treated as trustworthy. Neither do scientists personally reexamine the foundations of their discipline, check out atomic theory or the evidence for continental drift. They also trust machines. Scientific observation is more and more performed though the medium of complex machines, yet the design and operating principles of those machines are not exhaustively checked by each scientist who uses them. Even when scientists replicate their colleagues' or their opponents' experiments, they cannot be sure that the replication is precise in every detail (indeed, logically speaking, one might say that a replication can never be fully the same as the original, see Collins, 1985, pp. 83–84). At the forefront of the creation of new knowledge, no one may fully know what factors are going to be influential. Measurements are often being made at the very limits of sensitivity of apparatus or at the very edges of computing capacity, so minor differences in the configuration of equipment may play a big role.

Thus, though it is clearly correct to say that science has a skeptical character—advances in science are usually made by doubting what was believed before—it is at the same time true that science depends on trust. Further evidence for the importance of trust comes from two sorts of sources. First, one can understand the routine importance of trust by looking at the special circumstances of controversy. Under normal conditions, all of the things mentioned above (the functioning of apparatus, the reasonableness of peer reviewers, and so on) are usually taken on trust, but in a controversy each can be open to doubt. Long-held assumptions get called into question; the trustworthiness of other scientists and even of the peer review system itself can come to be doubted (see Collins & Pinch, 1979, pp. 239–253). Once this process begins, the possibilities for distrust and doubt begin to expand exponentially. Just as philosophers have tended to use the special circumstances of controversy as a laboratory for studying scientific argument, one can say that the study of controversies exposes the deep reserves of trust on which ordinary science is based.

The second source is historical. Shapin (1994) has made the case that the founding of key institutions of modern science in 17th- and 18th-century England was crucially dependent on newly consolidated conventions of trust and civility. The willingness to take as authoritative experimental findings whose production one had not personally witnessed, and to value those findings over learned teachings from traditional sources, depended on the acceptability of the word of the gentleman. Historically speaking, trust is the basis of the scientific community.

The second characteristic of the working of science I want to point to is the role of skilled judgment and interpretation. Although science is methodical, the ordinary life of the scientific community demands that people exercise judgment. Scientists have to decide which readings to ignore because of gremlins in the equipment; they have to decide which papers to read and whose interpretations of findings to take most seriously. Again, the fundamental importance of the operation of this aspect of science can be seen in controversies. Advocates on both sides in a scientific dispute typically see what the options are; all participants are familiar with the principal claims being made. But they evaluate them differently. They may differ in the weight they attach to the various kinds of evidence being presented, they may hold conflicting opinions about the reliability of the types of equipment people are using, they will typically differ in their estimations of the dependability of various scientists' works, or they may feel their own experimental or field-observation skills are underrated by their opponents. As in any area of skilled work, judgment is indispensable. As Thomas Kuhn (1977) emphasized:

> When scientists must choose between competing theories, two men fully committed to the same list of criteria for choice [such as simplicity or scope] may nevertheless reach different conclusions. . . . With respect to divergences of this sort, no set of choice criteria yet proposed is of any use. One can explain, as the historian characteristically does, why particular [scientists] made particular choices at particular times. But for that purpose one must go beyond the list of shared criteria to characteristics of the individuals who make the choice. (p. 324)

I have offered these two points in a neutral tone. The advance of scientific knowledge depends on the exercise of judgment and on the workings of trust. Expressed in this moderate way, the points may seem uncontentious, though I believe they have been overlooked by most commentators on the public's understanding of science. I am not saying that this is all there is to science, or that science boils down to mere judgment and trust. I am simply saying that trust and judgment are vital ingredients of scientific life and that their presence is important to the public's response to scientific expertise.

Why Do Trust and Judgment Matter to the Public Understanding of Science?

If it is accepted that trust and judgments are integral to the internal operation of science, these factors can help one understand difficulties with the public's reception of science. Take trust first. As discussed in the previous section, basic science has to proceed on the assumption of trust and on procedures that minimize distrust. For example, poorly thought-of articles are seldom ridiculed; rather, they are ignored. If trust is central to the maintenance of beliefs and the passing on of knowledge within science, then it is hardly irrelevant to the relationship between scientists' knowledge and the views of the public. It is not as though trust were an additional, optional element within the scientific community. The building of scientific knowledge depends upon it; the transmission of scientific belief depends upon it. And precisely because the trustworthiness of equipment, of assumptions, of scientists' work practices, and so on have to be taken for granted in the normal operation of science, such trustfulness is also central to the public acceptance of scientific knowledge. Or, to put it another way, a distrustful public can always find a pretext or grounds for doubting scientific knowledge claims, precisely because scientific agreements themselves depend on trust. However, as noted above, the public commonly encounters science in contexts in which trust in the institutions supplying the knowledge is already a contentious issue and where trust is quite possibly already endangered. In the majority of cases, the science that matters to members of the public matters to them because of the practical context within which the knowledge is to be used. In such instances members of the public have a keen interest in figuring out the trustworthiness of the sources of expertise.

For instance, people may worry about the hazards of a factory close to them or about the wisdom of disposing of waste by burning it in an incinerator near their homes. They do not encounter science in a contemplative or disinterested manner but in relation to practical projects and to the agendas of various companies, government departments, or public bodies. Scientific evidence is presented as part of a campaign to get people to eat beef, drink organic milk, adopt a specific diet, or use a certain toothpaste. Accordingly, in controversial and contentious matters, members of the public may already be of a mind to distrust the scientists and to ascribe ulterior motives to them. An extreme example is provided by the present European arrangements for environmental impact assessments (EIA), where the costs of the assessment have to be met by the company or person proposing the development. In effect, this provision means that the scientists performing the EIA are in the pay of the developers themselves. Though it is easy to understand why EIAs are organized in this way (otherwise, the taxpayer would have to fund all such investigations), such arrangements tend to encourage the public to be critical of any background assumptions, specialized jargon, conventional practices, or rules of thumb that scientists use. Sometimes, of course, citizens will have good grounds for distrusting scientists. But even when they do not have especially good reasons, a context of distrust ensures that some reasonable-looking grounds can be found. Furthermore, because science can often be encountered in such adversarial contexts as public inquiries and law suits, there is an interest in picking the other side's arguments to bits, in finding grounds for controversy even when they have subsided by common consent within the scientific community.

These problems with trust are magnified by difficulties over judgment. Particularly if scientists have publicly played up the methodical character of their knowledge, members of the public may not regard it as legitimate when scientists cannot exhaustively say why they regard a specific diagnostic test as adequate or why they accept statistical evidence at the $p = 0.0001$ level, or whatever. If science is methodic, then it should be possible to spell it out like an algorithm. But if such an explanation cannot be given (as in practice it cannot, because of the role of judgment), the public may suspect that the judgments used are tendentious. The issue of trust reenters. Again, issues of judgment can be exacerbated by courtroom procedures, as discussed below.

Moreover, policy-makers can sometimes allow problematic issues to be lent a spurious scientific character by having them delegated to experts. But the issues on which experts feel confident to comment may not resemble the issues as experienced by the public. Remarks about the safe limits of individual gases emitted from incinerators do not necessarily apply to the cocktails of gases released from industrial plants. Claims about the safety of agricultural chemicals sprayed on protective clothing under test conditions may not apply to the actual conditions in which some farm workers operate. Although scientists themselves are usually scrupulous about their statements, scientific assessments can be a guise under which untested practices are foisted onto the public. That scientific expertise has been invoked on "dodgy" occasions can then become a reason for routinely distrusting expertise even when it is brought to bear more sensitively.

The survey quiz questions used to estimate the public's understanding of science ask about more or less context-free science (see Evans & Durant, 1989, for example). But in everyday situations people have to use scientific information in a context-sensitive way. And this difference is what makes the usual attempts to assess the public understanding of science through quiz-type questions unrealistic. In sum, two central and enduring characteristics of academic, basic science are also the characteristics that make it suspect in practical circumstances. It is tempting for the scientific community to respond to these troubles by stressing the "hardness" of science and the scientific method. But any such response, however understandable, will tend to be counterproductive because science cannot be automatic and method-based, free of trust and judgment.

Furthermore, though pure, disinterested inquiry has always served as the exemplary model for scientific inquiry and lies at the core of quizzes probing the public understanding of science, the science with which members of the public have routinely to deal may equally well come from the commercial sector. Accordingly, the apparently benign idea that scientific understanding can be smoothly diffused to the public and that the public can readily accept scientists' claims because they are based on disinterested investigation can very easily become inverted to look like a charter for pulling the wool over the public's eyes. If not all scientific claims are disinterested, then there is clearly a danger that the diffusion of science will end up lending legitimacy to some questionable claims.

These reservations about the mainstream emphasis in the public understanding of science certainly have not been ignored. They have probably been taken most seriously in relation to public concerns over risk. The assessment of risks, whether of new technologies or from natural hazards, has long been an issue in science communication. In the field of risk communication, it is clear that the public has a strong interest in weeding out partial and tendentious claims and that knowledge and trust are closely related.[2] But once it is acknowledged that trust in the source of information about a risk or hazard is important to understanding people's response, the typical move among policy-makers and students of policy analysis has been to try to break the phenomenon of trust down into its various components.

For example, psychologists may attempt to model the procedures people appear to use in determining the trustworthiness of, say, the perceived

---

[2] I am grateful to an anonymous referee who helpfully drew this point to my attention. She or he argued that though trust may underlie the acceptance of knowledge claims, more knowledge commonly increases trust. Although I accept that this assertion can be true—greater knowledge about how exactly the ozone hole was detected may, for example, have led members of the public to treat the issue more seriously— knowledge-inspired increase in the public's trust in science does not seem to me to be an automatic reaction. In other words, more knowledge can increase distrust as well, as appears to have happened over nuclear power.

expertise and public interest orientation of the body. Or they may look at the credibility that the public attaches to different types of organizations or different communication media. (Scientists may, for instance, ask whether television news is regarded as more credible than official information leaflets.) The guiding assumption is that the resulting information can then be used to make scientific organizations more credible and to diminish the likelihood of irrational nonacceptance of their pronouncements. Analogously, where arguments about trust and credibility have been taken on board in relation to the public understanding of science, the typical response in the scientific community has been to supplement an interest in the public's understanding of science with a study of factors affecting public trust in science. To put it another way, confident in the correctness of their scientific views, science communicators see public distrust as a distortion, a problem to be overcome; they aim to find approaches that prevent science's signal being disrupted by the noise of distrust.

This response has both a practical and an intellectual drawback. First, it tends to imply that trust is only an issue when one is dealing with nonspecialized audiences for expert knowledge. However, trust is central to the business of science itself. Scientific research is not conducted by automata but by participants in a scientific community, a community bound by trust. Given that judgments about trustworthiness are so central to the practice of science, there is little hope that they can be eliminated from the processes by which the public comes to acquire and assess scientific knowledge claims. Trust is an indispensable component in the creation and passing on of scientific knowledge; it is not restricted to lay audiences for science, and it is not a feature that can be technically manipulated to promote high-trust conditions.

The second point is that trust and credibility are not fixed dispositions, of either individuals or institutions. Trust and credibility are the outcome of interactions and negotiations. This key point emerges from several recent qualitative studies, including Wynne's (1992) investigation of Cumbrian sheep farmers' responses to scientific advice given in the aftermath of the Chernobyl nuclear disaster. As a result of radioactive contamination, far-reaching restrictions were placed on the sale of livestock. Initial confidence that the official scientists understood the problem gave way to public skepticism as the quarantine period was progressively lengthened. According to Wynne's (1992) analysis, when farmers encountered the messiness of day-to-day science, when they saw how readings could vary over small distances and how difficult it was to get a stable figure for background radiation, the farmers revised their notion of scientific knowledge. This change was neatly captured in a story Wynne told about the live monitoring of sheep. One farmer, wrote Wynne, saw that just over ten sheep out of a sample of a few hundred failed the contamination test. They were too highly contaminated for release. Then the farmer recounted how the monitoring scientist "said, 'now we'll do them again'—and we got them down to three!" (Wynne, 1992, p. 293). Because the monitoring device had to be held against the rear end of the sheep and because, as the farmer noted, "sheep do jump about a bit," it was hard to get consistent readings. The farmer could see that what ended up as a fact about contamination started off as a messy and uncertain operation. For the farmer, the mystique and authority of other official data records began to evaporate, too. The credibility of expert opinion was revised, indeed renegotiated, during the farmer's encounter with scientific practice.

Other case-study analyses of the public understanding of science have indicated that members of the public are very capable of acquiring scientific knowledge and responding to the demands of technical debate when they are highly motivated to do so. Whether the stimulus is personal ill health leading to investigation of medical knowledge about one's condition (the determination of some HIV/AIDS sufferers and those in high-risk groups to understand and participate in medical trials is probably the best-known example), personal dietary choices leading to intense interest in nutritional science, or concern about an environmental hazard leading to engagement with officially recognized knowledge about contaminants, case studies have shown that members of the public (and often their legal representatives) can come to be skilled users of scientific knowledge and able participants in scientific argument.

Furthermore, in such cases it is not just that members of the public have "gotten up to speed," learning a reasonable amount of what scientific and technical experts already know, but that they have on several occasions helped extend scientific knowledge. For example, persistent public worries about nuclear safety, coupled with the work of concerned scientists, have led to closer attention to the subtle pathways through which radioactive contamination can be concentrated. Similarly, groups of patients who have insiders' knowledge of a disease or disorder have contributed to understanding how to manage their condition, particularly on how to manage it in the light of the varied and unpredictable demands of everyday family life. Scientific understanding has sometimes been enhanced also by the ability of citizens' involvement to break down existing disciplinary boundaries.

Overall, these insights from case studies in the public understanding of science reveal two main sorts of conclusions. First, they indicate that there is no one formula for transmitting scientific knowledge. The credibility of experts is, in a sense, always being negotiated and evaluated. A means of deploying expertise in one social context may not work in another, for public trust in expertise—a (perhaps *the*) central issue in the public understanding of science—cannot readily be routinized. Second, and more radical, it is not accurate or appropriate to regard the public understanding of science as a one-way street. Lay publics can be active participants in the generation of new knowledge and the overthrow of old scientific beliefs. The relationship between scientific expertise and the public is far more complex than is typically recognized in calls for public understanding that emanate from the scientific establishment.

Institutional Factors that Exacerbate the Misunderstanding of Science

In courts and other quasi-judicial contexts the difficulties faced by science are, as a series of case studies has indicated, magnified. This effect has been of particular significance in the United States, where constitutional arrangements allow official scientific rulings to be tested in court by the independent judiciary. The court-room scrutiny of science has proven to be surprisingly successful, not necessarily in the sense that justice has been advanced but that science has succumbed to legal interrogation. The fate of science in court can also be understood in relation to the factors of trust and judgment. As indicated early in the previous section, the circumstances of adversarial court-room interaction commonly lead to the suspension of trust. The opposing sides have an interest in throwing doubt on each others' credibility, even if they cannot replace the queried knowledge with positive answers of their own. Opposing legal teams concentrate on undermining the public credibility of their adversaries. The give and take of basic science, the assumption of reasonableness, is withdrawn. The problems with judgment, too, are often heightened by the adversarial process of legal examination. Scientists are typically called on in court to offer factual determinations. If it turns out that these "factual" statements hang on judgments and that, therefore, a different scientist might have judged the "facts" differently, this disclosure can corrode the expertise that allowed the scientists to be present in the first place. The opposing legal team therefore has a vested interest in exaggerating the role of judgment and counterpoising an individual's judgment against the ideal of purely objective, almost machine-like fact recognition.

The fact that scientific knowledge stems from the everyday work of scientists can be developed for the purposes of legal argument as well, especially when it is linked to the issue of trust in machines and in administrative systems. It is common to see much play made of the fact that scientific knowledge is founded on working practices. Like everyone else, scientists have to clean their experimental equipment, make adjustments to the detectors, and keep records of their actions; laboratory workers can be as slipshod as any other employees. These incidental features that seldom make it into formal scientific argument can be invoked in legal settings to undermine the apparent indubitability of science. Additionally, as Jasanoff (1990) has noted, the connection between scientific results and custom and practice can be queried (p. 202; see also Jasanoff, 1992). If a case turns on the toxicity of a substance whose dangers have been indicated by animal experiments, it can still be questioned whether the decision to use animal evidence is itself scientifically justified. Of course, in this modern version of an ancient paradox, tests cannot themselves all be independently tested nor can all methods be methodically checked.

The examination of such issues has not been restricted to U.S. courts. For example, in Britain, too, the legal system has exposed weaknesses in scientific reasoning. However, this probing has chiefly been limited to forensic science, the evidence from which has been questioned in a number of high-profile cases, including ones dealing with alleged terrorist bombers. It is, however, the persuasiveness of legal challenges to scientific judgments and particularly (as discussed later in this chapter) judgments about the regulation of harmful substances that has made the U.S. experience of such pivotal importance.

The centrality of this legal deconstruction raises a further interesting point. In many cases, critiques of the authority of science in public contexts are perceived to come from the political left. For example, the elitism of scientific medicine has been subjected to criticism commonly associated with a policy preference for preventive approaches. Equally, there tends to be an elective affinity between the left and an antinuclear stance, a connection less visible on the right. One might therefore think that the critique of scientific authority is associated with the left's critique of political authority. In fact, however, commerce and business have not been shy of using exactly parallel arguments when they have wished to combat policies sanctified by scientific evidence.

This willingness to appropriate the critique of scientific authority was made particularly clear to the U.S. Environmental Protection Agency, which introduced far-reaching reforms in its early years in the 1970s. These changes were met by lobbying from the business sector and by successive challenges through the courts from companies questioning toxicological evidence and claims about environmental harm. Whether coming from the left or from the deregulation-minded right, the arguments make a similar assertion. The scientific interpretation in question (e.g., the safety of a certain level of exposure to radiation or the toxicity of an industrial solvent) has been presented as being based on judgment, a judgment tendentiously exercised.

The Response of Scientists and the Aggravation of Problems with the Public Understanding of Science

From this analysis, it appears that the typical response of scientists to problems with public understanding tends to accentuate those problems because scientists have been (and are) encouraged to overstate the role of method. Since the origins of modern science, and particularly since the professionalization of science in the 19th century, scientists have pressed for more influence in society. They have presented themselves, not individually but as a body, as being in possession of expert knowledge that society needs in order to run itself better. Scientists advise the military, manufacturing companies, health officials, educationalists, social-welfare practitioners, and so on. To justify this expanding influence, scientists have stressed the special character of their knowledge. Although, as philosophers have lately come to acknowledge, no one has been able to spell out "the scientific method" in any detailed way, the idea that there is a scientific method has been widely promulgated. The benefits of this notion are clear. It sets scientists apart from other advisers and professionals, who can lay claim only to unmethodical knowledge.

Moreover, the idea that scientists are the best governors of science has been used to establish unparalleled autonomy for the scientific profession. Hundreds of millions of pounds are supplied by the government for expenditure on research each year, without—at least until recently—much in the way of public accountability. So, even though there are good reasons to be skeptical about the existence of *a* scientific method, scientists have had reason to promote it as the key to the special superiority of science.

Faced with challenges to the authority of their knowledge claims, the scientific community may find it tempting to respond to these troubles by stressing the hardness of science and the robustness of the scientific method. But any such response, however understandable, will tend to be counterproductive because science cannot be method-based, free of trust and judgment. Except under totalitarian circumstances, the application of scientific understanding to matters of public concern will involve some element of public participation and, accordingly, must depend on a measure of trust. Shutting the public out is likely to increase distrust and thus further corrode the practical authority of science.

The Changing Social Function of Science

The final point that has emerged from social science work on questions surrounding the public's understanding of science is that the changing societal role of science is itself a key variable, though one commonly overlooked. The scientific community's discussion of the public understanding of science tends to concentrate on what in principle are the timeless issues of scientific method and on unchanging core truths of science. By contrast, social scientists have pointed out that science's social role is itself changing in at least two ways that are significant for laypeople's approach to science. The first concerns the growing role of commercialism in research. Partly to deliver the assumed gains of market competition and partly to meet the rising costs of scientific work, research institutions are being obliged to become more commercial. For example, public research agencies are being privatized and business people are increasingly involved in the evaluation of basic research proposals. The stereotype of the disinterested scientist is now no longer representative of the circumstances under which the majority of scientists are employed. Consequently, an image of scientific knowledge as essentially disinterested is more and more implausible and, in a certain sense, inaccurate.

In his compelling and popular overview of the "short twentieth century," Hobsbawm (1994) drew attention to the extent to which contemporary science, even basic research, cannot avoid accusations of interestedness. Noting the contest for money and the involvement of political and commercial interests in the support of science, he observed that postwar science has not been value-neutral:

> as all scientists knew, scientific research was *not* unlimited and free, if only because it required resources which were in limited supply. The question was not whether anyone should tell researchers what to do or not to do, but who imposed such limits and directions, and by what criteria. (p. 556)

The co-opting of science by economic and political goals lay at the heart of Hobsbawm's analysis of the precarious situation in which one finds the ideal of science at the close of the 20th century. I consider his analysis to be correct, but only in part. His summary reveals what he overlooked:

> All states therefore supported science, which, unlike the arts and most of the humanities, could not effectively function without such support, while avoiding interference so far as possible. But governments are not concerned

with ultimate truth (except those of ideology or religion) but with instrumental truth. At most they may foster 'pure' (i.e. at the moment useless) research because it might one day yield something useful, or for reasons of national prestige, in which the pursuit of Nobel prizes preceded that of Olympic medals and still remains more highly valued. Such were the foundations on which the triumphant structures of scientific research and theory were erected. (Hobsbawm, 1994, p. 557)

His observation about the basis of the deal between science and the state is an important point, but I believe he neglected the extent to which the instrumental truths at which science and technology are aimed are themselves undergoing a change. In the immediate postwar decades, the primary social role of (and justification for) science was to increase productivity and competitive performance, whether economic, military, or medical. But, as noted by Beck (1986/1992), Roqueplo (1994), and others, people have begun to switch away from seeing science as a way of increasing production to seeing it as a means of handling risks and achieving regulation. Of course, much R&D is still aimed at innovative products and processes. But, to take an extreme example, scientists are now nearly as likely to be advising politicians on the health risks arising from BSE (mad-cow disease) as they are to be advising them on ways of increasing agricultural productivity. This growing regulatory role (sometimes referred to as the *expertization of science*) places increasing, and increasingly unrealistic, demands on science because questions are more likely to be publicly raised over trust and judgment in regulatory disputes than over the development of innovative products. If designers want racing cars to go faster or airplanes to carry more passengers, then there can be legitimate differences in the way that such performance is measured. One car may go faster flat out on the straight and another car may corner more quickly; one plane may carry large numbers of passengers on short hops, another plane may transport large numbers of passengers across continents. Both cars can claim to be fast, both planes claim to have a huge capacity. Very little hangs on proving, let alone proving to the public, which car is "really" fastest, especially because a racetrack will have both bends and straights.

However, in the matter of ruling out risks, there are pressures to prove which pesticide is safest, which disposal method for oil platforms is the least environmentally harmful, and so on. Moreover, these proofs have to be offered in public forums where various interest groups have a legitimate role. The tasks now facing science, as well as the context in which that task has to be performed, place new, more exacting demands on scientific knowledge.

Conclusion

At one level, the conclusion of this chapter is rather straightforward. It is that the public-understanding "industry" and most quantitative research on the public understanding of science operate with a restricted understanding of both what science is and what its role in contemporary society can be. Accordingly, an improved approach to the practical and intellectual issues surrounding the public understanding of science will demand a fuller understanding of key features of the scientific enterprise and of the context in which scientific research is currently carried out.

My argument is that analysts of science need to appreciate the role of trust and judgment not only in scientists' dealings with the public but also within the practice of the natural sciences themselves. Moreover, it is clear that the public's understanding of the political economy of science and of the regulatory role of scientific knowledge is a key additional element in shaping public attitudes to science and expertise. Social scientific work on the public understanding of science suggests that the public's understanding of the institutions and politics of science is a significant aspect of the public's overall attitude and responsiveness toward science and technology, yet it is an aspect that has been largely ignored in discussion of the public understanding of science in the official policy world. This neglect leads to the ironical position that the U.K. White Paper on science urges more public understanding of science but not of the policy process of which the White Paper is itself a part.

Though it is always too easy to suppose that there is something special about the present day, I believe I have given grounds for accepting that the science in public is today facing a uniquely difficult challenge. In some sense the points I have raised about the fragility of scientific knowledge are timeless. Trust and judgment have always characterized the scientific enterprise. But their persistence as issues nowadays relates to recent changes in the institutional challenges to science (in particular from the U.S. courts) and to alterations in the social role of scientific expertise. In addition, these corrosive tendencies have a self-perpetuating character. Once public trust in the scientific community is threatened, that very lack of trust jeopardizes the future acceptability of scientific expertise. In that sense, if my analysis is correct, there is—notwithstanding the exhortations of the Bodmer report and the associated publicity and dissemination activities of the public understanding of science movement—no easy way to reverse this decline.

Two things do suggest themselves, however. The first is simply increased recognition of the new regulatory role of science. The social role played by science has changed in recent decades, and increased public debate about the new demands on science can only be helpful to the public understanding of the role of expertise. Second, it may be worthwhile looking at ways of building public trust in the institutions of science. From 1979 to 1997, successive Conservative governments accustomed university scientists in the United Kingdom to the idea that business people should sit on the research councils. Though many commentators have expressed concern about this development and the implied commercialization of British science, these changes could be taken in another light. They could instead be seen as prefiguring a rise in public participation in science, with citizen participation not only on the research councils but, for example, on university research committees, too. I do not propose this grass-roots representation as an immediate solution but rather as an indication of trust-building institutional arrangements that may offer the best prospect for a socially respected science.

A final irony is that another residue of Conservative policy, the privatization of formerly public functions, may be tending to push in the direction I advocate. With the advancing commercialization of the National Health Service, for example, new scope has risen for doubts about how oriented its decision-making is to the public interest. As has long been the case, treatment on demand is not available to all; the resources are simply not adequate. This scarcity used to be managed by having long waiting lists with clinicians setting priorities for treatment. Quasi-privatized managers now realize that they themselves are increasingly held responsible for rationing health care. One way to handle this difficult problem is for them to involve patients or "customers" in decision making. There is evidence that a similar logic is even being considered for health-service expenditure on research. The desire to avoid sole responsibility for unpopular decisions becomes a spur to democratization. From such unlikely sources may come practical experiments in wider and more meaningful public engagement with science.

References

Beck, U. (1992). *The risk society: Towards a new modernity.* (M. Ritter, Trans.) London: Sage. (Original work published 1986)

Bodmer, W., & Wilkins, J. (1992). Research to improve public understanding programmes. *Public Understanding of Science, 1,* 7–10.

Collins, H. M. (1985). *Changing order: Replication and induction in scientific practice.* London: Sage.

Collins, H. M., & Pinch, T. (1979). The construction of the paranormal: Nothing unscientific is happening. In R. Wallis (Ed.), *On the margins of science* (pp. 237–270). Keele, England: University of Keele.

Evans, G., & Durant, J. (1989). Understanding of science in Britain and the USA. In R. Jowell, S. Witherspoon, & L. Brook (Eds.), *British Social Attitudes: Special International Report* (pp. 105–119). Aldershot: Gower.

Hobsbawm, E. (1994). *Age of extremes: The short twentieth century 1914–1991.* London: Michael Joseph.

Irwin, A., & Wynne, B. (Eds.). (1996). *Misunderstanding science? The public reconstruction of science and technology.* Cambridge, England: Cambridge University Press.

Jasanoff, S. (1990). *The fifth branch: Science advisers as policymakers.* Cambridge, MA: Harvard University Press.

Jasanoff, S. (1992). What judges should know about the sociology of science. *Jurimetrics, 32,* 345–359.

Kuhn, T. (1977). *The essential tension.* Chicago: University of Chicago Press.

Martin, S., & Tait, J. (1992). Attitudes of selected groups in the U.K. to biotechnology. In J. Durant (Ed.), *Biotechnology in public: A review of recent research* (pp. 28–41). London: Science Museum for the European Federation of Biotechnology.

Roqueplo, P. (1994). *Climats sous Surveillance: Limites et Conditions de l'Expertise Scientifique* [Climates under surveillance: Limits and conditions of scientific expertise]. Paris: Economica.

Royal Society of London. (1985). *The public understanding of science.* London: The Royal Society.

Shapin, S. (1994). *A social history of truth: Civility and science in seventeenth-century England.* Chicago: University of Chicago Press.

U.K. Government. (1993). *Realising our potential: A strategy for science, engineering and technology.* London: HMSO, Cm 2250.

Wynne, B. (1991). Knowledges in context. *Science, Technology, and Human Values, 16,* 111–121.

Wynne, B. (1992). Misunderstood misunderstanding: Social identities and public uptake of science. *Public Understanding of Science, 1,* 281–304.

Yearley, S. (1994). Understanding science from the perspective of the sociology of scientific knowledge: An overview. *Public Understanding of Science, 3,* 245–258.

Yearley, S. (1995). Misunderstanding the public: The public understanding of science movement. *Science, Technology and Innovation, 8*(6), 25–29.

CHAPTER 10

AGAINST LINEARITY—ON THE CULTURAL APPROPRIATION OF SCIENCE AND TECHNOLOGY*

Knut H. Sørensen, Margrethe Aune, and Morten Hatling

The Anatomy of Linearity

Most studies of science and technology in modern societies are production-centered. They focus on the production of facts and artifacts in institutions of innovation and scientific inquiry. Still, when on occasion one follows specific forms of knowledge or artifacts that are on their way into society, the perspective remains production-centered because the topic of inquiry has been defined in laboratory terms. Even within the field of the "public understanding of science," such bias seems very common (see, for example, Wynne, 1995). In this chapter we argue the need for a user-centered focus and explore the potential of such a shift by discussing three empirical cases based on theoretical and methodological developments in technology studies.

The preoccupation with linearity in science and technology studies has touched off a critique of the assumption that relationships between science and technology are simple and unidirectional. Such criticism is necessary because the so-called linear model of innovation has long been a highly efficient rhetorical device for emphasizing the economic potential of scientific research. Although the unequivocal conclusion of innovation studies has been that "the notion that innovation is initiated by research is wrong most of the time" (Kline & Rosenberg, 1986, p. 288), this mode of thinking has been difficult to change. As Steve Woolgar (1994) has stated, most alterna-

---

* An earlier version of this chapter was presented at the conference "Public Understanding of Science and Technology," Wissenschaftszentrum Berlin für Sozialforschung, November 30–December 2, 1995; at the Center for Cultural Studies of Science, Technology, and Medicine, University of California at Los Angeles, April 9, 1996; and at the seminar of the Science Studies Program, University of California at San Diego, May 6, 1996. The comments from the participants of these seminars are gratefully acknowledged.

tive views of the relationship between research and innovation still have a linear structure.

According to the linear model of innovation, there is a unidirectional link between basic science, applied science, and the development and marketing of a new product. Similar beliefs exist in the analysis of public understanding of science. In terms of simple communication theory, there is (a) a *sender* who (b) *codes* (c) a *message* that is to be (d) *decoded* by (e) a *receiver*. This arrangement is the basic anatomy of the linear, production-centered (or science-and-technology-centered) model found in studies on the public understanding of science. By making use of developments in cultural studies and theories of literature, researchers can apply this model in quite sophisticated ways to view both coding and decoding as contingent and embedded processes (see, for instance, Lewenstein, 1995). Similarly, Latour (1987) maintained that for all the rhetorical efforts of technoscientists, the fate of their claims about facts and artifacts is in the hands of their readers. Still, thinking in these terms, we can see many possible problems in the transfer of knowledge from science and technology to other localities of society because of the cultural embeddedness of coding and decoding and the lack of interest of senders and receivers in the process.

Scientific and technological knowledge may therefore be distorted and not always appreciated. However, communication theory does not provide much insight into the processes by which knowledge is reshaped, transformed, and eventually put to use in people's everyday lives. Even refined linear models of the public understanding of science and technology remain fundamentally flawed for a number of reasons. First, as mentioned, they reflect an underlying asymmetry between producers and users of knowledge. The producers are seen as active, essential, and defining. Users are seen as reactive and limited in the scope of their actions. Second, such models express an overly instrumental and rationalist view of scientific and technological knowledge. They presuppose the existence of generally shared goals, such as the goal of educating the general public. It is taken for granted that people need to know certain, often decontextualized, facts about nature.

Third, it is assumed in linear models that researchers and the public generally agree on what the relevant problems and fruitful concepts are. In addition, research results are believed to be broad enough to serve both innovation and ideology. Neither of these presumptions holds up under scrutiny. In fact, a major problem for modern science and technology is that their results are increasingly meaningless outside narrow specialist communities. The relevance of results to others must be constructed, and the work of creating relevance is demanding.

Finally, linear models produce stories with too-well-defined beginnings and endings. Narratives about public understanding of science like traditional narratives about innovation, appear to be too conventional in this respect. They let on that researchers may put "tracers" on facts and artifacts and follow them around in society. This characteristic of linear models may be meaningful, but it is basically an expression of producer-side bias: What producers make are the facts and artifacts worthy of pursuit.

The agenda of research on the public understanding of science and technology reflects an educational bias. The ideal is a classroom with good students who are eager and able to digest and reproduce packets of knowledge from teachers. Moreover, because good students are considered in this line of thinking to be those who have a positive view of science and technology, they are also seen to embody the norms emerging from the agenda of public understanding.

Appropriation and Agency

Medical anthropologist Emily Martin (1989, 1994) analyzed how medical concepts like menstruation and immunity are appropriated and transformed to give meaning and purpose within a local context. From a scientific point of view, such appropriation and transformation results in a distortion of medical facts and consequently in a failure to really understand the concepts. However, as Martin showed, most people do not appropriate scientific concepts in order to emulate the scientist or medical professional but to make sense of their own lives and relevant natural phenomena from within their own cultural framework.

The tension between people's efforts to act in and make sense of their everyday world and the authority and insights found in scientific and technological knowledge calls for appreciation and understanding. Superstition and misleading beliefs may be detrimental, but so may the practices of science and technology. A way to study appropriation and agency is needed.

The study of media and the technologies of everyday life has elicited an understanding of the appropriation of artifacts as *domestication* (Silverstone & Hirsch, 1992). Metaphorically, this concept conveys the need to "tame" facts and artifacts that are taken from a "wild" outside world and put into a domestic setting. Although facts and artifacts may embody scripts or programs designed to get users to comply with the producer's vision of how a piece of knowledge or equipment should be employed (Akrich, 1992; Latour, 1992), these scripts or programs represent an *affordance* at most, a suggestion that one way of acting or reacting be preferred (Pfaffenberger,

1992). There is no necessity to comply. Indeed, the script or program in a machine may be difficult to read, a situation that most people have experienced. Users need to act in order to integrate knowledges or artifacts into their everyday lives.

There are different ways to analyze the consumption of technologies. McCracken (1988) emphasized the necessity to construct meaning from artifacts in order to place them, mentally and culturally. People employ many rituals to this end: possession rituals, maintenance rituals, grooming rituals, and so forth. Silverstone, Hirsch, and Morley (1992) explicitly used the domestication concept to describe how artifacts are integrated into "the moral economy of the household," by which they mean to highlight the fact that there is an important normative as well as symbolic structuration of the home and its internal discourses. Their approach focused on the way households are both "economies of meaning" and "meaningful economies" and thereby highlighted the interaction between economic and symbolic transactions within the household and between the household and the outside world. Moral economies are negotiated spaces. Thus, facts and artifacts must be culturally appropriated if they are to function and make sense.

However, this process of appropriation is not a simple integration of technology into a cultural setting. To domesticate an artifact is to negotiate its meaning and practice in a dynamic, interactive manner. This negotiation implies that technology as well as social relations are transformed. To use the concept of domestication as an analytical tool is to emphasize that the cultural appropriation of an artifact is a multidimensional process. The artifact must be—

1. acquired (either bought or in some other way made accessible);
2. placed (i.e., put in a physical space, a mental space, or both);
3. interpreted so that it is given meaning within the household or a similar local context of identity and given symbolic value to the outside world; and
4. integrated into social practices of action.

Strategies of domestication thus take place in three main dimensions: (a) practical, (b) symbolic, and (c) cognitive. In the practical dimension, domestication entails a pattern of usage. How shall the artifact be employed? What are the practical implications of a given set of knowledge? Symbolic efforts are about the production of meaning and the relationship between meaning, identity, and the public presentation of self. Cognitive work pertains to learning about an artifact or to the intellectual appropriation of new knowledge. However, facts have more than just cognitive qualities. From our point of view, facts may be domesticated in a manner similar to that used with artifacts. To emphasize the practical and, in particular, the symbolic qualities of facts may make it easier to understand how facts may or may not be appropriated.

One could argue that the domestication of facts is different from the domestication of artifacts. Facts do not have the visible and material qualities of artifacts, and they are not objects of use, technologies are. While keeping such potentially important differences in mind, we explore the idea that the domestication of science and the domestication of technology often go hand in hand. For example, knowledge may piggyback onto artifacts because they are made relevant to local actors through efforts to appropriate, say, a particular piece of equipment.

Although domestication can be carried out by institutions and other collectivities, it is also performed by individuals or groups in various settings, such as households or workplaces, which are the focus of this chapter. Regardless of the setting, however, domestication should be studied as a contingent process. It depends on local resources, structural or global intersections, and juxtapositions of local and global strategies. An analysis of domestication must be sensitive to local conflicts as well, for domestication is not a process free of friction and resistance. Gender, class, ethnicity, and age may be relevant analytical categories for framing and understanding domestication.

What is constructed through domestication may be understood as micronetworks of humans, artifacts, knowledge, and institutions (Sørensen, 1994). A home, a drive to work, or electronic homework can serve as examples of micronetworks. Whereas a home is obviously a complex system of relations, a drive to work is experienced by most commuters as something simpler. It can be analyzed as a network consisting of a car; a system of roads and other infrastructure; a driver; his or her skills; and his or her knowledge about infrastructure, traffic rules, and destination. To function within this network, the driver draws upon symbolic, practical, and cognitive resources and produces symbolic, practical, and cognitive effects that result in an observable style of driving, a pattern of use, and identity (for a more detailed analysis, see Sørensen & Sørgaard, 1994). The heterogeneous outcomes of domestication may also be understood as something analogous to a cyborg (Haraway, 1991), a hybrid or quasi object (Latour, 1993), or a monster (Law, 1991).

In order to explore domestication as an approach to the "public understanding of science and technology," we present three narratives that highlight different aspects of the appropriation process. The first narrative is about energy use, the understanding of energy, and energy conservation in Norwegian households. This presentation serves as a basis for looking at the relationship between the appropriation of science and the appropriation of technology. In the second narrative, which emphasizes the cognitive aspects of domestication, we explore the social integration of personal computers (PCs) into homes and examine the related process of knowledge acquisition. The third narrative shifts attention from homes to workplaces. In order to look at strategies for shaping workers' future domestication of computer systems, we analyze the way designers of such systems try to configure (Woolgar, 1991) users.

Contingencies of Understanding Energy, Energy Use, and Energy Conservation in the Home

Energy conservation is important for a global policy of environmental sustainability. This message is conveyed to Norwegian households, but public policy does not promote energy-saving as such. Rather, the goal is to improve the economic efficiency of energy use. Consumers are advised to review the costs and benefits of their energy use in order to find ways to increase the level of their energy conservation. Norwegian consumers know that the country has large reserves of relatively cheap energy, much of it in the form of hydroelectric power. This kind of energy is perceived as clean, at least with respect to emissions of CO_2 and other gases. In addition, Norway has a cold climate. To find out how Norwegian households manage their use of energy under such conditions, extensive interviews were conducted with members of 30 Norwegian households about their views on energy, their uses of energy, their values and commitment to sustainability, and their lifestyles and ways of life. (See Aune, 1998, which provides a more detailed account of the research.)

To comply with public policy on energy conservation, one needs quite a broad understanding of physical, economic, and technological bodies of knowledge. A simple example of the level of complexity involved is found in Kempton's (1987) research on thermostats. He argued that there are two different folk models of how thermostats work. The first, the feedback theory, is the scientifically correct one: Thermostats turn off heating when a given temperature is reached. The second, called the valve theory, assumes that the higher the thermostat is set, the more energy will flow from it. In terms of control theory, the latter view is incorrect.

Kempton's argument was that many laypeople find the valve model more adequate, and thus more functional, than the correct feedback model when it comes to guiding their management of home heating. The reason for this departure from science is that scientists and laypeople use different criteria to evaluate theories. However, the pertinent question is whether the understanding of the thermostat is the relevant issue. For people primarily interested in comfort, as most Norwegians are, the main task is to achieve a comfortable temperature quickly without overheating the house. Thus, the main challenge is not the understanding of how thermostats work but the far more complicated problem of the production and diffusion of heat under conditions of turbulence and heat loss. The person who relies on a thermostat at a fixed setting to produce a comfortable indoor temperature may be making a mistake, as when the amount of turbulence and heat loss is great because of cold and windy weather. Whether the thermostat is understood correctly is of minor relevance. The main problem facing the user is to manage what in scientific terms is the complex interplay of physical and engineering thermodynamics and physical and engineering fluid dynamics.

Thus, heating a house has more in common with engineering than science. The challenge is to produce a satisfying result, not a correct analysis or an optimal solution (Constant, 1984). In addition, people in a household have to make sense of the economics of their energy use, adapt to policy measures and current ideology concerning the use of energy, and manage a cultural integration of energy use that can meet the different needs that household members have, say, with respect to indoor temperature. This task of finding an expert solution calls for a large multidisciplinary team. Most households do it on their own.

In the terms used in this chapter, what the members of the households do is domesticate their houses in order to transform them into homes, and maintain them as cozy, comfortable dwellings. In the process the regimes of indoor climate, showering and bathing, lighting, and so forth have to be negotiated and renegotiated. As one of the informants explained: "[W]e have different needs in the family. Two persons think it's OK to have a rather cold indoor climate, and two need to have it warmer. We regulate that by use of pullovers and blankets." The emerging pattern of materially embedded actions influences the use of energy, but in most households the consumption of energy is not an issue. Energy as such is uninteresting. It is invisible to the household and is thus consumed indirectly, as a consequence of other acts of consumption.

It has been shown that increasing the frequency of metering, billing, and information from the utility companies can make energy use more visible and of greater concern in Norway than is currently the case (Kempton & Layne, 1994; Wilhite & Ling, 1992). Norwegian households seem to be pretty well informed about the main characteristics of their energy consumption. Knowledge is lacking about neither the amount of money spent on energy (electricity) nor the distribution of energy for main items such as heating, warm water, and lighting. An improved billing system probably serves more as a reminder of expenditures than as a provider of new knowledge.

Home is not just a haven in a heartless world; it is also a project. Between 1973 and 1983, 42% of Norwegian households invested in one or more energy conservation measures, and 32% continued to do so between 1984 and 1990. In technical terms, in other words, the energy conservation quality of Norwegian houses improved considerably. However, in most cases energy conservation is not practiced as an end in itself. Rather, it piggybacks upon general home improvements and increased standards of comfort. Homeowners insulate their walls or buy new double- or triple-glazed windows to avoid draughts and to have an indoor temperature they regard as more comfortable. Triple-glazed windows are advertised with pictures of people in T-shirts and shorts in the middle of the winter, a depiction signifying increased levels of comfort rather than reduced electricity bills.

These investments do not mean that people disregard the costs of energy consumption. They do, however, give priority to indoor comfort. Informants explained that they used to freeze when they were children. They remember having to wear sweaters to keep warm in the house. They do not want to experience this discomfort and inconvenience again. However, knowledge about the costs of energy use does not necessarily lead people to calculate their consumption or investments in relation to the use of energy. In fact, economic arguments are often used against some clearly profitable energy-saving equipment, probably because it is unclear whether the equipment lends itself to appropriation without excessive effort.

Thus, most Norwegian households construct everyday life in ways that conflict with the official goals of a sustainable environmental policy. It is not that these goals are unperceived or misunderstood or that people are ignorant of basic economic, technological, or scientific aspects of energy use. Official policy is countered by other messages that emphasize the need to reinforce and increase the material standard of living. It is also countered by the perception of national energy abundance. Even more important, dominant symbolic systems do not seem to accommodate energy conservation well. Curbing one's energy use does not carry an unambiguously positive meaning. On the contrary, to many people it denotes being cold and having to dress in a way perceived as uncomfortable.

Most Norwegian households appropriate the official energy conservation agenda by transforming it into a collective, national, or even international responsibility rather than an individual one. From the household point of view, energy conservation policy, knowledge, and technologies are domesticated by chance, not by design. They are, so to speak, appropriated by proxy, in connection with concerns that are more important to the households.

Because we found a varied pattern of domestication strategies pursued by Norwegian households, we were unable to detect a system of local energy conservation that reflected in any reasonable way the politics, science, or technology of energy conservation as it is defined at a global level. Even if feedback loops were to be introduced into a linear model of the transfer of knowledge and artifacts, it would still not be possible to make sense of the diversity and transformative capacity of energy cultures among Norwegian households. They must somehow be understood in their own terms.

To summarize, the micronetworks that people construct in terms of energy consumption revolve largely around everyday life in general. Energy is not even a means to an end but a means to make other means work. It is invisible and indirect and thus complicates learning and knowledge acquisition. In dealing with energy issues, people act and understand pragmatically, perhaps because standard, popular scientific knowledge is not very helpful. The negotiations that people must conduct within their households to juxtapose needs, values, artifacts, and money, take place primarily within a discursive framework different from that of schoolbook physics.

Bringing Computers Home—Or Coping with Presence

The advent of the PC meant that computers became available for home use. But what could be the role of PCs in the home? What would be the purpose of bringing a computer into the home?

When Turkle (1984) conducted her pioneering study of computer cultures in the early 1980s in the United States, her informants were the vanguard of future computer users. The patterns she identified were manifold and without any clear tendency toward standardization. People, it turned out, had a variety of different styles of using and talking about computers. Turkle described how her informants were exploring their own identities and transforming the meaning of their activities and selves. In particular, Turkle emphasized different ways of relating to computers. She identified two main kinds of relationship to PCs in the home: one that focused on the seemingly magical abilities of computers, and one that stressed the need for transparency, that is, the need to understand the computer. Early British research on computer cultures emphasized computers as a hobbyist phenomenon, although belief in the future importance of the technology was an additional important motive to buy one (Haddon, 1992).

To explore how the household as a unit domesticated the PC through negotiations, a study of computers in Norwegian homes was carried out in 1991 (for details, see Aune, 1992; see also Aune 1996). Involving 23 conversations with a total of 42 informants and sometimes group interviews with families, the approach differed from Turkle's partly because of the emphasis on microsocial negotiations between household members and partly because of the 1991 study's greater concern for the cultural embeddedness of the appropriation of computers.

Aune (1996) identified several distinct types of domestication strategies employed by the PC users who were questioned. The most common type was the *extender,* a person (usually a male) who had bought a PC in order to extend work into the home. This arrangement made it easier to do overtime work at home rather than at the office. The extender's relationship to the PC was mainly instrumental; it was a tool to do a job. Instrumentality was even more pronounced with female extenders than with male extenders. This type contrasts with the *explorer,* whose relationship with the PC was more playful and exploratory. Trying out software, hardware, or both, he (there were only males in this category) would spend a lot of time figuring out what the computer could do and making additions that extended its use. At times, this *hobbyist* approach even led to products intended for sale.

Among young boys, the personal computer was primarily a machine for games. A majority of these participants in the study nevertheless retained an instrumental relationship to their computers. The PC was sometimes fun to play with. It could also be helpful for doing schoolwork, and occasionally it was used in conjunction with drawing and other software programs. A few boys found PCs especially gratifying and spent substantially more time and energy than their peers in front of the screen. They would often connect to other boys in an underground culture of Amiga users, cracking protection codes of games or, above all, making so-called demos to show off their skills in combining the graphical and sound capabilities of their computers.

Computer literacy has been a catchword in the calls to prepare for the information society. The argument is that everyone is going to need computer skills. Computer literacy is a concern in the politics of information technology. However, in social studies of computing, there has been a surprising lack of interest in studying how people learn to use computers. Computer skills are an interesting case illustrating transfer of scientific knowledge (computer science) whereby an artifact (the computer) may support learning, in particular because of the assumed ability of computers to facilitate self-instruction. In addition, the attribution of competence and noncompetence could be a very important aspect of the domestication process.

Formal training in computer use is not included with the purchase of home PCs. Some extenders and explorers in our study (primarily the males) had computer skills, which in most instances they had picked up on the job or through their education. In the majority of the households studied, the male responsible for buying the machine thus had at least some knowledge about computers. In other words, the domestication process started from a gendered attribution of skills: men as competent, women as noncompetent.

In principle, there are many ways to develop skills in the use of personal computers. Learning by trial and error is very important, but the users in our study also acquired their knowledge by reading manuals and computer journals; using instruction programs; and getting instruction and help from other users, such as family, friends, or colleagues. A few of the respondents took courses as well, mainly in word processing. Most of the participants combined several methods for acquiring skills. Four major learning strategies were observed:

1. *Experimentation* emphasized learning as integral to the use of computers. For the practitioners of the experimentation strategy, the slogan was that using is learning, that there is no use without learning, but that it is also important to learn from the experiences of others. Learning by trial and error was thereby combined with learning from other users and with experimentation with programs copied from computer journals. Explorers and Amiga hackers typically belonged in this category.

2. *Tinkering* was a more pragmatic approach to learning than experimentation and was mainly used by extenders. The basic idea was that computer use is based on previously acquired knowledge. To these participants in the study, the slogan was that learning is using, that there is no learning without use, and that an individual begins to learn to use a computer if he or she has a pragmatic need to know about new aspects of the software being used. Trial and error was the most important type of learning for these informants, although many of them also said that computer journals were important sources of information. They read, however, about general features of computers or software, not about particular programming sequences.

3. *Analysis* was common among young men with sound formal knowledge about computing. They wanted a general understanding of a given piece of software, how it should be used to solve problems, and how eventual failures might be remedied. To achieve this, they read manuals and made efforts to analyze problems formally. They sometimes employed a trial-and-error approach as well.

4. *Training* was the preferred strategy among the women and children. Their use was pragmatic and was related to such tasks as word-processing, school work, or games, and they learned the minimal skills necessary to

carry out their tasks. They learned by going through careful and supervised trial and error, asking for instructions, and using self-instruction software or help functions available on the machine. A few also attended courses.

The emerging picture of people's approaches to computers is more complicated than Turkle's distinction between magic and transparency. Although a striving for transparency in the third group ("analysis") was evident, none of the other groups emphasized any "magic" qualities. In fact, all four learning strategies mainly exhibited different pragmatic relations. Even the more expressive explorers, using experimentation as their method for learning, measured skills according to one's ability to perform on the computer.

This pragmatism is related to the way computers are domesticated. Home PCs are not objects for their own sake. They are acquired for work, education, play, or some other purposes. Their domestication entails the establishment of a relationship between the PC and an emerging set of activities. The cognitive aspect of this relationship is the learning of some basic computer skills, but the main point is to learn to use the PC as a tool. This learning is embedded in previous skills and the relevant areas of practice.

To summarize, our findings on the social integration of PCs and the related process of knowledge acquisition are similar to those in the section on energy use. People in everyday settings do not acquire knowledge in a disembodied manner. Their concern is the management and practice of everyday tasks. Knowledge about energy and computers has only pragmatic relevance and must be combined with other kinds of knowledge if it is to be applied.

Nonetheless, the PC as a material artifact figured prominently in the stories our informants told about how they learned. This was most clearly observed in learning that took place through the use of help features and other self-instruction software. However, when the informants explained how they employed the strategy of learning by trial and error, nearly all of them described practical and cognitive interaction in which the computer played a crucial role in the process of finding out how it worked in relationship to a particular user. Clearly, knowledge about computers is acquired in a way mediated by the machine. In most cases, the user may not have an understanding of computers as such but may have knowledge about sequences of events that elicit an intended result. Thus, computers possess a kind of contingent and local instructional quality. They do not instruct by means of some inner essence waiting to be discovered but by means of a complex set of enabling and disabling mechanisms that interacts with users, bringing out the users' different "styles" of computer usage.

The contingencies of learning are, of course, very important. We have already hinted at influences that effect the acquisition of computer skills, and gender and age are related to the way PCs become domesticated. The jobs people have are also important in many aspects of computer usage. In the next section, we look at a different sort of contingency related to the way software engineers attempt to stage the domestication of computers in the workplace.

The Construction of User Knowledge in the Design of Computer Systems

The domestication[1] of workplace technologies happens in a social terrain different from that of the household. The domestication of technologies at the workplace, just like the domestication of technologies in the home, is not free from conflicts and efforts to control outcomes. But that similarity does not mean that the two processes are the same, for the social dynamics of the workplace are different from those of the home. New technologies are usually acquired by management in order to achieve some goal, and even if the process of implementation may not be as rational as economists claim, managers are usually able to influence the outcome by, say, controlling tasks and the information and training offered to employees. The terms of the transfer of knowledge are consequently different from those discussed in the previous two sections of this chapter.

On the other hand, there are important similarities. First, users need to be able to domesticate new technologies at work as well as at home. Unless an artifact can lend itself to integration practically, symbolically, and cognitively, it will not be used. Second, in both settings an artifact's use is embedded in an asymmetric relationship between the assumed expertise of its designers and the assumed nonexpertise of its users. There exists an idea of correct and incorrect usage that is predefined as it were through design, production, and marketing. In principle, users are supposed to acquire knowledge from the designer in order to be able to use the artifact in the correct manner. Of course, this expectation reflects the notion of linearity embedded in the standard understanding of knowledge transfer.

As shown in the previous two narratives, users often act differently than they are supposed to do. This difference may be related to the fact that

---

[1] Etymologically speaking, domestication is related to the home, the domestic setting. In this section, however, we use the word as an analytical concept that transcends the physical meaning of *home*. Domestication in this context designates a particular understanding of cultural appropriation that may occur in any social setting where people may transform artifacts in order to make them "their own" (Lie & Sørensen, 1996).

designers and users in those two cases occupy separate worlds with little or no communication between them. With closer interaction, this mutual exclusiveness could be different. If designers and users were to collaborate on the construction of a particular technology, designers would have a better idea of what users want, and users would acquire a better understanding of how designers think the artifact or system is intended to be used. Examining the design practice behind computer systems would thus yield strategic insight into this issue of worker participation, for there is general agreement that workers possess knowledge about the work processes that designers need in order to make new systems. When asked about the need for such participation, the surveyed Norwegian computer systems designers unanimously agreed that user involvement is crucial (Hatling & Sørensen, in press).

The following analysis of systems design practices is based on extensive interviews with 63 practitioners and academics active in the field and a postal survey of Norwegian companies doing systems design (see Hatling & Sørensen, in press, for more details).

From an abstract sociological perspective, the challenge for computer systems designers is to create such representations of the information-processing in a given organization as will provide the basis for increasing the delegation of tasks to the computer. The task of these designers is to inscribe user actions into the system (Akrich, 1992; Woolgar, 1991). One point of view is, in our opinion, that designers are setting the stage for users' domestication of the new system, including the possibilities for users to learn about the system and its logic. Another point of view is that designers depend on the knowledge that workers have about their workplace and the local practices of information-processing.

These diverging interpretations of the systems designer's role result in difficult paradoxes. Users (workers) are perceived as a very important source of procedural and organizational knowledge. In some sense, their participation in the efforts to design a new system is mandatory. Without access to worker's knowledge systems, designers cannot do their jobs adequately. This reciprocity should in principle invite a kind of collaboration, an exercise in participatory design based on an exchange of knowledge. If that cooperation were to be achieved, computer systems designers would learn about local practices, and workers could learn about the possibility of using computers to improve the quality and/or efficiency of local information processing.

Still, we learned from the interviews that it is important to designers to emphasize the difference between their expertise and the skills of workers. This distinction is made by understanding workers' skills as a form of knowledge that is demarcated by its concrete, local qualities. One of the informants stated that

> Often, they [workers] have problems in relating to issues at an abstract level. They find it difficult to relate to specifications. They almost have to see concrete examples of how it is going to be. . . . They are not able at all to formulate their needs in a precise manner. Often, the situation is such that we have to take the task upon ourselves and present solutions to which they may say yes or no.

In their accounts, designers also constructed users (workers) as conservative and unable to recognize possibilities for change. Thus, the interaction set up between designers and workers was very asymmetric. Although both parties possessed valuable knowledge, only one form of knowledge was seen by the informants as progressive and capable of creating possibilities for change. In other words, closer interaction between the designers and users depicted in this narrative continued to reflect in general the social construction of expertise in modern societies. Experts were assumed to possess the rational knowledge that is critical to innovative design and to know what ideally should be transferred to nonexperts in order to enable them to perform in their roles as users or citizens. This view was described by one informant in the following manner:

> It is fun to look at users' specifications because very often they are so extremely hung up in beliefs about what's possible. When users talk, they are not very visionary. What else should they be? Of course, they cannot talk about things they haven't seen.

Such points of view should not be dismissed as mere arrogance on the part of experts. Systems designers and workers have different kinds of knowledge and experience. In addition, the practices of systems designers are embedded in a larger social context where the methods and resources available impose important constraints on their practices. One could also argue that these designers are faced with the challenge of domesticating the methods they apply in order to perform their tasks. Most of the methods used to domesticate technology in Norway originally came from the United States. Developed in a different culture, they may need cultural translation in order to work within the Norwegian context.

The problem lies in the way the differences between designers and workers were managed and how those differences, particularly their cognitive dimension, affect the domestication processes of workers. Computers are too flexible for the sociotechnical configuration of a new computer system to be achieved through technical inscriptions alone. Managerial instructions as well as training are needed to make users comply with given procedures. This prerequisite usually sets the stage for a traditional exercise in the linear transfer of knowledge so that workers can be educated to perform as envisaged by the designers of the new system. But this image usually differs from the self-perception of workers, and the efforts made by designers to compartmentalize the workers' influence do not foster a productive learning environment. When workers' concerns about issues of employment security and quality of work life are dismissed as misguided, when participation is reduced to questions about the layout of screen images, the divide between designers and users is reproduced even in efforts of participatory design. One should not wonder if workers in this situation find it legitimate to follow their own views and interests when domesticating a new system.

In summary, conflicting representations of reality may be a basic constituent of the cultural appropriation of kinds of knowledge as well as of technologies. Even when computer systems designers and users interact, it is possible that their preconceived ideas about each other will prevent mutual understanding of skills and needs. We have shown this to be the case for designers, and we may safely assume that workers often construct designers in a similar, only reversed, manner.

Linear Transfer versus Linear Appropriation?

This chapter has explored the concept of domestication as a theoretical and methodological alternative for studying issues related to the public understanding of science and technology. Traditional research on the public understanding of science and technology was criticized for being production-centered, that is, for defining issues from the perspective of engineers and scientists and analyzing knowledge-crossovers in terms of linear transfer. A further critical note is the observation that in the traditional approach to the public understanding of science and technology, public understanding is evaluated, whereas scientific understanding is not. This imbalance makes it difficult to grasp the ways in which scientific and technological knowledges are appropriated.

We argue that one needs to employ a different methodology. The concept of domestication is such an alternative. Above all, it focuses on the public rather than on science and scientists, inverting the questions and issues on the traditional agenda of research on the public understanding of science and technology. This shift forces one to see how the appropriation of knowledge is embedded in practices related to a well-defined sociotechnical ensemble, such as constructing a home or a work environment (a user's point of view) and in a complex of many different knowledges and artifacts (a scientific or technological point of view).

In a sense we can say that scientists' efforts of linear transfer of knowledge become the object of the public's efforts of linear appropriation, but the linearities are of different origin and nature, and they do not match. As a whole, this situation of production and reproduction of knowledge is definitely nonlinear. A methodology flexible enough to cater to nonlinearity is needed.

To some extent, one may draw a comparison with studies of how engineers work. Often, engineers may not be particularly concerned about having a correct scientific understanding of technology or about making scientifically correct applications. Their concern is mainly to make something "work," and to this end they make pragmatic use of available material and intellectual resources (Bucciarelli, 1994; Constant, 1984; Levold, 1990).

As is evident from the two first narratives presented in this chapter, when many people have an understanding of a phenomenon that is incorrect from a scientific point of view, they do not experience this discrepancy as a problem. Unless their actions fail to elicit the desired results, most homeowners or computer users do not bother to search for a scientific understanding of their tasks. This lack of concern is probably due to an absence of a simple and unidirectional relationship between scientific knowledge (at least of the sort that is accessible to laypeople) and the physical aspects of human action. Laypeople are seldom, if ever, offered an understanding of technology or nature that provides complete and unambiguous insight and guidance for action. Even in the case of PCs equipped with self-instructional software, knowledge cannot be assumed to piggyback upon technology. When knowledge is domesticated, when it is locally embedded and embodied, it is made relative to local culture and practice. Formal, codified knowledge is important in quizzes but is not important in the same way in real life. Failure to take account of that distinction is the primary mistake of surveys on the public understanding of science and technology.

We do not argue that all knowledge is of equal value or validity. One can provide sound arguments for supporting agendas of scientific and technological literacy, at least as part of an enlightened culture. However, in order to have a better understanding of the conditions of producing such literacy, one needs to be concerned with how facts and artifacts may or may not be domesticated. This realization is a major methodological message of our paper.

Further, domestication is a process through which people produce micronetworks of material and nonmaterial, human and nonhuman elements. They make use of new and old knowledge and artifacts, accepting some, discarding others. These networks are contingent on local resources, structural or global intersections, and juxtapositions of local and global strategies.

Domestication is therefore pragmatic but also a potential issue of conflict. New knowledge or technology must be domesticated in order to be integrated into local culture, but not all new knowledge or technology is domesticated, and different people domesticate different kinds of knowledge and technology.

Conflicts arising from domestication are related to—

1. different local interests and roles (gender, class, ethnicity, age);
2. the relationship between hegemonic, global codes versus local, possibly oppositional codes of interpretation and evaluation; and
3. the relationship between, on the one hand, the efforts of designers and producers of knowledge to configure use through inscriptions, prescriptions, and programs and, on the other hand, users' creativity and ability to counteract programs.

Issues of scientific and technological literacy should not be addressed without such concerns in mind. This caution is also an important methodological point.

Because micronetworks have no universal features, except for their formal nature as an ensemble of human and nonhuman parts with practical, symbolic, and cognitive qualities, standardized methodologies cannot be used to explore them. In the first two narratives in this chapter, there was no single, homogeneous, well-defined relationship to energy and energy-saving technologies or toward computers. The empirical observations lead to an emphasis on differences, say, in terms of cultures of energy use or in terms of learning strategies in relation to computers. Neither the artifacts nor the scientific knowledge behind these artifacts could shape human action or thinking in an unequivocal manner. Thus, the analysis of domestication has to be sensitive to the possibility of such differences as well as to the social and cultural differentiations of taste, use, and availability of artifacts and knowledge that represent some of the contingencies of domestication.

Nevertheless, the concept of micronetworks is important for the field of research on the public understanding of science and technology. Too often, understanding is perceived only in cognitive terms. When scientific knowledge is taken not as a set of disembodied truths but as locally embedded discourse, one discovers the need to analyze its practical and symbolic aspects as well. The case of the home computer suggests that there are important cognitive challenges to the computer user, but even so computers become integrated into practical and symbolic patterns of routines, placement, and division of labor. Thus, computers become part of a micronetwork where learning is needed to keep the network together. The emphasis is somewhat different in the case of energy use. Few, if any, of the informants experienced problems of understanding energy and energy issues. They are not looking for knowledge about energy as such, and the information that they have has a contradictory relationship both to practical and symbolic aspects of their everyday life. They construct micronetworks where learning about, say, new energy-saving technologies is not seen as important. This lack of interest in learning blocks the domestication of such technologies.

By exploring the public understanding of science and technology through the concept of domestication, one basically insists that analysis should focus on (a) the public, rather than on science or technology; (b) symbolic and practical issues, not just cognitive ones, (c) the study, not just the observation, of cultural embeddedness; and (d) analysis that is sensitive to conflicts, differences, and differentiations between people. The resulting methodology is well known in science and technology studies as one that is qualitative, action-oriented, semiotically informed, and thus flexible.

In our first two narratives, many users had difficulties when they domesticated available knowledge. Some of these problems were cognitive, for some knowledge is difficult to understand. But the major obstacle was the lack of concern for practical and symbolic challenges among knowledge suppliers. Our third narrative suggested why that lack of concern is such a problem. Designers (and by implication, scientists) are so concerned with their own knowledge and skills and with how they are different from users that they are unable to communicate with users in a productive manner. Ultimately, the problem of the public understanding of science and technology is a problem of democracy.

References

Akrich, M. (1992). The de-scription of technological objects. In: W. Bijker & J. Law (Eds.), *Shaping technology/building society* (pp. 205–224). Cambridge, MA: Massachusetts Institute of Technology (MIT) Press.

Aune, M. (1992). *Datamaskina i hverdagslivet. En studie av brukeres domestisering av en ny teknologi* [The computer in everyday life: A study of how users domesticate a new technology] (Report 15/92). Trondheim, Norway: Centre for Technology and Society.

Aune, M. (1996). The computer in everyday life. In M. Lie & K. H. Sørensen (Eds.), *Making technology our own? Domestication technology into everyday life* (pp. 91–120). Oslo: Scandinavian University Press.

Aune, M. (1998). *"Nøktern eller nytende?" Energiforbruk og hverdagsliv i norske husholdninger* [Energy consumption and everyday life in Norwegian households] (Report No. 34). Trondheim, Norway: Center for Technology and Society.

Bucciarelly, L. L. (1994). *Designing engineers.* Cambridge, MA: Massachusetts Institute of Technology (MIT) Press.

Constant, E. (1984). Communities and hierarchies: Structure in the practice of science and technology. In R. Laudan (Ed.), *The nature of technological knowledge* (pp. 27–46). Dordrecht: Reidel.

Haddon, L. (1992). Explaining ICT consumption: the case of the home computer. In R. Silverstone & E. Hirsch (Eds.), *Consuming technologies: Media and information in domestic spaces* (pp. 82–96). London: Routledge.

Haraway, D. (1991). *Simians, cyborgs, and women.* London: Free Association Books.

Hatling, M., & Sørensen, K. H. (in press). Social constructions of user participation. In K. H. Sørensen (Ed.), *The spectre of participation: Technology and work in a welfare state.* Oslo: Scandinavian University Press.

Kempton, W. (1987). Two theories of home heat control. In D. Holland & N. Quinn (Eds.), *Cultural models in language and thought* (pp. 222–242). Cambridge, England: Cambridge University Press.

Kempton, W., & Layne, L. L. (1994). The consumer's energy analysis environment. *Energy Policy, 22,* 857–866.

Kline, S. J., & Rosenberg, N. (1986). An overview of innovation. In R. Landau & N. Rosenberg (Eds.), *The positive sum strategy: Harnessing technology for economic growth* (pp. 275–306). Washington, DC: National Academy Press.

Latour, B. (1987). *Science in action.* Milton Keynes, England: Open University Press.

Latour, B. (1993). *We have never been modern.* Cambridge, MA: Harvard University Press.

Latour, B. (1992). Where are the missing masses? The sociology of a few mundane artefacts. In W. Bijker & J. Law (Eds.), *Shaping technology/building society* (pp. 225–258). Cambridge, MA: Massachusetts Institute of Technology (MIT) Press.

Law, J. (Ed.). (1991). *A sociology of monsters: Essays on power, technology and domination.* London: Routledge & Kegan Paul.

Levold, N. (1990). *Ingeniørarbeidet—bransjeskapt eller lokalt konstruert?* [Engineering work—Shaped by industry or a local construction?] (STS working paper no. 12/90). Trondheim, Norway: Centre for Technology and Society.

Lewenstein, B. V. (1995). Science and the media. In S. Jasanoff, G. E. Markle, J. C. Petersen, & T. Pinch (Eds.), *The handbook of science and technology studies* (pp. 343–360). Thousands Oaks, CA: Sage Publications.

Lie, M., & Sørensen, K. H. (Eds.). (1996). *Making technology our own? Domesticating technology into everyday life.* Oslo: Scandinavian University Press.

Martin, E. (1989). *The woman in the body.* Milton Keynes, England: Open University Press.

Martin, E. (1994). *Flexible bodies: Tracking immunity in American culture from the days of polio to the age of AIDS.* Boston: Beacon Press.

McCracken, G. (1988). *Culture and consumption: New approaches to the symbolic character of consumer goods.* Bloomington, IN: Indiana University Press.

Pfaffenberger, B. (1992). Technological dramas. *Science, Technology, and Human Values, 17,* 282–312.

Silverstone, R., & Hirsch, E. (Eds.). (1992). *Consuming technologies: Media and information in domestic spaces.* London: Routledge.

Silverstone, R., Hirsch, E., & Morley, D. (1992). Information and communication technologies and the moral economies of the household. In R. Silverstone & E. Hirsch (Eds.), *Consuming technologies: Media and information in domestic spaces* (pp. 15–31). London: Routledge.

Sørensen, K. H. (1994). *Technology in use: Two essays on the domestication of artifacts* (STS working papers 2/94). Trondheim, Norway: Centre for Technology and Society.

Sørensen, K. H., & Sørgaard, J. (1994). Mobility and modernity: Towards a sociology of cars. In K. H. Sørensen (Ed.), *The car and its environments: The past, present and future of the motorcar* (COST social sciences, COST A4 vol. 2, pp. 1–32). Brussels: Commission of the European Communities (CEC), Directorate General XII.

Turkle, S. (1984). *The second self: Computers and the human spirit.* New York: Simon & Schuster.

Wilhite, H., & Ling, R. (1992). The effects of better billing feedback on electrical consumption: A preliminary report. *1992 Proceedings of the ACEEE Summer Study on Energy Efficiency in Buildings, 10,* 173–175.

Woolgar, S. (1991). Configuring the user: The case of usability trials. In S. Law (Ed.), *A sociology of monsters: Essays on power, technology, and domination* (pp. 57–99). London: Routledge.

Woolgar, S. (1994). *Rethinking the dissemination of science and technology* (CRICT Discussion Paper no. 44). London: Centre for Research in Innovation, Culture and Technology (CRICT), Brunel University.

Wynne, B. (1995). Public Understanding of Science. In S. Jasanoff, G. E. Markle, J. C. Petersen, & T. Pinch (Eds.), *The handbook of science and technology studies* (pp. 361–388). Thousands Oaks, CA: Sage Publications.

PART FOUR

INFORMING THE PUBLIC AND DEBATING SCIENCE AND TECHNOLOGY

INTRODUCTION

Meinolf Dierkes and Claudia von Grote

In these final chapters the critical review of the traditional approach to studying the public understanding of science and technology comes full circle to center on the function of the politicohistorical context. It is an aspect that commanded particular attention in the first part of the book because of the metatheoretical question of the degree to which the history of the relation between science, society, and the public has shaped the public understanding of science and technology as an object of research. The political context as a space for public debate is considered again, but this time the purpose is to examine the way in which the public understanding of science and technology is articulated—that is, how the forums and forms of scientific and technological discussion work—and what it means for knowledge and the popularization of knowledge and trust.

Analytical interest centers on the assertion that raising the level of knowledge (i.e., informing the public) has a positive influence on attitudes, a statement based on the prevailing assumption that knowledge and attitudes are positively related. However, no clear empirical confirmation of such a direct relation has yet been offered, and experience with public debates on science and technology has shown that the relation between knowledge and attitudes varies greatly. For those reasons, the research focus is no longer the knowledge that one conveys by informing people but rather that activity's underlying social system consisting of those who impart the knowledge, the contexts in which they do so, the function acquired by such processes, and the role that the media and experts play in them.

Peters (chapter 11) and Jelsøe (chapter 12) emphasize that the investigation of factors determining public attitudes toward science and technology calls for much more than analysis of how much the public knows about them. Both authors reason from the perspective of risk communication and technical assessment as they address the question of the conditions under which the public gains trust in the persons and institutions passing on information and making decisions. Bucchi (chapter 13) devotes attention to a specific channel for the formation and expression of public opinion: the media. But in contrast to the familiar question of whether and how the press influences the public, the issue at the center of Bucchi's interest in the newspaper coverage of science and technology is why scientists seek communication through that medium.

In chapter 11 Peters examines the didactic assumption that follows from the cognitive-deficit model in the study of the public understanding of science and technology—the expectation that imparting more knowledge will improve the attitude toward science and technology. The goal of his two empirical studies is to measure the relations between knowledge and attitude at a given time after a phase of controversy (in this case, Chernobyl) and to investigate those relations at two different times in order to ascertain whether local controversies over the construction of an incinerator change the local public's attitudes as a function of information flow. The results of Peters's studies, which contradict the assumption that a universal positive relation exists between knowledge and attitude, draw attention to technology-specific factors and a host of empirical effects that bring other interrelations to light. In particular, Peters looks at (a) the connection between the credibility of the information sources and the acceptance of information and (b) the difference between the expert's and the lay public's concepts of assessment, such as the concept of having control over a given technology. He therefore considers it too narrow to restrict the study of the public understanding of science and technology to the public's knowledge about science, technology, and scientific methods.

Jelsøe picks up on precisely that view in chapter 12. He stresses much more explicitly than Peters that it is inappropriate to discuss the topic as though the role of information were context-free. In his opinion the analysis must be broadened to include the circumstances under which information is given and the assumptions that the public entertains about the relation between science, scientific institutions, industry, and political institutions. Key questions in this regard are how and in which political context information is regulated and what prompts the public to assimilate information.

Historical analysis of the technology debates in specific countries allows the significance of information to be ascertained as an element of such broadened discussion and activities. Jelsøe reconstructs the critical controversies on biotechnology of the last 25 years in Denmark in light of experience with the discussion about nuclear power. The building of trust in regulatory processes through the use of discursive procedures is a key point in explaining the willingness of a critical public to accept a given technology.

Whereas media analyses often take up the question whether press coverage forms attitudes against science and technology, such as by sensationalizing scientific and technological topics, Bucchi asks the question the other way round: What function does the scientific community give to the press? He therefore asks why an abstract scientific finding like the ripples in the cosmic background, which is virtually irrelevant to everyday life, managed to trigger the avalanche of press reports that it did in England. His "case analysis" of media reporting is an investigation into the role that mass-media communication itself acquires for the treatment of such scientific topics. He means the role of giving a scientific paradigm a discursive framework that is less constrained than a forum of experts. As Bucchi points out, scholars can use public communication to attract and coordinate researchers from a wide variety of approaches and can open up academic disciplines as well as delimit topics of discussion.

CHAPTER 11

FROM INFORMATION TO ATTITUDES? THOUGHTS ON THE RELATIONSHIP BETWEEN KNOWLEDGE ABOUT SCIENCE AND TECHNOLOGY AND ATTITUDES TOWARD TECHNOLOGIES*

Hans Peter Peters

One hypothesis often held by researchers and by people engaged in practical communication with the public is that pessimistic risk judgments, which lead to fears or so-called *acceptance problems* of modern technologies, are caused by insufficient knowledge about the sources of risk and their impacts. Differences between risk judgments arrived at by laypeople and risk judgments arrived at by experts are often attributed mainly to knowledge differences between the two groups. The more scientific and technological knowledge laypeople have, so the argument goes, the better they should be able to assess technologies in a rational way and come to conclusions similar to those of technological experts. Furthermore, if conflicts between experts surface, they are often explained away by distinguishing between "real" and "self-appointed" experts.

Because modern technologies and many environmental problems are characterized by complexity and intransparency, this hypothesis that lack of knowledge leads to negative attitudes has at least some initial validity on the surface. If one accepts the hypothesis, then information campaigns and the like seem to offer a probable cure for acceptance problems and fears about the environment. By closing the knowledge gap, one expects to diminish the differences between the risk judgments, attitudes, and fears of laypeople and those of experts. As a consequence, a matter-of-fact approach for providing information to the public is seen as the key to solving disagreements between the lay public and scientific experts. This line of reasoning seems to have guided the design of the 1992 Eurobarometer 38.1 survey on the public

* I thank Philip C. R. Gray for his comments on an earlier draft of this chapter and for his editorial help in making my English appear less clumsy than it otherwise would have been.

understanding of and attitudes toward science and technology (see Institut National de Recherche Agricole [INRA] & Report International, 1993). It is an approach to which I take exception, a response I explain by examining a particular perspective on the relationship between the public on the one hand and science and technology on the other—the expectation that knowledgeable people will have a positive attitude toward science and technology and will therefore accept technologies that are recommended by experts.

Certain assumptions underlie the technocratic way of thinking conveyed by the 1992 Eurobarometer 38.1. First, it is assumed that risk perception and attitude formation are decisively influenced by scientific and technological knowledge. The second assumption is that the most significant difference between laypeople and experts with respect to the evaluation of technologies consists in the amount and quality of factual information available to inform opinions.

As demonstrated in this chapter, both assumptions are invalid. The discussion and empirical data presented in the following pages do not pertain to science per se but to controversial technologies, particularly nuclear power and waste incineration. The information examined is therefore not knowledge about science in a narrow sense but rather information about technologies, technological projects, and their impacts.

The question of how knowledge is related to attitudes has a scientific and a practical dimension. Scientifically, the relationship between knowledge and attitudes is instructive for what it can reveal about how attitudes toward science, technologies, and technological projects are formed. Practically, this question is important because increasing the knowledge level of the population is often seen as a method for "improving" public attitudes.

The hypothesis that a lack of knowledge is the main cause of acceptance problems is criticized in this chapter on empirical and theoretical grounds. This criticism is presented in four steps:

1. The implicit assumptions that a positive relationship exists between knowledge and attitudes toward science and technology are reconstructed, with the Eurobarometer survey being used as an example.

2. A number of hypotheses about the relationship of knowledge to attitudes are discussed. Empirical results from two studies are presented in order to illustrate, rather than to prove, some points of my criticism. I dispute the assumption that there is a consistent correlation, or even a simple causal relationship, between knowledge about a scientific or technological innovation and the attitude held toward it. Some critical remarks about the Eurobarometer study are made.

3. I argue that the divergent opinions that the promoting expert and the skeptical member of the public have on controversial technologies need not stem from the knowledge gap between those groups. Experts and laypeople differ in more ways than their scientific and technical competence alone.

4. In order to draw conclusions for further research from the critique provided, I sketch some promising research issues that should receive more attention than has been given thus far in studies on the public understanding of and attitudes toward science and technology.

Relations between Knowledge Level and Attitudes

Implicit Assumptions

The Eurobarometer survey on the public understanding of and attitudes toward science and technology may serve as an example illustrating a tendency to apply the above-mentioned technocratic perspective as an analytical model to the study of laypeople's perceptions and decision-making. The authors of the report on the Eurobarometer study are careful not to stress the link between knowledge and attitudes too strongly; after all, it appears to be quite weak and inconsistent in the Eurobarometer study itself (INRA & Report International, 1993, p. 82). However, the selection of basic scientific knowledge as the primary variable in the study of attitudes toward science and technology does say a great deal about the designers' implicit theories.[1] The statement that the "understanding of science and technology nowadays requires (a) a basic knowledge of scientific facts and (b) an understanding of the process or methods of science for testing our models of reality" (INRA & Report International, 1993, p. 19) is certainly correct. But social and political knowledge is also but crucial for an understanding of science and technology within society.

Knowledge of this kind includes information about the complex relationships of science and technology with industry and governments; the institutional arrangements in which science and technology development are carried out; the influence that professional communities, industry, and politics have on research programs and individual scientists; the conflicts between scientific authority based on claims to truth and the lay public's demands to

---

[1] Because the designing of a Eurobarometer study is a complicated process in which conditions and information required by the Commission of the European Communities may clash with the interests of the researchers preparing the questionnaire, this criticism is addressed to a somewhat amorphous target group.

participate in decisions leading to technological change that affects the conditions under which one has to live; the relationship of science and technology with ethics and culture; and the question of who controls technological development and its applications and for what purposes. Why did the Eurobarometer survey and many other studies not cover these areas of knowledge, which are at least as important as a knowledge of scientific facts and methods are when it comes to forming reasonable opinions and attitudes and making rational decisions about science and technology?

Empirical Data

There are studies that have found positive relationships between the level of knowledge about science and attitudes toward science or toward applications based squarely on science (e.g., Grimston, 1994; McBeth & Oakes, 1996, p. 424; Zimmerman, Kendall, Stone, & Hoban, 1994, p. 74). Quite a number of studies, however, illustrate that such relationships are varied, usually weak, and rather complicated (nonmonotonous). Take, for example, two sources showing the tenuous nature of the relation between knowledge of and attitudes toward science: (a) Hennen & Peters's (1990) analysis of the nuclear power debate that spread throughout Germany after the 1986 Chernobyl disaster, and (b) Wiedemann, Schütz, and Peters's (1991) study of a local controversy about the planned construction of a waste incinerator in Aldenhoven (a small municipality near Aachen in North Rhine–Westphalia). In these two projects, members of the general population (Hennen & Peters) and of the local population (Wiedemann et al.) were surveyed with an instrument that included questions eliciting their scientific knowledge (knowledge items) and questions about their attitudes toward science (attitude items). Additionally, the local population involved in the incinerator study was surveyed in November 1989 and again from November through December 1990, meaning that the authors were able not only to analyze the relationship between knowledge level and attitudes at a given time but also to show the change that occurred in that relationship between the first panel wave and the second.

In the Chernobyl study (Hennen & Peters, 1990), the level of knowledge was measured by means of three questions with which respondents were asked about the number of nuclear power plants operating in West Germany, the proportion of national electricity consumption met by nuclear power plants, and the dominant type of nuclear reactor used in West Germany. Each question had four answer categories (in addition to *don't know*), one of which was correct. An index "knowledge level" representing the sum of the interviewee's correct responses was calculated (theoretical value range 1–3).

In Wiedemann et al. (1991) three open-ended questions tested respondents' knowledge about the type of waste to be processed in the planned waste incinerator, the companies involved in the project, and the agency responsible for the licensing procedure. Because the plant was designed to process more than one type of waste and because more than one company was involved in the project, eight correct responses were possible. The knowledge index was calculated in the same manner as for Hennen and Peters (1990). Because respondents rarely gave more than two correct responses, values of 3 to 8 were recoded as index value 3. Hence, the value range of the knowledge index was likewise 1 through 3.

Attitudes toward the respective technology were measured by means of a single question. In the Chernobyl study respondents were asked to select one of five possible stances on the future use of nuclear power in Germany, ranging from *immediate withdrawal from nuclear power* to *use as much nuclear power as possible*. Similarly, respondents in the incinerator study were asked whether they were in favor of the plant, opposed to it, or undecided. (A fourth option—that it did not matter to the respondent whether the incinerator plant was built—was rarely chosen and is excluded from the analysis reported in this chapter.)

Figures 11-1 through 11-3 show the mean levels of knowledge for each attitude group in Hennen and Peters (1990) and Wiedemann et al. (1991). From these three bar graphs, it is evident that there is no universal relationship between knowledge level and attitude. Figures 11-1 and 11-2 show a nonmonotonous relationship between these two variables. That is, the gradient from most positive attitude to most negative attitude is not paralleled by a gradient from most knowledge to least knowledge. In Figure 11-1 (the Chernobyl study), for example, the respondents with the most positive attitude toward nuclear power had the highest level of knowledge about it, but the respondents with the second most negative attitude had the second highest level of knowledge about this technology. Figures 11-2 and 11-3 show that the form of the relationship can shift dramatically over time. The first wave of the incinerator study shows a U-shaped relationship between knowledge level and attitude. (Respondents who were undecided had less knowledge than either the proponents or the opponents.) The picture that emerged from the second wave twelve months later reveals a monotonous pattern. The direction of the relationship runs counter to the hypothesis that a high knowledge level should go hand in hand with positive attitudes: Opponents had significantly more knowledge than either the "undecided" respondents or the proponents.

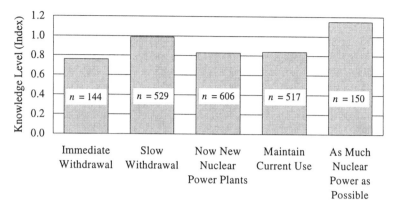

Local Attitudes toward Use of Nuclear Power

Figure 11-1. Post-Chernobyl empirical relationships in Germany (1987) between level of knowledge about and attitudes toward the use of nuclear power. The data are from *"Tschernobyl" in der öffentlichen Meinung der Bundesrepublik Deutschland—Risikowahrnehmung, politische Einstellungen und Informationsbewertung* [Chernobyl in the opinion of the Federal Republic of Germany—Risk perception, political attitudes, and information assessment] (Report No. Jülich-Spez-551), p. 23, by L. Hennen and H. P. Peters, 1990, Jülich, Germany, Research Center Jülich.

In the incinerator study, an analysis of the changes in knowledge level and attitude between the groups of the first wave and the groups of the second wave shows that the undecided respondents of the first wave (i.e., those with the lowest knowledge level) acquired much more knowledge by the second survey than did the groups with clear-cut attitudes. Furthermore, undecided respondents who learned more than average were more likely to develop negative attitudes than those who learned less than average. (The latter subgroup tended to remain undecided.) Learning more about the project during the period from the first to the second survey had a medium negative correlation with attitude change $(r = -.19, p \leq 0.1)$. That is, those who obtained more knowledge about the project were more likely to develop negative attitudes than those whose knowledge level remained constant.

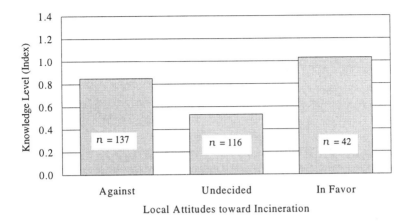

Figure 11-2. Empirical relationships between knowledge level about and attitudes toward construction of an incineration plant as measured in the local populations close to a planned site (first wave results, November 1989). The data used in the graph are obtained by a re-analysis of the original survey data conducted by the author. The methodology of the survey is described in "Information need concerning a planned waste incineration facility," by P. M. Wiedemann, H. Schütz, and H. P. Peters, 1991, *Risk Analysis, 11*, pp. 229–237.

Consistent with a recent British study (Evans & Durant, 1995), the empirical data presented here do not support the hypothesis of a universal positive relationship between knowledge level and attitude. What kind of relationship does exist seems to depend on a number of factors particular to the respective technology or technological project. In the following paragraphs, I introduce hypotheses about the relationship of attitudes and knowledge and briefly outline how these hypotheses might be used to explain the empirical results reported by Hennen and Peters (1990) and Wiedemann et al. (1991).

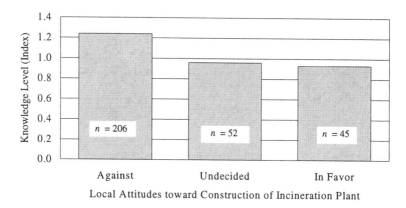

Figure 11-3. Empirical relationships between knowledge level about and attitudes toward construction of an incineration plant as measured in the local population close to a planned site (second wave results, November–December, 1990). The data used in the graph are obtained by a re-analysis of the original survey data conducted by the author. The methodology of the survey is described in "Information need concerning a planned waste incineration facility," by P. M. Wiedemann, H. Schütz, and H. P. Peters, 1991, *Risk Analysis, 11,* pp. 229–237.

Hypotheses about the Relationship between Knowledge and Attitudes

It is quite evident that knowledge influences opinions in many cases. Under certain circumstances new knowledge may completely alter opinions and attitudes without there being much influence exerted by personal or situational factors. But in most cases, particularly when claims to knowledge are controversial and uncertainty is evident, it may be difficult to know whether to believe new information, and it may therefore be transformed by laypersons into subjective knowledge. Furthermore, indisputable facts may be interpreted, evaluated, and weighted differently. With complex issues (and most controversies over science and technology *are* complex), there is little direct relationship between facts and attitudes. Hence, the kind of relationship between knowledge and attitudes may in some situations be the reverse of that postulated in the hypothesis that more knowledge leads to more positive attitudes toward science and technology. Attitudes may influence attention structures and attention levels, in turn affecting information-seeking and, finally, the level and structure of a person's knowledge.

A number of possible effects produce the variety of empirically observed relationships between knowledge level and attitudes. Some of these effects tend to produce a monotonous relationship (high knowledge level associated with a clearly positive or clearly negative attitude), some lead to a U-shaped relationship (the higher the knowledge level, the more extreme the respondent's attitude). I first discuss a number of effects that might cause monotonous attitudes toward science and technology.

Biased selection of knowledge items (methodological artifact). The observed relationship between knowledge and attitudes may be heavily dependent on the kind of questions that are asked about knowledge and hence may be a methodological artifact. Rieder (1995, p. 4) convincingly argued that it is useful to distinguish between different areas of knowledge and that proponents and opponents in scientific or technical controversies differ with respect to the areas in which they possess expertise. If proponents and opponents use different kinds of knowledge to form and defend their attitudes, then it is crucially important to follow a systematic procedure that ensures a selection of knowledge questions that is equally fair to both sides. In the Chernobyl study two of the three knowledge questions were related to positive aspects of nuclear power (reliable energy supply), knowledge that is more likely to be considered relevant by proponents of nuclear power than by its opponents. Retrospectively, it becomes clear that the knowledge index was biased against the opponents of nuclear power.[2] It is therefore not surprising that the group of respondents with the most positive attitude toward nuclear power received the highest score on the knowledge index and that the group with the most negative attitude received the lowest score.

Again, the 1992 Eurobarometer study on public understanding of science and technology can be used to illustrate this argument. In that survey a similar methodological bias may have influenced the ranking of nations with respect to their corresponding populations' public understanding of science. A look at the individual knowledge items (INRA & Report International, 1993, pp. 152–164) shows that the ranking of nations differed from item to item. For example, British respondents scored much lower than respondents from Greece and Portugal on questions about astronomy (questions 56–57), and Greek and Portuguese respondents gave more incorrect answers to all other questions than did the British. These results show that the ranking of nations is quite sensitive to the distribution of knowledge items across different fields of science and to their use in the calculation of the summary

---

[2] Neither Hennen and Peters (1990) nor Wiedemann et al. (1991) were aware of this methodological problem when they conducted their surveys. Their main concern was to formulate knowledge questions that had a single, indisputably correct answer and that seemed to be relevant to opinion formation in one way or another. These two criteria alone were difficult to meet.

index. An index including many astronomy questions will favor Greek and Portuguese respondents; one with few astronomy questions will favor British respondents.

Hutton (1996) called attention to the relationship between school science curricula and the public understanding of science. According to him, members of the public will be particularly knowledgeable in those fields that were given priority during their secondary education. Hutton's observation, together with the plausible hypothesis that the school science curricula also influence survey-question developers in their judgment of which knowledge is relevant, may explain why Great Britain was ranked among the top group of nations with respect to the public understanding of science (INRA & Report International, 1993, p. 38). It may have to do with the fact that most knowledge questions in the 1992 Eurobarometer survey were developed and selected by Anglo-Saxon researchers who probably (and unconsciously) tended to select items that reflected their country's school science curricula.

An even more obvious cultural bias in the 1992 Eurobarometer survey is found in item 60, in which the experimental method was rashly equated with the scientific approach per se (INRA & Report International, 1993, p. 42). The respondents were asked to select one of three possible strategies for finding the reasons behind frequent breakdowns of a machine: (a) talking to the machine's operators, (b) using scientific knowledge about the material used, and (c) testing different materials by means of an experiment. Only the experimental approach was considered the correct answer by the researchers (INRA & Report International, 1993, p. 42). British empiricism, which is held in particularly high esteem in Great Britain and countries with similar cultures, certainly underlay the formulation of this question and the assessment of the responses.

By contrast, theory-driven and observational ways of going about the task might be regarded as "scientific" more often in such countries as France and Germany. For example, a qualified engineer commissioned to find the cause of the machine's breakdowns would probably not consider experimentation as the primary point of departure. Before embarking on a costly series of experiments, he or she would most likely inspect the machine's design in order to see whether it satisfied established principles of engineering. Furthermore, it would be a good idea for this engineer to start the investigation by talking to the operators of the machine, who, as eyewitnesses of the breakdowns, might be able to suggest possible reasons for them. Although it is true that empirical proof of the cause(s) of the failures would ultimately require experiments, theoretical explanations might also be possible (and more economical), and practical solutions might be contributed by the machine's operators without there being a full experimental analysis of the problem.

It seems unlikely that the initially nonexperimental procedure for solving a problem is any less valid in Britain than elsewhere. The implication of this unlikelihood is that the people who formulated item 60 of the 1992 Euro-barometer drew upon a culturally determined idealization rather than upon differences in practice. The responses to this question therefore only reveal that in some countries the experimental approach is more strongly associated with the scientific method than is the case in countries where theory-driven or observational approaches, which are also well-established in science, exert greater influence on what people think of as scientific method.

If the knowledge levels of different groups (e.g., different attitude groups or different cultural groups) are to be compared, then survey items that are intended to measure people's knowledge must be selected through a system-atic procedure that takes into account the possibility that the emphasis given to different fields of knowledge may differ from one group of respondents to the next. When analyzing what knowledge proponents and opponents of technical projects know, one must therefore first inquire about the kinds of "facts" that are used by the two groups to support their respective arguments. In intercultural comparisons of scientific literacy (such as the Eurobarometer survey), knowledge items should be selected only after an analysis of the science curricula taught in the schools of all the countries included in the study.

To summarize, normative judgments about what constitutes the proper domain of knowledge for a well-informed or scientifically literate citizen inevitably influence the process of constructing knowledge scales. It is there-fore necessary to mount a critical discussion of the procedure by which knowledge items are selected and of the normative assumptions underlying that selection.

Asymmetrical availability of information sources. Which information sources people use when forming opinions about particular issues partly depends upon the perceived credibility of those sources. Moreover, the credibility ratings of information sources in controversies about technology are correlated with the perceived similarity of sources' and recipients' atti-tudes (Peters, 1992). If the information sources on one side of a controversial issue are more accessible than those of the other (e.g., because of different public-relations strategies pursued by the actors or because of biased use of sources by mass media), people who initially agree with the side that is more present with information in the public will be likely to learn more about the controversial project than will people who disagree with that information, for the former group will find the available sources more credible and hence will use them more extensively.

This process was probably involved in the changes that occurred in proponents' and opponents' knowledge between the first wave and the second wave of the incinerator study. During that period, local public discussion was dominated by opponents of the project. The activists against the incinerator plant (the construction of which was favored by most of the local population) were far more often represented in the local newspapers than the proponents were. The companies working on the project used a public-relations strategy that focused on a small group of local opinion leaders and avoided communication with the public at large. Furthermore, opponents were more prepared to demonstrate their opinion publicly than proponents were and talked more often with friends and relatives about the project. Hence, opponents of the project found more opportunities to inform themselves through credible information sources (mass media and interpersonal sources) than did proponents of the project, for whom the publicly available information sources were less credible.

Differing approaches to forming attitudes. Detailed information on the characteristics and positive and negative impacts of a technology or a project is particularly crucial for attitude formation and decision-making that are based on cost–benefit analysis. However, as Mazur (1985) argued, there are other, not necessarily irrational, ways to form attitudes and make decisions, ways that entail a need for detailed scientific and technical information. For example, some technologies may be perceived to conflict with nonnegotiable ethical principles. This eventuality is likely with particular medical techniques, such as organ transplants, abortion, or artificial insemination with the sperm of anonymous donors. In some cases attitudes and decisions may be determined by cues evoking fear, disgust, or other strong emotion rather than by arguments based on fact. Ideologies or hidden agendas might cause advocates or adversaries of a technology to instrumentalize their defense of or opposition to that technology for purposes unrelated to its immediate advantages or disadvantages (purposes such as demonstrating against nuclear power as a way of protesting capitalism or the form of government; supporting nuclear power as a means of achieving or preserving power for one's nation). Finally, attitudes might be formed by strong personal interests. In that case, the attitude-formation process might be dominated by only one motive (e.g., economic profit, political power, or minimization of personal risk), with other aspects seeming irrelevant.

The utility of detailed information varies considerably among people, who pursue different approaches to form their attitudes. People are therefore also likely to differ with respect to information demand, information-seeking behavior, and knowledge level. Because the particular attitude-forming approaches that individuals apply seem to be associated with attitudes, knowledge level may also vary with attitude. Proponents of a given technology, for example, may derive their attitudes primarily from cost–benefit calculations, whereas determined opponents may respond more to ethical arguments, ideological orientations, and emotions. (That is not to say that opponents are the only ones motivated by ethical considerations, ideological orientations, or emotions.)

In the Chernobyl study, the attitude group expressing categorical rejection of nuclear power (*immediate withdrawal*) does not fit into the picture of a U-shaped relationship suggested by the overall picture of the results. This group differed sharply from the moderate opposition group (*slow withdrawal*) with respect to knowledge but diverged only slightly on most other variables, a fact suggesting that the two groups experienced different attitude-formation processes. Perhaps emotional motives (fear of a nuclear disaster in Germany similar to that in Chernobyl) played a much greater role for the group consisting of nuclear power's adamant opponents, whereas the moderate opposition group may have based its attitudes more on the cost–benefit calculations of abandoning the use of nuclear power.

The effects mentioned so far tend to produce a monotonous relationship between knowledge level and attitudes. The effects discussed in the next section may lead to a U-shaped relationship.

Attitude extremeness and knowledge level correlated with "involvement." It is basic knowledge in social psychology that involvement—a variable indicating the subjective relevance of a problem—is usually related to extremeness of attitudes. High involvement in something leads to more extreme attitudes (positive or negative) than does low involvement, which is more often associated with indifferent attitudes (Mackie & Gastardo-Conaco, 1988).

Petty and Cacioppo (1986), in their influential model of elaboration likelihood, postulated a specific relationship between involvement and the attitude-formation approach. They distinguished between a "central" and a "peripheral" route to persuasion (see especially p. 3). Taking the central route, recipients intensely process persuasive messages and examine the quality of arguments. Taking the peripheral route, recipients only superficially process messages and react on the basis of simple "cues" (p. 18). According to one of the main hypotheses of the model, high involvement is one of the factors leading to the central processing of messages, and low involvement leads to peripheral processing. It is evident that central processing requires and creates more substantial knowledge than peripheral processing.

Hence, attitude extremeness and knowledge level are both thought to be correlated with involvement. This hypothesis implies a statistical correlation between attitude extremeness and knowledge level, even though a direct causal link between these two variables may not exist. A monotonous correlation between attitude extremeness and knowledge level produces a U-shaped functional relation between attitude direction (e.g., for or against) and knowledge level. Such a relationship is expressed clearly in the data of the first wave of the incinerator study and, though less so, in the data of the Chernobyl study.

Defending one's stance and persuading others as motives for knowledge acquisition. There is another mechanism that might also lead to a U-shaped component in the observed relationships between attitude and knowledge level. Forming one's own opinion might not be the only, or even the most important, motive for acquiring knowledge. A strong motive for information-seeking may be the intention to persuade others of one's opinion or attitude. Because persuasion requires argumentation, a person who anticipates having to defend his or her stance, and/or to convince others of it, might prepare for this encounter by acquiring knowledge.

The need to defend one's attitude as well as the intention of persuading others of it are very likely related to the conviction behind one's attitude. People whose attitudes are extreme as compared to those of the majority have a greater need to justify their opinions and have a more difficult time doing so. Furthermore, the intention to persuade others is most likely correlated with involvement. Because extremeness of attitude is correlated with involvement, one may expect people with relatively extreme attitudes to be more eager to persuade others and hence to prepare for discussions by intensely seeking and processing information. The incinerator study provides evidence supporting this hypothesis. Proponents and opponents stated more often than undecided respondents that they would probably try to convince others in private discussions and would probably advocate their opinions publicly (Wiedemann et al., 1991).

The shape of the observed functional relationships may reflect the joint effect of several of the above-mentioned influences, which may reinforce each other or cancel each other out. In general, then, there is no reason to expect a positive correlation between knowledge level and attitudes toward science and technology, and even less of a reason to expect a change from negative or neutral attitudes to positive ones as a result of information campaigns.

Broadening the View on the Relationship
Between Experts and Laypeople

Possession of special knowledge is certainly a major feature that distinguishes between experts and laypeople, and one often encounters clear-cut differences of opinion and attitude, with laypeople occupying one side of the divide and communities of experts on certain technologies occupying the other (Rothman, 1990). At first glance, it may seem plausible that the knowledge gradient is the reason for attitude differences. But as the previous section shows it is empirically and theoretically unjustified to expect strong and consistent links between people's knowledge about and attitudes toward science and technology. There is no unequivocal relationship between knowledge level and attitudes. Hence, attitude differences between laypeople and communities of experts cannot be attributed to knowledge differences.

If the knowledge gradient is not the crucial factor for attitude differences between experts and laypeople, where should one look for an explanation of those differences? It seems necessary to focus attention and research efforts on factors other than just knowledge. In the following paragraphs I discuss a number of other dimensions that may cause disagreements between the expert and the layperson and may influence attitudes toward technologies more than factual knowledge does. Table 11-1 summarizes some of the hypotheses about potentially relevant differences between experts' and laypeople's perceptions and attitude-formation processes. The mental-model approach, together with constructivist analyses of hidden values, assumptions, and frames, may be used to identify these differences.

It can be assumed that laypeople use different—not necessarily less reasonable and sophisticated—criteria (i.e., values) than experts do when evaluating the merits and drawbacks of technologies. They may take into account possible effects that do not appear in the expert's cost–benefit analysis because they are extremely hypothetical, hard to quantify, or presumably irrelevant. Furthermore, the relation that laypeople have to technology with respect to control is completely different than that of experts. The dominant view among experts who produce and operate technology is that technology is a tool, that humans use technology to achieve their goals. For example, Hubert Markl, a leading German science manager, has likened the relationship between humans and technology to that between a bird and its wings (Markl, 1993). Just as the bird is not afraid to use its wings, humans should not be afraid to use technologies. The concept of technology as an extension of an individual's natural organs lies behind such rhetoric. Because scientists are part of the professional community developing and controlling these technologies, they find this perspective plausible. Laypeople, however, who

Table 11-1. Perceptions and Attitude Processes: Hypotheses about Potentially Relevant Differences Between Scientific Experts and Laypeople

Dimension of difference	Scientific experts	Laypeople
Systems of knowledge and rationalities	Special knowledge and perspectives	Common sense and everyday perspectives
	Scientific concepts and language	Everyday concepts and language
	Appraisal of knowledge with respect to truth	Appraisal of knowledge with respect to usefulness
Relevance structures and values	Scientific and technical aspects dominant in mental models	Social aspects dominant in mental models
	Precise and narrow problem definition	Vague and broad problem definition
	Balancing of costs and benefits on the societal level	Balancing of costs and benefits on the personal level
	Technical and economic efficiency as an important value	Fairness as an important value
Perceptions and preferences of control	Preference for administrative and technical control	Preference for control by the individual
	High trust in science and technology	Limited trust in science and technology
	Technology as a tool	Technology as a threat to environment

hardly participate in the control of technologies, are more likely to take another view. To them, technologies are part of the uncontrollable and potentially dangerous environment against which protection is necessary. Whether one regards science and technology as a tool or as a set of constraints that must be adapted to is probably a crucial factor in attitude formation. An important difference between laypeople and experts, therefore, is their respective passive and active relationship to technology.

Another aspect that may influence the willingness of laypeople to accept or reject the opinions and attitudes offered by experts has to do with the social relations between the two groups. Usually, this aspect is subsumed in the label *trust and credibility*. It is not sufficient, however, to understand credibility as a unidimensional concept that guides the rejection or acceptance of information. What one thinks about the persons who generate and disseminate scientific information may determine not only which information figures in an individual's attitude formation but also how that information is used.

The questionnaire used in the incinerator study (Wiedemann et al., 1991) included questions asking respondents to indicate their agreement or disagreement with statements about experts (see Table 11-2, my translation). Three aspects were selected for this exploratory survey about perceptions: (a) the social distance between experts and laypeople (items 1 & 2), (b) the objectivity of experts (items 3 & 4), and (c) the ambiguity of scientific knowledge (items 5 & 6). Such aspects may be thought of as part of more comprehensive mental models of science and technology.

The results of this survey as shown in Table 11-2 demonstrate that about half of the population (in the context of the mentioned controversy about an incineration plant) felt a social distance between themselves and the experts. Even more people doubted that experts act independently from interests and provide objective analyses. But most interestingly, a large majority of the population acknowledged that even if experts did follow the scientific method, science would often not lead to clear answers. For example, conflicting stances taken by experts are not always explained as conscious unilateral or mutual manipulation or error but rather are perceived as the "two sides of a coin," that is, as dependent on individual perspectives. This lay epistemological view is consistent with constructivist notions of science and is perhaps more mature than the naive positivist epistemologies of many scientists.

A descriptive analysis of lay beliefs about the epistemology of science and the social role and vested interests of experts is intriguing enough in itself. It becomes even more so if one compares subgroups of respondents, different types of experts, and different fields of public debate with respect to such beliefs. Furthermore, such analysis may enable one to determine the influence that different concepts of science and technology and different beliefs about scientists and experts exert on attitudes toward science and technology.

It is a truism that technologies are sociotechnical systems that are developed not by scientific innovation alone but rather by the interaction of science, politics, industry, and consumers. Even when assessing the risk of a technology in a narrow sense, one has to consider not only the details of its technical design but also the socioeconomic context of its implementation and use. It is important to know who sets the safety standards according to which procedure, who checks to see that they are observed, who develops and applies the technologies, and which interests are involved. The public acceptance of these systems depends as much on their sociopolitical aspects as on their technical aspects. In Germany, for example, public attitudes toward nuclear power were as much affected by a "social accident" (a bribery scandal within the German nuclear power industry) as by the Chernobyl disaster (Peters, Albrecht, Hennen, & Stegelmann, 1990). Any analysis of technology acceptance must therefore take into account the sociopolitical context of technology development and use.

Table 11-2. A Local German Community's Perceptions of Experts, 1990

Survey items	Agreement (in %)
1. Experts are interested only in their special field and do not care about the real concerns of the public.	56
2. In their work experts are oriented to the needs of the public. After all, many of them have children and a family themselves.	55
3. Expert statements are based on scientific analyses and are therefore objective.	46
4. Experts aren't really independent but rather back up the opinions of those who pay them.	81
5. As long as experts remain on scientific ground, no controversies can arise between them.	26
6. Science doesn't always provide unequivocal results. Experts therefore are often at variance with each other.	94

Note. $n = 313$. The data used were obtained by a survey of the population in the neighborhood of a planned incineration plant in North Rhine–Westphalia, Germany (second wave, November–December 1990). The methodology of the survey is described in "Information needs concerning a planned waste incineration facility," by P. M. Wiedemann, H. Schütz, and H. P. Peters, 1991, *Risk Analysis, 11,* 229–237.

The arguments presented thus far are not in any way meant to discourage science, educational institutions, and the actors involved in controversies from providing scientific and technical information to the public. On the contrary, the public should have easy access to such information and to all other kinds of information that are considered useful. As far as attitude formation and decision-making is concerned, however, scientific information is unlikely to be as crucial to laypeople as it is to experts. Laypeople use this information, if they use it at all, according to their own rationalities. As long as these rationalities are not known, the impacts that increased scientific or technical literacy have on attitudes are all but unpredictable. After all, scientific or technical facts do not speak for themselves; they have to be given meaning in the context of an interpretative frame.

Conclusions and Suggestions for Further Research

This chapter started with criticism of popular beliefs that a close relationship exists between scientific literacy and positive attitudes toward modern science-based technologies. Empirical evidence and theoretical arguments contradicting those beliefs were stated in the sections that followed.

Three conclusions are drawn from this line of reasoning. First, the links between scientific and technological knowledge and attitudes toward science and technology are usually very complex and vary from one subject field to the next over time. Second, technologies are sociotechnical systems, so the field of competence relevant for their assessment is not limited to scientific and technological knowledge. It includes other areas of expertise as well (e.g., economic, political, sociological knowledge). Third, differences between lay and expert assessment of technologies are not necessarily the consequence of inferior scientific and technological competence. They might stem from the fact that lay judgment models used to assess advantages and disadvantages of technologies, and lay positions on the control of technologies, often differ from the corresponding judgment models and positions of experts. When exploring the determinants of public attitudes toward technologies, one must therefore do much more than identify the lay public's level of scientific and technological knowledge. I close this chapter by outlining three suggestions for the direction in which future studies on people's attitudes toward science and technology might be broadened.

Exploring Subjective Lay Theories about Science and Technology

An analysis of people's frames of reference and mental models is crucial to an understanding of the ways people use information to form opinions and attitudes. These cognitive structures determine whether and which factual knowledge is relevant for attitude formation. It is therefore most important to analyze laypeople's frames of reference and mental models with respect to technologies. While the exploration of subjective theories is primarily the domain of qualitative research, quantitative survey research can be used to determine the relative frequency of their occurrence in different segments of the population, in different countries, and with respect to different technologies. There is already a rich body of sociological and psychological literature based on case studies of scientific and technological controversies (e.g., Jungermann, Rohmann, & Wiedemann, 1991; Nelkin, 1992) and on analyses of subjective theories about technologies and their impacts (e.g., Bostrom, Morgan, Fischhoff, & Read, 1994; Daamen, van der Laans, & Midden,

1990) that could be used to develop more sophisticated surveys for the purpose of analyzing the cognitive determinants of attitudes toward science and technology.

Broadening the Concept of Public Understanding of Science and Technology

It is far too limited to equate an understanding of science and technology with the possession of scientific and technological knowledge and an understanding of scientific methodology. There are equally relevant questions to study. For example, what do people assume and know about science policy, the relationship between science and industry, the institutions of the science system, and the regulation of scientific activities and science-based technologies? What do people think about the motives and institutional constraints of scientists and technology developers? Which criteria do people use to decide about the truth of factual claims? How does scientific authority, which is based on the socially recognized claim of privileged access to truth, accommodate democratic beliefs that legitimacy is only gained by consent of the affected?

Public Esteem for and Access to Expert Information in Public Controversies

If science and technology are not neutral to values and interests (as least in controversial fields), it is important to know how members of the public define their relation to experts. Do they feel that their concerns are being neglected by established experts, and do they therefore consider counterexperts as "their" experts? Do members of the public expect scientific experts to be judges settling conflicting factual claims to (or even policy options on) scientific authority, or do they think instead that experts are just instrumentalized by those with economic or political power? What role does the public assign to experts in political decision-making? Pursuing this line of inquiry would lead to a useful paradigm shift away from "public understanding of science" and toward "public assessment and utilization of scientific expertise and experts."

References

Bostrom, A., Morgan, M. G., Fischhoff, B., & Read, D. (1994). What do people know about global climate change. 1. Mental Models. *Risk Analysis, 14,* 959–970.

Daamen, D. D. L., van der Laans, I. A., & Midden, C. J. H. (1990). Cognitive structures in the perception of modern technologies. *Science, Technology, and Human Values, 15,* 202–225.

Evans, G. A., & Durant, J. R. (1995). The relationship between knowledge and attitudes in the public understanding of science in Britain. *Public Understanding of Science, 4,* 57–74.

Grimston, M. C. (1994). Public opinion surveys in the U.K.. *Nuclear Europe Worldscan, 14*(7–8), 98.

Hennen, L., & Peters, H. P. (1990, January). *"Tschernobyl" in der öffentlichen Meinung der Bundesrepublik Deutschland: Risikowahrnehmung, politische Einstellungen und Informationsbewertung* [Chernobyl in the public opinion of the Federal Republic of Germany: Risk perception, political attitudes, and information assessment] (Report No. Jül-Spez-551). Jülich, Germany: Research Center Jülich.

Hutton, N. (1996). Interactions between the formal U.K. school science curriculum and the public understanding of science. Public Understanding of Science, 5, 41–53.

Institut National de Recherche Agricole (INRA), & Report International (1993, June). *Europeans, science and technology: Public understanding and attitudes* (EUR 15461). Brussels: Commission of the European Communities.

Jungermann, H., Rohrmann, B., & Wiedemann, P. M. (Eds.). (1991). *Risikokontroversen. Konzepte, Konflikte, Kommunikation* [Controversies about risk: Concepts, conflicts, communication]. Berlin: Springer.

Mackie, D. M., & Gastardo-Conaco, M. C. (1988). The impact of importance accorded an issue on attitude inferences. *Journal of Experimental Social Psychology, 24,* 543–570.

Markl, H. (1993, October 20). Zu klug und doch nicht klug genug. Über die Angst vor dem Wissen und die Angst vor der Angst [Too clever, yet not clever enough: On the fear of knowledge and the fears of fear]. *Frankfurter Allgemeine Zeitung,* p. 36.

Mazur, A. (1985). Bias in risk-benefit analysis. *Technology in Society, 7,* 25–30.

McBeth, M. K., & Oakes, A. S. (1996). Citizens' perceptions of risks associated with moving radiological waste. *Risk Analysis, 16,* 421–427.

Nelkin, D. (Ed.). (1992). *Controversy: Politics of technical decisions* (3rd ed.). Newbury Park: Sage.

Peters, H. P. (1992). The credibility of information sources in West Germany after the Chernobyl disaster. *Public Understanding of Science, 1,* 325–343.

Peters, H. P., Albrecht, G., Hennen, L., & Stegelmann, H. U. (1990). 'Chernobyl' and the nuclear power issue in West German public opinion. *Journal of Environmental Psychology, 10,* 121–134.

Petty, R. E., & Cacioppo, J. T. (1986). *Communication and persuasion: Central and peripheral routes to attitude change.* New York: Springer.

Rieder, B. (1995, June 5). "Über mehr Wissen zu positiveren Einstellungen"!? Zur Tragfähigkeit einer populären "Devise" kommunikations- und bildungspolitischer Konzepte im Lichte der Meinungs- und Einstellungsforschung ["Via more knowledge to more positive attitudes"!? On the soundness of a popular "motto" for concepts of communication and educational policy in the light of opinion and attitudes research]. *Reflexion,* pp. 1–8. (Newsletter of Ri*QUESTA, In der Hainlache 42, D-68642 Buerstadt).

Rothman, S. (1990). Journalists, broadcasters, scientific experts and public opinion. *Minerva, 28,* 117–133.

Wiedemann, P. M., Schütz, H., & Peters, H. P. (1991). Information needs concerning a planned waste incineration facility. *Risk Analysis, 11,* 229–237.

Zimmerman, L., Kendall, P., Stone, M., & Hoban, T. (1994). Consumer knowledge and concern about biotechnology and food safety. *Food Technology, 48*(11), 71–77.

CHAPTER 12

INFORMATION'S ROLE IN THE INTRODUCTION AND SOCIAL REGULATION OF NEW BIOTECHNOLOGIES[*]

Erling Jelsøe

The public understanding of science and technology is considered in light of the debate about new biotechnologies in Denmark. After a review of this debate and a discussion of surveys on public perceptions of biotechnologies, I conclude that a particular kind of "critical acceptance" of new biotechnologies exists in Denmark. It is characterized by a widespread critical attitude in the population and by a degree of consent to and trust in the social regulation of biotechnologies. As argued in this chapter, this response can be understood generally as a reflection of certain traits of Danish political culture and specifically as a result of the controversies over new biotechnologies in the 1980s. Information per se did not play a major role in those processes, only as an element of the overarching debate and political discussion. Regulatory activities, too, played an important role in the rise of critical acceptance. Thus, the impact of information depends on social and historical context. Information itself must be seen in its relation to such wider processes of social and political debate and to the learning processes of the various groups and institutions involved.

In the Eurobarometer surveys on science and technology, there has been a strong focus on the significance that knowledge has for attitudes about science and technology. Various measurements of knowledge, the construction of knowledge variables, and cross-tabulations of these variables with measures of attitudes have been a characteristic feature of these surveys. This stress on knowledge appears to stem primarily from an underlying assumption that knowledge and attitudes are positively correlated.

[*] This chapter is a result of preliminary work in relation to a study, "Biotechnology and the Danish Public," supported by the Danish Research Council for the Social Sciences (Grant No. 9500780). The Danish study is part of a concerted action of the European Commission (B104/CT95/0043), "Biotechnology and the European Public." The other members of the Danish research team are Jesper Lassen, Arne Thing Mortensen, Helle Frederiksen, and Mercy Wambui Kamara.

Parallel emphasis has also often been placed on the need to enlighten the public about "the facts" of risks and opportunities entailed by science and technology, the assumption being that this information will increase the popular acceptance of new technologies as expert views about their associated risks become known to the public. This stance has been disputed and criticized frequently by social scientists and others studying the public understanding of science and technology. Although it has been quite persistent among decision-makers and experts within science and technology, there are probably not many of them who still defend it.

The problem of knowledge points to some of the social interests and contradictions connected with the public understanding of science and technology. The question of information is thus also a question of the democratic aspects of science and technology policies. In short, who is providing the information? What kind of information is it? In what social and political context is that information embedded? And what function does the information have in establishing the conditions for the population to influence the decision-making processes related to science and technology?

The linear view of the function of information—that information will influence attitudes and, in turn, lead to a more "rational" view of science and technology—is paralleled by a simple model in which information fosters an informed debate and leads to decisions that are more democratic than would otherwise be the case. According to this model there is little problem with democracy, only a technocratic difficulty with improving the efficiency of the information activities and making them sufficiently goal-oriented. Of course, the basic questions of what makes people receive information, how they use it, and so forth are ignored in this simple view. Experience shows that information, if it is to be meaningful, must relate to the needs or experience of the people concerned and that there must be at least some possibility that the information will be useful, say, by improving people's ability to influence decisions that they feel are important to them.

One important implication that this observation has for information about science and technology is that information must be seen as an element of overarching debate and political discussion. There is no such thing as objective information, of information free of social and political context. The social and political context comprises a number of topics: actual regulatory initiatives, the history of regulation, and the broader question of national political culture; earlier cases of information campaigns, public debate, or both; the influence and socioeconomic significance of industry and agriculture within the field of technology in question; the role of experts and other professional actors; and the role and experience of grass-roots organizations.

In the following pages I try to illustrate some of these points by using the debate about new biotechnologies in Denmark as an example. The focus is broad, so I will not go deeply into all the above-mentioned questions about the social and political context. I describe the development of the debate and of the information campaigns. I relate that development to the history of the regulation of new biotechnologies in Denmark and to some of the most important events in the industrial utilization of gene technology. I also mention earlier cases of information campaigns and touch on the question of political culture, but not in any detail. Despite such delimitations, this approach seems sufficient to illustrate my point that there is a complex interplay between information and a number of contextual circumstances and that, consequently, no simple or context-free model for the role or function of information about science and technology is likely to be adequate. I emphasize the context of information activities perhaps even more than the activities themselves. Furthermore, the focus is on the *public* debate and on the dissemination of information to the general public. In a long-term perspective it is not unlikely that, for instance, changes in school education may change the public's readiness to receive and make use of information about science and technology.

Biotechnologies in Denmark

In recent years much of the discussion about the introduction of new biotechnologies in Denmark and the reactions of the public have centered on an apparent paradox. It relates to the 1991 and 1993 Eurobarometer findings on what Danes think about biotechnology and genetic engineering (Eurobarometer 1993):

– The expectations that Danes had of the improvements that genetic engineering might be able to make in their way of life were very low compared to the EU average.
– Danes had the highest perception of environmental and human-health risk associated with applications of biotechnology and genetic engineering.

At the same time, however,

– The Danes' high degree of support for research in biotechnology and genetic engineering was close to and a little above the EU average, respectively.
– Danes had the EU's highest level of objective knowledge about biotechnology and genetic engineering.

There are a number of methodological problems connected with the way the questions were formulated.[1] If the 1991 and 1993 Eurobarometer findings are accepted at face value, they give rise to at least two considerations. First, it is worth noting that low anticipation of positive effects of genetic engineering and a high perception of risk go hand in hand with a high level of knowledge about biotechnology and genetic engineering. Second, and more important, the Danes' relatively high degree of support together with high risk perception and low anticipation of positive effects must be a result of social experience and learning processes.

General explanations for the results of the 1991 and 1993 Eurobarometer surveys in Denmark have been given. It is ventured, for example, that the population in Denmark has a relatively high level of trust in government (Cantley, 1992; Institut National de Recherche Agricole [INRA] & Marlier, 1993). The character of the Danish debate—a broad popular debate instead of a media-battle between industrialists, scientists, environmentalists, and

---

[1] The Eurobarometer findings on risk perception and on support were arrived at through a battery of questions about different areas of research in biotechnology and genetic engineering. First, the descriptions of these areas were characterized by frequent use of positive words, such as *improve, useful, better quality*. In addition, most of these descriptions presented the perspectives of new biotechnologies in a way that was likely to be regarded as quite positive by the respondents.

Second, it is also remarkable that herbicide resistance, which until 1993 had been the single most important area of development in plant biotechnology, was not mentioned, whereas improvement of food quality, which had been of limited importance until 1993, was addressed in two items.

Third, none of the negative consequences that might result from genetic engineering, the study of which had occupied many scientists, were brought up at all. One example was the ecological consequences of releasing genetically modified plants into nature. This positive bias in the descriptions of the areas is probably one of the reasons for the high overall support expressed in the Eurobarometer findings. Of course, it is less obvious that changing the descriptions of the areas would change the ranking of the Danish answers relative to the EU average. My discussion in this chapter is based on the assumption that the *ranking*, but probably not the absolute level, of the Danish answers expresses a certain degree of realism.

Fourth, the questions about knowledge were basically textbook questions, and it is at best unclear what the responses indicate about the interviewee's capacity to form an opinion about specific applications of biotechnology and genetic engineering. Experience seems to show that this capacity is often underestimated by experts because of the lack of science-based knowledge among laypeople.

Lastly, most respondents cannot easily answer the very general question about whether genetic engineering will improve "our way of life." Only those respondents who are unambiguously for or against genetic engineering have little difficulty with such an item. Furthermore, the expression "our way of life" is vague in this connection. The responses to this question can only be a very crude indication of a general orientation. It is totally unclear how different respondents find a balance between specific attitudes toward various aspects of biotechnology and genetic engineering.

the like—is also pointed out (Teknologinævnet, 1994). I suggest that there are more complex developments behind the situation expressed through the Eurobarometer results. This explanation calls for a historical approach focusing on both the debate and the policy processes related to biotechnology and genetic engineering and on more general traits of the public discourses about science, technology, and environmental issues in Denmark.

To underscore the importance of understanding the role of information in the wider activities of debate and political discussion, I try to identify changes in the public discourses about biotechnology and genetic engineering as a basis for the periodization used in the following text. Periodizations are always somewhat problematic. One can argue about the specific time and circumstances when changes occur and about the criteria for the periodization. Interestingly, other attempts to make periodizations on the basis of different analytic approaches do not differ very much from my presentation. Some of the significant shifts in the patterns of development are characterized by changes in many dimensions, such as the political framework, industrial activities, and the intensity of the debate. Therefore, a periodization based on public discourses about new biotechnologies may resemble one based on the number of articles on the subject in the media (as in Teknologinævnet, 1994) or one based on shifts in the phases of the innovation circle (as in Baark & Jamison, 1990).

No Regulation—and Limited Debate: The Development until 1985

The Danish situation during the 1970s was characterized by very limited public debate about new biotechnologies. The Asilomar conferences in 1973 and 1975 gave rise to a handful of articles in Danish papers but no real debate. The articles on the subject addressed the academic audience.

However, the intense U.S. debate on the safety issue after the moratorium on genetic engineering in 1975 did not pass unnoticed by researchers in Denmark with an interest in the new biotechnologies. As early as 1976 four of the Danish research councils[2] appointed a Registration Committee whose purpose was to register and monitor the research projects on genetic engineering in Denmark. The Registration Committee was also expected to come

---

[2] The Danish research councils are public bodies under the Ministry of Research. They have three functions: To advise other public authorities and institutions, to fund research projects for which financial support has been applied, and to promote specific research areas.

forward with proposals for a more permanent organization of public control over genetic engineering in Denmark. No rules or regulatory framework existed, but the Registration Committee followed the NIH guidelines adopted in the United States in 1976 and brought the moratorium to an end.

By setting up the Registration Committee, the research councils apparently anticipated a debate in Denmark similar to the one in the United States (Toft, 1985a). In 1977 the Registration Committee published its first report, which contained a number of recommendations for the regulation and control of genetic engineering in Denmark. The most important recommendation was a proposal that a more permanent commission be established to assume responsibility for these tasks. However, such a commission was not set up. Not until six years later, did in 1983, the Minister of the Interior appoint a committee to prepare a proposal for such a commission (see below).

Notifying the Registration Committee of genetic engineering projects was voluntary, and the information about the projects was confidential and thus not publicly accessible. This arrangement was criticized but did not give rise to much debate until about 1982, when the dominant discourses on biotechnology and genetic engineering changed. Before 1982 it seemed as though there were no real conflicts about the subject. The debate about risk dominated the media coverage but gradually ebbed toward the end of the 1970s as distance to the moratorium increased and as research in genetic engineering became a more established activity than it had been.

The issue that probably attracted most attention in this period was genetic engineering on humans. Especially around 1980 a debate about test-tube babies was sparked by trials with *in vitro fertilization* at the national hospital in Copenhagen. Even though this area of medical science is not an example of gene technology, the debate in the media was extended to include visions of a "brave new world" with genetic modification of human characteristics.[3] Bjorn Elmquist, a member of the Danish Parliament (MP) from the Danish liberal party, directed a number of questions in Parliament to the Minister of

---

[3] This extension of the debate made a degree of sense because the introduction of in vitro fertilization was seen by many people as a first step toward a development with much wider implications. On the other hand, there was also a tendency in the public debate and in public perceptions, especially in the early phases of the debate, for various aspects of genetic engineering to become interwoven. Modification of genes was seen as a major issue that can be illustrated with examples from all fields of application. A funny demonstration of this tendency's influence on the debate was given when one of the members of the Registration Committee commented on the 1980 debate about human genetics, saying to a newspaper reporter: "Gene manipulation is more than test tube babies" (Christensen, 1980). He felt that the debate about human genetics had become a barrier to the possibilities of promoting genetic engineering in agriculture and industry.

the Interior in 1980. Upon hearing that the Minister saw no reason to take any initiatives, Bjorn Elmquist made a proposal to the European Council for setting limits on the development.

This debate about human genetics launched activity that included public hearings, regulation, and the establishment of a council of ethics in 1987. In many respects this development paralleled that related to the use of genetically modified organisms (GMOs) in agriculture and industry. In the rest of this chapter I focus on GMOs in industry and agriculture.

During the 1970s there was also a more theoretical and critical debate about issues such as reductionism and ecology, in which the paradigmatic foundations of the new biotechnologies were discussed. This debate, which was pursued chiefly in a journal for critical scientists (*Naturkampen*), reached a turning point in 1981 when its leading protagonists began to emphasize the positive possibilities of the new biotechnologies, say, as a means solving environmental problems and replacing synthetic chemical products with less poisonous and less energy-consuming products of biological processes.

In summary, public discourses on biotechnology were confined to minor groups of intellectuals and experts until the beginning of the 1980s. There were no profound conflicts over the development, and researchers seem to have been successful at anticipating debate about genetic engineering and avoiding regulation. In this context little information about genetic engineering was given to the broad public, and no public initiatives to increase that input were taken.

The Debate about Nuclear Energy in the 1970s

During the 1970s there was very intense debate and public information activity as well as political conflict related to another field of technology, nuclear energy. After the Danish power companies made plans for nuclear power plants in Denmark in the early 1970s, an NGO called the Organization for Information about Nuclear Power (OOA) was formed in January 1974. The OOA started debate and information campaigns about nuclear power and embarked on political campaigns against it. It launched its campaigns together with the "sun badge," *Nuclear power—no thank you!*, which became internationally known. The debate was further intensified when a Swedish nuclear power plant, *Barsebäck*, began operating only 16 miles from Copenhagen and its 1.5 million people in 1975. Numerous demonstrations and other collective manifestations against nuclear power attracted up to 50,000 people in the second half of the 1970s.

The Danish government and most of the political parties in Denmark, on the other hand, wanted to promote nuclear energy as an alternative to fossil fuels after the energy crisis. In November 1974 it was decided to run a government-funded public information campaign about energy problems. The main responsibility for appointing the members of the independent body that was to run the campaign, the Energy Information Commission, lay with the Danish Joint Council for Youth and Adult Education (Dansk Folkeoplysnings Samråd), an umbrella organization for various organizations undertaking popular youth and adult education (or "people's enlightenment"). The information activities had the purpose of making it possible for all viewpoints to be expressed and should include the question of nuclear energy. The Energy Information Commission issued a series of books and minor publications and gave financial support to local arrangements such as panel discussions or information activities (Mikkelsen & Olesen, 1991).[4]

This public information campaign only added fuel to the debate, which was already intense. In 1976 the Danish Government decided not to continue funding the campaign, asserting that there was no longer need for it. The government was not ready to take any decision on nuclear energy, however, and it became increasingly clear that a majority of the Danish population was against the use of nuclear power plants. The issue was not settled until 1985, at which time a majority of Parliament finally decided that nuclear power should not form part of Danish energy planning.

This outcome of the debate about nuclear energy was an important part of the experience with information and debate about new technology when biotechnologies became an issue in the 1980s. Despite the government's attempts to establish a consensus in the population through the information campaign, the context for the debate was set by broader activities, including growing popular resistance and the campaigns initiated by the OOA. During the 1970s the climate of the nuclear debate was very polarized.

Industrial Interests Change the Scene: The Development from 1982 to 1985

In the early 1980s the debate about new biotechnologies broadened. Some of the Danish biotechnological companies began actively articulating their

---

[4] The expenditures over the two years of the information campaign's existence totaled 2.8 million Danish Crowns (U.S. $0.5 million), of which about 50% went for publications and about 20% for support of local debate, support was given to 456 local arrangements.

interests in production based on the new biotechnologies. In a panel together with two MPs at a meeting of the Danish engineers' organization in April 1981, the research director of the leading Danish company, Novo, complained about a lack of qualified biotechnologists. He called for a more active public policy regarding the priorities of higher education in that field of science. Reported in the press, his statements gave rise to debate. It had also been noticed that Novo, one of the world's leading producers of insulin and enzymes, had begun research on gene technology and was assumed to be under pressure because the American company Genentech had recently reported having tested genetically modified bacteria that could make human insulin ("Gensplejserne," 1981). Likewise in 1981, the Council of Technology, which under the Ministry of Industry was responsible for funding a major part of publicly financed industry-related R&D, set up a group for genetic engineering, whose members went on to propose a number of initiatives for promoting research and education in that field.

By the end of 1982 the Registration Committee had received the first production notification. It was from Novo, which wanted to start pilot production of an enzyme, maltogene amylase, by means of GMOs. Respecting the principle of confidentiality, the Registration Committee did not inform the public about this intention. The public learned of it from a notice in Novo's 1982 annual report. It divulged few specifics about the production plans, but given the great secrecy that enshrouded all activities of genetic engineering in Denmark at that time, the notice in Novo's annual report could be regarded as a strategy of relative candor. It has been assumed that the disclosure was Novo's way of seeking to test the public's reaction to its plans (Toft, 1985b).

Even though public debate about biotechnology was still limited, the noticeable interest that Danish industry showed in biotechnological development and production elicited discussion in the newspapers and in meetings of academic societies. A few politicians began to participate in the discussions, and one social democratic politician who wanted to promote biotechnological development in Denmark advocated public control and regulation as well as technology assessment as a way of gaining knowledge about the consequences of such development (Hilden, 1981).

When the Council of Technology established the initiation group for genetic engineering later in 1981, the weekly journal of the Danish engineers' association (*Ingeniøren*) ran an editorial criticizing the lack of public regulation:

> The debate about nuclear power should have indicated that the feeling or lack of feeling of public security is a very concrete factor to consider if technology is to be utilized by research and industry. We must, therefore, find as soon as

possible means for social control that simultaneously meet this demand for security and draw on the most recent experience in the United States so that Danish industry will not be burdened with unnecessary constraints in its attempts to utilize a new technology that should be a perfect match for a highly educated society in a country poor in raw materials. ("Gensplejsning og tryghedsbehov," 1982; my translation)

This explicit reference to the nuclear debate was no surprise at the time this editorial was written, for the voices were still loud, and the consequences for the decisions on nuclear power in Denmark must have been clear to most of the participants. Later in the 1980s the influence of the experience with the nuclear issue became more implicit than explicit.

The decision to set up a committee to prepare the framework for regulating new biotechnologies in Denmark as proposed by the Registration Committee six years earlier was not taken until 1983, however. Danish industry was against any specific regulation of biotechnological production, and most biotechnological researchers were against regulation of research altogether. The committee—Indenrigsministeriets gensplejsningsudvald, the Gene-Splicing Committee of the Ministry of the Interior (my translation)—which consisted mainly of government officials from a number of ministries together with a few researchers, was supposed to finish its work by the end of 1984, but its report, *Gene technology and safety,* did not appear until mid-1985. This body, which had been asked to investigate existing laws and to make any necessary suggestions for new legislation on genetic engineering, recommended no fewer than three new parliamentary acts to cope with various aspects of the field. *Gene technology and safety* (Gene-Splicing Committee, 1985) was characterized by a large number of minority statements from the representatives of various agencies, a fact that indicated the rivalry between the ministries. The Ministry of the Environment and the Ministry of Industry in particular were at odds over the need for regulation.

This traditional handling of the regulation issue by the state bureaucracy was soon overtaken by events. In November 1984 Novo applied for approval of a fermentation plant for the industrial production of human insulin, other peptide hormones, and enzymes by means of GMOs. Because there was no regulatory framework for handling such an application, it was submitted by Novo to county authorities, as stipulated by existing environmental legislation.

The county authorities, however, had no competence for dealing with such an application, so they required assistance from the Ministry of the Environment. Novo found the subsequent procedure of approval too lengthy, and during the summer of 1985 the company launched a press campaign to put pressure on the authorities. One of Novo's directors, Kim Hueg, criticized the approval procedures and stated in several newspapers that Novo might move its production facilities to another country, such as Japan or the United States (Hansen & Kjøgx Pedersen, 1985; see also Toft, 1985b). At the same time, he wrote an extensive article in one of the leading Danish newspapers and gave a television interview (TV Avisen, July 7, 1985) in which he described the perspectives of the new biotechnologies very positively (Hueg, 1985; Toft, 1985b). Lastly, the population in the vicinity of Novo's production facilities received a company brochure promoting the new biotechnologies and providing information about Novo's recent initiatives concerning safety measures at its production facilities. In short, Novo's strategy shifted from an initial mixture of threats and sales promotion to a much more forward style of quite open communication with the public.

This time, however, a Danish environmental NGO called NOAH entered the public discourse within this field. Formed in 1969, NOAH had played an important role in starting the environmental debate and in heightening general social awareness of environmental problems in Denmark. During the 1980s the organization had acquired status as the Danish section of Friends of the Earth. A few weeks after Hueg's article appeared in *Politiken,* a representative of NOAH, Jesper Toft, wrote a reply to Kim Hueg, pointing out that there was a lack of knowledge about the risks associated with GMO-based production (Toft, 1985c). Novo responded by inviting representatives of NOAH to an internal meeting in order to establish contact with the company's critics (Toft, 1985b).

NOAH played a more active role in a second case of launching GMO-based production in Denmark, a process that began in 1985 when Nordisk Gentofte applied for approval to produce human growth hormone. Nordisk Gentofte had not adopted the same principles of openness as Novo, and before the application was submitted Nordisk Gentofte's activities in genetic engineering were not known to the public. Neither was it known that the company had already engaged in pilot production of growth hormone.

In this case, too, the procedures laid down in environmental law were followed. Approval by the Greater Copenhagen Council, which acted as the authority in this instance, came within only two months. In a press release NOAH called for signatures of local citizens to be compiled as required for filing a complaint with the Danish Environmental Agency. Because the Council's decision was the first approval for production based on GMOs, NOAH especially wanted to make clear whether the Agency would stand by its earlier statement that a risk assessment should be conducted before the use of GMOs in production could be approved.

A considerable number of citizens signed NOAH's petition. Impressed by this local response, which the company had not expected, Nordisk Gentofte decided to arrange a public information meeting to explain its plans to the community. This meeting was characterized by a traditional one-way communication strategy, with company representatives spending most of the evening explaining the new technology. There was time for only a few questions toward the end of the meeting. Dissatisfied with this experience, some of the participants decided to arrange a new meeting, one in which company representatives, independent experts, NGO representatives, and politicians should participate in a debate. In a radio interview about the information meeting, the director of Nordisk Gentofte said that the company would move its production facilities to another country if the approval procedure became excessively long.

NOAH was a major actor in sparking Denmark's intense debate about new biotechnologies in the second half of the 1980s. From 1982 to 1985, however, there were still relatively few participants. Criticism of state bureaucracy and confrontations between industrialists and environmentalists were the typical components. Information was not much of an issue for the public authorities. By contrast, the biotechnological companies that were ready to begin production based on GMOs adopted very different communication strategies. Nordisk Gentofte sought to avoid any communication with the public as long as possible. Novo followed a strategy of relative openness. Pressed by their competitive situation, both companies tried to pressure authorities into speeding up the approval procedures and threatened to shift their investments to other countries if approval became too troublesome to obtain in Denmark.

From 1982 to 1985 the public discourse about biotechnologies changed. There arose new issues revealing that the interests of industry, state bureaucrats, and environmentalists were divided. Elements of these discourses still prevail, but in the second half of the 1980s they were transformed by widened public debate and by changes in the character of the political processes bearing on the new biotechnologies.

Broad Popular Debate: The Period from 1985 to 1990

Despite the proposals for new legislation that were contained in the report from the Gene-Splicing Committee, the document did not prompt any immediate initiatives from the government. The Danish conservative government was not highly in favor of specific legislation in this field. Furthermore, it seemed that the various ministers in the government perpetuated the conflicts that had existed among the committee's members (Arnfred & Lindgaard Pedersen, 1987).

Nothing happened until February 1986, when an MP from the Socialist Peoples' Party (*Socialistisk Folkeparti*) raised a debate in Parliament. Together with spokesmen from a broad spectrum of parties, he proposed a resolution that passed Parliament. This resolution directed the Minister of the Environment to make proposals for an act on conditions for producing and using GMOs (Folketingstidende, 1986, cols. 6363–6406). The minister did so and the Act on the Environment and Gene Technology was passed by Parliament in June 1986. It is said to be the world's first law on genetic engineering.

Even though the act had been elaborated in close contact with the relevant interest groups, including representatives from industry, its final wording was more restrictive than the biotechnological industry had wanted. The act contained a formal ban on the deliberate release of GMOs into nature[5] (but authorized the Minister of the Environment to give approval in special cases). Approval would be suspended pending review of the complaint by the competent authority. Unlike the existing Act on the Environment from the 1970s, the 1986 Act on Environment and Gene Technology entitled a number of NGOs to act as plaintiffs. Industry specifically opposed all these features in the new act. The legislation also stipulated that a decision taken by the authorities had to be publicly promulgated together with the conditions for approval, a provision adopted from the Act on the Environment.

Thus, it was the Danish Parliament that took the initiative. Furthermore, the Minister of the Environment agreed with Parliament that it should have an account of the deliberations before the first approval of deliberate release of GMOs into nature so that the principle aspects could be discussed. All these decisions demonstrated the active role of Parliament in the making of the legislation on genetic engineering, which, in turn, was an expression of the level of politicization of the issue and the degree of anxiety in the population. Two contributing factors, however, were that the conservative government was a minority government and that there was a "green majority" of social-liberal and socialist parties in Parliament.

In the parliamentary debate the Minister of the Environment, who favored the legislation, stated:

---

[5] *Deliberate release* of GMOs denotes production that implies that GMOs will have to be released into nature as a consequence of the character of the production, as when genetically modified plants are grown in a field or if genetically modified bacteria for cleaning up oil-spills are spread in the sea. By contrast, *contained use* of GMOs occurs when GMOs are used in closed production systems, as when genetically modified bacteria produce enzymes and other proteins in large fermentation tanks in a factory.

It is very important that we have this debate in this Body and have the opportunity to discuss the implementation of the new technology in society. I also think it is important that there is a debate in public. Many people are interested in genetic engineering and in the consequences that the new technology can have for all of us. (Folketingstidende, 1986, col. 6363)

This statement demonstrated the growing recognition among politicians at that time of the need for information and debate about new biotechnologies. A first attempt to provide such information was seen in 1985 when the Ministry of the Environment published 3,000 copies of a booklet entitled *Environment and genetic engineering* (Danish Ministry of the Environment, 1985) in connection with a video also produced and distributed by the ministry. It emphasized the need for public debate and was presented as an introduction to such an exchange, not as the official viewpoints of the Ministry of the Environment. The booklet focused on the potential importance of genetic engineering for Danish industry and agriculture. A specific issue it addressed was the prospect of solving environmental problems by means of new biotechnologies, but it also stressed the risks associated with genetic engineering and the need for regulation and risk assessment.

Environment and genetic engineering was criticized by NOAH for not discussing which applications of the new biotechnologies would be most profitable to Danish industry and therefore most likely to be implemented. The omission, as NOAH stated, gave the impression that the environmentally friendly applications would play a more important role than one would expect from an economic viewpoint. In addition, the booklet was faulted for not directing enough attention to ecological risks and for not including a sufficiently serious discussion about the lack of knowledge about risks. NOAH called for resources with which to provide much deeper knowledge about risk and safety issues (Toft, 1985b). Of course, this criticism, which reflects drawbacks in the way the perspectives of genetic engineering were communicated by the ministry, also points to some of the seemingly almost unavoidable weaknesses in this kind of one-way communication by a public authority.

The Broad Popular Debate

Large-scale public support of information and debate did not exist until 1987, when an R&D program for biotechnologies was passed by the Danish Parliament after lengthy negotiations between the parties (Arnfred & Lindgaard Pedersen, 1987). Basically, it was a technology program for promoting biotechnological R&D in Denmark. At the insistence of the green majority in Parliament, the Minister of Education was obliged to allocate 20 million Danish Crowns (about U.S. $3.3 million) out of 500 million Danish Crowns (about U.S. $83.3 million) for information activities and technology assessment over the four-year period from 1987 to 1990. These sums were not impressive compared to similar programs in other countries, but they were substantial in relation to the size of Denmark and to what had been spent in other areas of R&D. Novo's annual R&D budget for 1986 alone was around 400 million Danish Crowns (about U.S. $66.6 million).

The money was to be administered by the Board of Technology (*Teknologinævnet*), which was established in 1985 as an independent institution responsible directly to Parliament. Its purpose was to follow and initiate comprehensive assessments of technological development's possibilities for and impacts on society and citizens and to support and encourage public debate on technology. The establishment of the board was prompted by the previously cited debate about nuclear power in Denmark in the second half of the 1970s and by the expectations of a comparably strong debate about information technology and biotechnology in the 1980s (Klüver, 1995). By handing responsibility over to the Board of Technology, Parliament had direct contact with the body in charge, placing it outside government control. Several MPs argued that it was important to ensure an accrual of knowledge in the population so that the public could participate actively in the debate about development in biotechnologies (Arnfred & Lindgaard Pedersen, 1987).

Parliament specified that the information activities be organized in cooperation with Dansk Folkeoplysnings Samråd. Another condition was that the NGOs were to become involved in the information activities and debates and were to receive funds from the biotechnological program.

The funds for which Dansk Folkeoplysnings Samråd was responsible was spent primarily on support of local arrangements of debate or information, which accounted for more than half of the 6.4 million Danish crowns (around U.S. $1 million) that it received over the three-year period from 1988 through 1990. Far fewer resources were spent on information material in the campaign on technology than had been the case in the campaign on nuclear energy (in which information material had been the subject of most of the discussion). The information material contained only very basic information about biotechnology and, of course, information about the possibility of obtaining support for local arrangements about new biotechnologies. There were very few restrictions on the possibility of receiving support, and from 1988 through 1990 658 arrangements, with more than 37,000 participants, received funding (Mikkelsen & Olesen, 1991).

The Board of Technology granted money for numerous activities from 1987 through 1990, including technology assessment, information and debate publications, and conferences. As stipulated by Parliament, support was also given to an NGO: NOAH. NOAH organized a series of seminars with key persons from Denmark and other countries, providing the most recent information about current issues related to the application of new biotechnologies (e.g., GMO animals, gene technology and the Third World, patents on life, and bovine growth hormone [BST]). This effort resulted in a series of publications that became influential despite the limited number of copies printed (1,000), for they were used by opinion leaders as sources. NOAH also wrote an information paper about gene technology, which was passed on to farmers.

Another important activity initiated by the Board of Technology was a series of consensus conferences. Inspired by the American consensus conferences on medical issues, the Board of Technology developed a Danish model that emphasized the involvement of lay persons, a commitment expressed by the fact that its assessment panel consisted of laypeople (Grundahl, 1995; Klüver, 1995). The first of these conferences took place in 1987 and addressed the subject of gene technology in industry and agriculture. The conference attracted a great deal of media attention and was attended by most members of the parliamentary committee responsible for the contact between the Board of Technology and Parliament. The final document by the lay panel, written after two days of hearings with a panel of experts, was generally very critical toward gene technology. For ethical reasons, the members of the panel unanimously wanted a total ban on genetic engineering on animals. There was also a majority for a ban on the deliberate release of GMOs into nature. Generally, the panel felt there was a lack of knowledge especially about the ecological impacts of deliberate release.

The industrialists in the expert panel openly expressed their disappointment with the final document after its presentation on the last day of the conference. The document confirmed the impression that the 1986 Act on the Environment and Gene Technology was in keeping with a widely shared critical attitude toward gene technology. Politically, one outcome of the conference was that the Danish Parliament decided not to fund animal-gene technology projects under the R&D program for biotechnology.

More important perhaps, the consensus conference demonstrated that laypeople could not only participate in the debate but also make informed decisions about what they thought should be done in the field of gene technology. In this way it contributed to the general impression that the debate about gene technology was no longer a matter solely for experts and elite groups.

Danish Surveys on Public Perceptions of Gene Technology

In 1987 the first surveys on public perceptions of gene technology were conducted.[6] The earliest one (Jensen & Thing Mortensen, 1987) was run in the first months of the year for the Danish Food Agency to prepare for an information campaign that the agency intended to launch. This survey combined focus-group interviews and a few interviews with individuals, some with ordinary citizens and some with representatives of consumer and environmental groups. Overall, the interviews revealed a critical attitude toward gene technology, though various degrees of skepticism were discernible. Most of the respondents expressed uncertainty and limited knowledge about the subject. Another general finding was a critical attitude toward the authorities. "We are more skeptical than the authorities," as one of the interviewees said, delivering the words that became the title of the report on the survey. In the report the authors concluded that a fundamental distrust of the decision-makers' willingness and ability to regulate and control stood behind much of the negative or skeptical attitudes toward gene technology. The authors went on to warn the Danish Food Agency against basing its information on experts' statements and viewpoints. They recommended an information campaign designed to support and qualify advocacy of viewpoints that could already be found in the population (Jensen & Thing Mortensen, 1987). Although reasons unrelated to the results of this qualitative survey prevented the launch of the information campaign planned by the Danish Food Agency (A. Thing Mortensen, personal communication, November 1995), the study attracted keen attention in the Danish debate, not least because of its clear indication of the population's critical attitudes toward the authorities.

The survey was conducted at a time when general experience with the social control of gene technology was still influenced by the controversies surrounding the applications by Novo and Nordisk Gentofte and before the information activities of the biotechnological program had been initiated. But the skepticism about the authorities also reflected rather general experience with environmental regulation and control. The environmental authorities had shown lack of trustworthiness in the eyes of the population in a number of cases during the first half of the 1980s. When the information practices of Denmark's Environmental Agency were criticized in the Danish

---

[6] All the surveys mentioned in this section consistently used the term *gene technology* in the interviews and questionnaires. By contrast, the 1993 Eurobarometer survey compared the significance that the use of the terms *biotechnology* and *genetic engineering* had on attitudes.

press and elsewhere in the aftermath of the Chernobyl disaster in 1986, an expert group was appointed to deal with the subject.

Another investigation (Borre, 1990a, 1990b) consisted of four consecutive quantitative surveys (September 1987, February 1988, May 1988, and May 1989) with samples ranging from 670 to 1,512 respondents.[7] The results indicated growing support for gene technology from September 1987 to May 1988. The group classified as supporters, about 20% of the respondents in 1987, increased from 6% to 7%, and there was a decrease of similar magnitude among the opponents. The largest group, which consisted of respondents with *neutral* opinions about gene technology, remained the same, about 62%. The classification was made on the basis of the answers given to four questions expressing viewpoints on gene technology. In this period the debate about gene technology was intense, and Danish television added to the general debate by airing a number of broadcasts on the subject. The 1989 survey indicated that the *neutral* group had grown to 67% of the respondents, a finding that Borre (1990a) interpreted as indicating that polarization of attitudes toward gene technology had lessened. The results also showed a quite critical attitude toward certain aspects of gene technology. In April 1989 40% of the respondents were against modification of plant genes and 68% were against modification of the genes of cows and pigs, whereas very few were against modification of the genes of bacteria. Furthermore, an item about trust showed that 67% of the respondents had little or no trust in the government and much higher trust in researchers and environmental movements (Borre, 1990a, 1990b).[8]

The general trend toward less polarization that Borre claimed to see may be disputed for methodological reasons because his classifications do not provide a sufficiently differentiated understanding of what really happened from September 1987 to May 1989. The large group of respondents with *neutral* opinions, which changed little (from 62% to 67%) over the 20-month

---

[7] The respondents of the survey in May 1988 had also been part of the survey in September 1987. The survey in February 1988, occasioned by a series of broadcasts about biotechnologies on Danish television in January 1988, was conducted by the Danish TV and radio corporation. The other surveys were funded by the Board of Technology as part of its information and technology assessment activities (Borre, 1990b).

[8] This result, which is consistent with the findings of Jensen and Thing Mortensen (1987), is not easily comparable to the findings of the 1993 Eurobarometer survey, in which 45% of Danish respondents said they trusted their public authorities to "tell the truth about biotechnology/genetic engineering." Borre (1990a) asked his respondents how much they trusted a number of participants in the debate about genetic engineering and mentioned the government, not public authorities. The Eurobarometer's result is interesting in itself, however, and I discuss it briefly later.

period of the surveys, makes it difficult to conclude that significant changes really occurred. Nevertheless, politicians and others responsible for the funding of the information activities saw the results of the survey as a confirmation that the strategy behind the promotion of the broad popular debate had been right.

The Continued Debate about Regulation

Parallel to the popular debate about biotechnologies, considerable effort was being made by the Danish biotechnological industry to influence politicians to change the 1986 Act on the Environment and Gene Technology. In 1987 the Association of Biotechnological Industries was formed as an interest organization for the industry. In early 1988 it issued a book containing a review of the regulation of biotechnologies in a number of countries as well as the OECD and the EC. According to this review, the general international trend was toward loosening regulation. Moreover, most countries regulated new biotechnologies under existing legislation rather than under a specific act as in Denmark. In the conclusion it was pointed out that the international scale of biotechnology is a reason why production facilities tend to be located in countries where conditions for approval are most favorable (Fink & Terney, 1988). Novo and Nordisk Gentofte similarly went to the press and, as they had done a few years earlier, warned that if the 1986 Act on the Environment and Gene Technology were not made less restrictive they would move their production facilities to another country. A number of biotechnological researchers likewise criticized the act. At the political level the Minister of Education promised that he would try to get the act changed.

The result of these pressures was that the Minister of the Environment proposed a number of changes in the act in January 1988. Large-scale trials with the contained use of GMOs were to be permitted without prior approval, cell hybridization was no longer to be covered by the act, and the power of complaints to suspend approval to manufacture biotechnological products was to be abolished, though only for contained use, not for deliberate release as the industry had wanted. The proposal led to further debate and a public hearing by the Board of Technology. When the act was passed in Parliament in May 1989, cell hybridization had been retained in the act and the power of complaints to suspend approval of production involving the deliberate release of GMOs into nature was upheld.

In the media the political debate continued, but after the amendment of the act the focus was on deliberate release. In December 1988 the Danish sugar factories sought approval to field test two different genetically modified sugar beets. In May 1989 the Minister of the Environment submitted his account of the approval to Parliament, and after a number of additional questions had been addressed to the minister, the Parliamentary Committee on Environment and Planning consented. This procedure demonstrated what probably had been clear from the beginning—that the formal ban on deliberate release in the 1986 act was not really a ban but rather a basis for a careful case-by-case assessment of all aspects of the risks associated with the release of GMOs. In conjunction with the intense debate, however, these careful procedures demonstrated that biotechnology was, in fact, subject to regulation. The discussions about the sugar beet took place not only as a media battle but also as one of the issues for debate at the numerous public meetings that were sponsored by the Board of Technology.

Summary for the Period from 1985 through 1990

The debate about biotechnology and genetic engineering was initiated in the mid-1980s, and the number of actors swelled as the decade continued. Because of the priority given to a broad debate, many people became involved. The great number of opportunities for the participants to influence the debate and the flow of information seemed to depolarize viewpoints somewhat and to foster willingness to listen to others. This outcome was due to the decentralized and nondirected character of the interaction and to the explicit involvement of NGOs like NOAH, which may have enhanced the credibility of the debate and its sponsors. The implementation of the 1986 Act on the Environment and Gene Technology, especially the careful procedures for approving the manufacture of products of gene technology, may have led to increased trust that authorities did exercise control over the biotechnologies in question.

Unlike Denmark's nuclear debate, the broad popular debate was not initiated by Parliament until after the passage of the Act on the Environment and Gene Technology and its restrictive regulatory procedures. This timing may have contributed to the more open and less polarized exchange. At the same time, of course, politicians were able to avoid committing themselves to a specific position that would have made decision-making even more difficult than it had been before the debate. The agenda for most of Denmark's political parties was to promote the use of new biotechnologies within the country. Achieving that goal was considered essential because of the economic importance of the Danish biotechnological industry and because of the country's agricultural production. (The economic interests in nuclear energy were, after all, less vital, especially when energy prices began to decrease after 1980, a factor that contributed to the outcome of the nuclear debate in Denmark.) The effort to forge social consensus on this point was apparent in both the formulation of the 1986 Act on the Environment and Gene Technology (see also Toft, 1995) and in the subsequent decision to fund the information campaign. On the other hand, there is no indication that the population's critical points of view on biotechnology and genetic engineering weakened from 1985 through 1990. However, viewpoints became specifically related to the various applications rather than to new biotechnologies in general. Such a differentiation of viewpoints was also seen in the outcome of the consensus conference and the surveys.

A number of factors contributed to the specific character of the development from 1985 to 1990. One of them was the Danish Parliament, which, among other things, reflected the emergence of a "green majority." Parliament's active role added to the democratic character of the policy processes more than highly bureaucratic control would have done.

The Status in the Mid-1990s

By 1990 and 1991 the debate about new biotechnologies and their social regulation had begun to decrease. Nevertheless, a good deal of discussion was heard about the adoption and subsequent implementation of EC Directives 90/219 and 90/220 in Danish legislation. Directive 90/220 was subject to particularly intense scrutiny, for it meant an end to the ban on deliberate release of GMOs and entailed several other changes in the 1986 Act on Environment and Gene Technology. The directive was seen as a weakening of control over gene technology. This debate shifted some of the attention from the Danish context to the European Community (EC), which in the early 1990s (and as of 1993 as the European Union [EU]) increasingly became the "villain" in the debates about biotechnologies.

A consensus conference in 1992 on transgenic animals, which was arranged by Denmark's Board of Technology, once again led to a clear statement against the use of gene technology on animals in agriculture. In addition, both the Danish authorities and the biotechnological industry have frequently emphasized their support for the principles of the existing regulations on gene technology. Novo Nordisk (the result of the merger between Novo and Nordisk Gentofte) has publicly criticized several attempts to weaken the EU regulation as advocated for especially by German industry (see, for instance, Terney, 1995). Novo Nordisk maintains that the present regulation, as implemented in the Danish legislation, is satisfactory to the company and preferable to the social and political conflicts and instability that would be likely to ensue if the regulations were watered down and public distrust were to result. The experience under the present regulations is explicitly referred to by the company as the cause of its present viewpoints, which are clearly different from those that had been expressed by both Novo and Nordisk Gentofte in the 1980s.

When it comes to GMO-based production, Novo Nordisk's communication strategy has been to inform the public and the authorities quickly about irregularities, such as an unintended discharge of GMO-laden waste water into the environment. The company thus consciously seeks to avoid any disclosures in the press about its production practices. It has also arranged a number of information and discussion meetings with representatives from NGOs in Denmark and at its export markets. Such meetings must, of course, be seen as two-way communication events giving the company information of strategic value about the viewpoints of the NGOs.

For the government's part, the ongoing attempts to amend and relax EC Directive 90/220 have been criticized by the Danish Minister of the Environment. Speaking in February 1996 at a public hearing arranged by the Council of Technology[9] to review genetic engineering being conducted on plants, he also warned against changing the regulation in any way that could foster distrust and uncertainty in the population.

The broad popular debate that characterized the period before 1990 has lost much of its intensity. The kind of critical acceptance of new biotechnologies that was described above as one of the findings of the Eurobarometer surveys of 1991 and 1993 perhaps summarizes the present situation quite well. Furthermore, discussions about biotechnologies and genetic engineering are focusing increasingly on specific issues related to the subject, not about its general foundations.

The situation in the mid-1990s may be characterized by the existence of a kind of social contract as demonstrated by the behavior of industry and the authorities referred to above. If industry and authorities observe certain rules (both legislated and unwritten), there will be a willingness in the population to accept a number of applications of biotechnologies and to discuss the prospects of their further development. If the rules are violated too clearly or too frequently, distrust and lack of acceptance may be the result. It must also be remembered that within some of the most critical areas, such as the use of gene technology on animals, genetically engineered foods, and gene therapy on humans, the first practical applications are still in the developmental stage or pending approval. When such approval is granted, as will happen soon with some applications, especially within the food area, the debate about

---

[9] In 1995, the Act on The Council of Technology gave the Board of Technology permanent status and, among other changes, changed its name to The Council of Technology.

new biotechnologies in Denmark may take a new turn. It is not likely that the term acceptance will be just as appropriate for, say, transgenic animals as it is for the present use of bacteria in fermentation plants. A rather critical debate about labeling, which became the focus of lengthy discussions prompted by the proposal for the EU Order about Novel Foods in 1995, seems to indicate great wariness about allowing the market to be flooded with products that might contain genetically engineered foods without people knowing it.

Conclusion

The development of the Danish debate about new biotechnologies and their social regulation reveals the broad character of the social processes that led to critical acceptance in the early 1990s in Denmark. Time series based on the static attitude measurements compiled through the Eurobarometer surveys may reveal some aspects of the development in attitudes about science and technologies. But the interplay between public debate, attitudes, and decision-making processes described in this chapter is essentially dynamic in nature. The impact of information must be seen as part of wide processes of social and political debate and decision-making, and that context is difficult to grasp through static measurements.

The development might be explained by more complex factors like Danish political culture, which is often associated with the kind of consensus-seeking evident in some of the development during the 1980s. Indeed, I do not dispute that consensus-seeking is an element of political culture in Denmark. But Danish politicians do not have a magic tool with which to avoid social conflicts and opposing interests. Acceptance (whether critical or not) is not a result of consensus-seeking policies. Instead consensus-seeking policies may facilitate a development toward acceptance if social conditions for achieving acceptance are otherwise present. Those social conditions include information—not information as such but as an element in a broad popular debate about the regulatory initiatives taken during the 1980s. The social conditions for achieving acceptance also include the broad experience of the nuclear and environmental debates, which show that political decisions about new technology may in fact be influenced by the population.

Conversely, it might be concluded that the debate about biotechnologies in Denmark has been "functional" (see Bauer, 1995) in the sense that politicians, regulators, and industrial actors have developed rather clear sets of rules specifying which arguments are important, how to inform the public, and so forth.

Lastly, trust in government or public authorities has been highlighted as an explanatory factor of attitudes toward new biotechnologies and their social regulation. As can be seen from the data presented in this chapter, evidence of this viewpoint is somewhat contradictory. It seems to be clear that trust in the government to regulate new biotechnologies was not great in large groups of the population during the 1980s (and that such skepticism was influenced partly by experience with the nuclear and the environmental debates). This attitude may have changed in the beginning of the 1990s, but it is a specific outcome of the social processes in the field rather than an independent explanatory factor. As mentioned above, it may also be that the critical focus has shifted toward the EU because an increasing number of regulatory initiatives are being taken at the EU level and must be negotiated within the EU system, with all its complicated, lengthy procedures.

References

Act on the Environment and Gene Technology. (1986, June 4). Lov nr. 288.

Act on the Council of Technology. (1995, June 14). Lov nr. 375, Lovtidende 1995, 1685–1687.

Arnfred, N., & Lindgaard Pedersen, J. (1987). Folketinget og bioteknologien [The Parliament and biotechnology]. *Samfundsøkonomen, 6,* 33–38.

Baark, E., & Jamison, A. (1990). Biotechnology and culture: The impact of public debates on government regulation in the United States and Denmark. *Technology in Society, 12,* 27–44.

Bauer, M. (1995). Towards a functional analysis of resistance. In M. Bauer (Ed.), *Resistance to new technology—Nuclear power, information technology, biotechnology* (pp. 393–417). Cambridge, England: Cambridge University Press.

Borre, O. (1990a). *Befolkningens holdning til genteknologi II, Kommunikation og tillid* [Public attitudes toward gene technology, II: Communication and trust] (Teknologi Nævnets Rapporter 1990/4). Copenhagen: Teknologi Nævnet.

Borre, O. (1990b). Public opinion on gene technology in Denmark 1987 to 1989. *Biotech Forum Europe, 7,* 471–477.

Cantley, M. (1992). Public perception, public policy, the public interest and public information: The evolution of policy for biotechnology in the European Community, 1982–92. In J. Durant (Ed.), *Biotechnology in public: A review of recent research* (pp. 18–27). London: Science Museum for the European Federation of Biotechnology.

Christensen, B. (1980, January 29). Genmanipulation er andet end reagensglasbørn [Gene manipulation is more than test-tube babies]. Interview with Kjeld Marcher. *Land og Folk* (Copenhagen).

Danish Ministry of the Environment. (1985). *Miljø og gensplejsning* [Environment and genetic engineering]. Copenhagen: Miljøministeriet.

Fink, K., & Terney, O. (1988). *Sådan reguleres genteknologi* [How gene technology is regulated]. Copenhagen: Foreningen af Bioteknologiske Industrier i Danmark.

Folketingstidende. (1986). Forhandlingerne i Folketingåret 1985–86, columns 6363–6406.

Gene-splicing Committee of the Ministry of the Interior (Indenrigsministeriets gensplejsningsudvalg). (1985). *Genteknologi og sikkerhed* [Gene technology and safety] (Afgivet af indenrigsministeriets gensplejsningsudvalg, betænkning nr. 1043). Copenhagen: Direktoratet for Statens Indkøb.

Gensplejserne [The gene-splicers]. (1981, April 28). *Dagbladet Information* (Copenhagen), p. 12.

Gensplejsning og tryghedsbehov [Genetic engineering and need for security]. (1982, December 2). *Ingeniøren* (Copenhagen), editorial.

Grundahl, J. (1995). The Danish consensus conference model. In S. Joss & J. Durant (Eds.), *Public participation in science: The role of consensus conferences in Europe* (pp. 31–40). London: Science Museum.

Hansen, A. M., & Kjøgx Pedersen, I. (1985, June 24). Fabrik føler sig dårligt behandlet [Company feels that it has been treated badly]. *Politiken* (Copenhagen), p 8.

Hilden, J. (1981, May 22). Gensplejsning og forskningsstyring [Genetic engineering and government of research]. *Dagbladet Information* (Copenhagen), p. 6.

Hueg, K. A. (1985, July 4). Industriel gensplejsning i Danmark [Industrial gene-splicing in Denmark]. *Politiken* (Copenhagen), section 2, p. 5.

Institut National de Recherche Agricole (INRA), & Marlier, E. (1993). *Eurobarometer 39.1—Biotechnology and genetic engineering: What Europeans think about it in 1993* (Report DG XII/E/1). Brussels: Commission of the European Communities.

Jensen, L., & Thing Mortensen, A. (1987). *Vi er mere skeptiske end myndighederne, En rapport til Levnedsmiddelstyrelsen om folks viden om og holdninger til genteknologi* [We are more sceptical than the authorities: A report to the Food Agency about people's knowledge about and attitudes toward gene technology]. Roskilde Universitetscenter: Institut for uddannelsesforskning, kommunikationsforskning og videnskabsteori.

Klüver, L. (1995). Consensus conferences at the Danish Board of Technology. In S. Joss & J. Durant (Eds.), *Public participation of science: The role of consensus conferences in Europe* (pp. 41–51). London: Science Museum.

Mikkelsen, K., & Olesen, J. E. (1991). *50 år med dansk folkeoplysning* [Fifty years of Danish youth and adult education]. Copenhagen: Dansk Folkeoplysnings Samråd.

Teknologinævnet (1994). Konklusioner fra programmet indenfor temaer [Conclusions from the program within themes]. In *Evalueringsrapport fra bioteknologitemaet* (pp. 172–200). Copenhagen: Teknologinvænet. Unpublished manuscript.

Terney, O. (1995). 20 år med genteknologi: erfaringer og perspektiver [Twenty years of gene technology: Experience and perspectives]. Fredericksberg: Foreningen af Bioteknologiske Industrier i Danmark.

Toft, J. (1985a). *Genteknologi—konsekvenser for miljøet ved anvendelse af gensplejsede mikroorganismer* [Gene technology—Environmental impacts of using genetically modified microorganisms]. *NOAH's Moderne Tider (No. 2)*. Copenhagen: NOAH's Forlag.

Toft, J. (1985b). *Kampen om genenerne: Industriel gensplejsning i Danmark* [The battle over genes: Industrial genetic engineering in Denmark]. *NOAH's Moderne Tider (No. 3)*. Copenhagen: NOAH's Forlag.

Toft, J. (1985c, July 24). Vort Miljø og industriel gensplejsning [Our environment and industrial gene-splicing]. *Politiken* (Copenhagen), section 2, p. 5.

Toft, J. (1995). *Seeking a broad-based consensus: A study of the implementation of EC Directive 90/220*. Roskilde: Roskilde University Library.

TV Avisen. (1985, July 7). Gladsaxe, Denmark: Radio Denmark.

CHAPTER 13

A PUBLIC EXPLOSION: BIG-BANG THEORY IN THE U.K. DAILY PRESS*

Massimiano Bucchi

> Imagine doing just what the Big Bang did/
> The whole world/Knew it was loaded
> Wave bye-bye 'cause it ain't never coming down now.
> The Breeders (1993), "I just wanna get along."
> On *Last Splash*. 4AD Records.

At 10:00 a.m. on Thursday, April 23, 1992, David Whitehouse from the British Broadcasting Corporation (BBC) Radio News informed his listeners that a NASA satellite, the Cosmic Background Explorer (COBE), had detected "ripples" in the universe's cosmic background radiation (CBR) and that this finding would provide conclusive evidence that the universe as we know it today began with an enormous explosion known as the big bang. On the following morning, all of the U.K.'s serious newspapers (*The Independent, The Times, The Guardian,* and *The Daily Telegraph*) and some of the tabloids featured front-page stories on COBE. For several weeks, the topic was one of the hottest around, at least in the British media, and became the subject of editorials, commentaries, and humor pieces. George Smoot, the physicist leading the COBE team, was offered several million dollars to write a book about the discovery (Smoot & Davidson, 1993; see also Chown, 1993; Rowan-Robinson, 1993). Widespread publicity was given to Smoot's remarks that he and his colleagues had found the oldest structures ever seen in the early universe: "primordial seeds of modern-day structures such as galaxies, clusters of galaxies . . . If you are religious, it's like seeing God" (Miller, 1994, p. 449). These and other comments, such as those by Stephen Hawking, who claimed that the finding was "the discovery of the century, if not of all time" (Hawking, 1992, p. 6; see also Hawkes, 1992a,

---

* A different version of this essay has appeared as a chapter in my book, *Science and the Media* (London: Routledge, 1998). I am grateful to the publisher for granting permission to publish this verson.

p. 1), and by cosmologist Michael Turner, who believed that "they have found the Holy Grail of cosmology" (as quoted in Miller, 1994, p. 450), doubtless attracted public interest in the discovery and fed a more general debate about science, religion, and their mutual relation in the human quest for truth.

What was the story all about? Was it just another case of sensational or spectacular science as exemplified by Stanley Pons and Martin Fleischmann's precipitate public claim to have discovered cold fusion (for background see Bucchi 1996a; Lewenstein, 1992, 1995; Sullivan, 1994)? Not really. Although Smoot and his team skillfully managed contact with the public arena through press releases and press conferences from the very outset of the COBE enterprise, nobody ever charged them with fraud or blamed them for hastily seeking publicity before the scientific community had had time to review the findings carefully. (Smoot and his colleagues did, however, hold a press conference without having an official publication of their final results at hand—just as Pons and Fleischmann had done—just a few hours after briefing the American Astronomical Society in Washington, DC. How did a rather esoteric, abstract finding having no apparent practical significance make it so big in the media? Explanations have been proposed that emphasize the role of media practices and the element of public resonance (e.g., Miller, 1994). According to such explanations, the attention given to COBE by the media, particularly those in the United Kingdom, stems from the fact that the media is already sensitized to such issues by ongoing debates about science and its achievements. While I acknowledge the importance of these factors, it can be easily argued that the role of scientists in making science public is, from that perspective, largely considered unproblematic. Media dynamics are seen as the only available explanation for the communicative relationships between science and the public.

In this chapter I try to focus on the other side of the process. What were the conditions that made it possible and appealing for the COBE scientists to go public? What consequences did their decision have for the scientific debate? Why was this instance of going public not chastised as in the case of cold fusion? In other words, how was science actually made in public in this case?

The theoretical guidelines of this attempt are given mainly by those studies ("continuity" models) whose authors have challenged the standard views on public communication of science by arguing that the public level of the communication of science may be important to the process of shaping the core scientific debate (e.g., Biezunski, 1985; Clemens, 1986; Cooter & Pumfrey, 1994; Fleck, 1935/1979; Hilgartner, 1990; Lewenstein, 1995; Shinn & Whitley, 1985). Such studies have shown that scientists often make active use of the public realm to acquire and exchange information, generate support for their theories, and seek legitimation for their institutions. The process of communicating science is therefore seen as a *continuum* of expository stages in which differences inevitably occur because of the dissimilar contexts and styles of communication and reception. Boundaries between "pure" knowledge and popular discourse cannot be sharply drawn even though they are often used by scientists to define and preserve their own authority over knowledge. Communication occurring at the "lowest" levels of the continuum, namely, public communication, can therefore influence specialist communication.

Unfortunately, most of these studies fail to address in detail the relationship between the public level and the specialist level. The research is based on the assumption that communication about science is filtered through a series of stages (e.g., intraspecialist, interspecialist, and pedagogical) that culminate in public exposure that reinforces the status of scientific actors and results (e.g. Collins, 1987; Whitley, 1985). The public is thereby relegated to a passive role quite similar to the one assigned it to in the canonical account: the role of amplifying scientific debate as defined by experts.

I have argued that this assumption is reasonable only when science communication proceeds smoothly through its usual stages (e.g. publication in the scholarly literature, then inclusion in university textbooks, followed by treatment in the popular media (Bucchi, 1998). Sometimes, however, some scientists are not willing to follow this procedure. Instead, they may be prone to "deviations" (Cloître & Shinn, 1985, p. 55), to interaction directly with the public, bypassing one or more stages of the scientific community's established sequence of communication. When the usual channels are circumvented, the more subtle contributions that the public can make can be obscured if one focuses only on outcomes when studying the public communication of science. The public should not be seen solely as an external, taken-for-granted, monolithic entity in which already dominating scientists and results are further consecrated. In order to understand the occasions when public communication acquires its greatest significance for the development of scientific debate, one need try to answer a crucial question: Under what conditions does the public communication of science take place through deviation rather than through standard channels (which could reasonably be called the popularization *modality*, despite the term's ideological burden)?

Cases in which scientists have turned to the public, as in COBE's revelation of evidence for the big bang, should be studied with an eye to "critical situations" that involve negotiation at different boundaries both within and outside the scientific community (see Bucchi, 1996b, 1998; on the concept of boundary work and demarcation strategies in science communication, see Gieryn, 1983, 1995; Holmquest, 1990). My proposal has two main implications. (a) Being related to specific conditions of core scientific debate, deviation to the public cannot be dissociated from such debate. This stance differs from that proposed in the so-called "canonical account of the communicational relationships between science and the public" (Shapin, 1990, p. 991).[1] In the canonical account, media actors are the ones entirely responsible for recklessly pushing certain issues into the public sphere. (b) Deviation to the public does not simply perpetuate the directions that scientific debate follows among specialists, as the continuity models suggest. Instead, deviation interacts with them in complex ways. In my view, the idea that boundary negotiation is what links core scientific debate with the public communication of science meets the need to acknowledge both the significance the public has for scientific discourse and the relative autonomy and specificity of that discourse.

A Brief History of the Universe

In 1927 Georges Lemaître, a Belgian mathematician and Catholic priest, hypothesized that the universe might have come about through the gradual expansion of an original "cosmic egg," a primeval atom (atom primitif) that divided itself into smaller and smaller atoms by virtue of a strong radioactive process. These ideas, however, were either heavily criticized or ignored outright by the prominent scientists of that period, including Einstein, who always refused to discuss the model that Lemaître had worked out. It took time for Lemaître's ideas to be coupled with the observations made by Edwin Hubble, the American astronomer who in 1929 discovered that other galaxies were moving away from the Milky Way galaxy at a speed directly proportional to their distance from it. (This discovery is known as Hubble's law and is expressed by a value called the Hubble constant [H_0].) However, in an address to the Mathematical Association in Cambridge, England, Eddington (1931) expressed the "philosophical repugnance" (p. 450) of the very notion that the universe had a beginning. Not until the late 1940s did George Gamow and his colleagues elaborate on Lemaître's notion with their concept of a singularity, a situation characterized by extreme density, excep-

---

[1] Other expressions with equivalent meaning have been used, such as "dominant view" (Hilgartner, 1990) and "diffusionist model" of public communication of science (Cooter & Pumfrey, 1994).

tionally high temperature, and zero dimension that, upon exploding, could have started the expansion process.

Ironically, the term *big bang* was coined in 1950 by British astronomer Fred Hoyle during his radio program, *The nature of the universe*. It was Hoyle who in 1948, together with two Austria refugees, Thomas Gold and Hermann Bondi, had elaborated an alternative model for the evolution of the universe, the so-called steady-state theory, and he had used the expression big bang to *ridicule* the idea that the universe could have originated from a primordial explosion. Despite Hoyle's intention, however, big bang became the standard term for referring to this theory of the universe's origin and expansion, and it met with increasing success in the following years.

The experimental result that is often said to have settled the matter between the two competing models was the discovery of CBR by Arno Penzias and Robert Wilson in 1965. Working as researchers for Bell Telephone Laboratories, the two scientists had experienced problems with eliminating a background noise from their reception of a satellite transmission. Later measurements led them to identify it as the radiation that Gamow, Alpher, and Herman said to be what remained of the intense radiation produced by the big bang. Still, this theory left some problems unresolved. In particular, it was difficult to explain the regularity that scientists observed in the universe. Because light (and therefore no other type of information) would not have had the time to travel from one region of a forming universe to another in the event of the hypothesized initial explosion, how have such regions ended up with the same temperature and expansion rate? In 1981 an American physicist, Alan Guth, proposed a solution that accommodated these and other difficulties: the idea of an "inflationary" universe. According to Guth, the universe, in its first phase, would have passed through a period of accelerated expansion. This line of thought would allow one to conceive of an original region so small that light would have been able to pass through.

COBE Takes Off

According to big-bang theorists, CBR should have a spectrum in the shape of a bell curve, which is typical of heat radiation. To investigate this conjecture, scientists have devoted extensive effort to obtaining an accurate measurement of the radiation. In 1987 a group of researchers from the University of California at Berkeley and from the University of Nagoya in Japan used an orbiting spectrometer and detected background radiation that differed somewhat from what they expected, for it had a number of spectral distortions. Physicists and astronomers wrestled with various interpretations of this observation until 1989, when NASA launched COBE, a satellite for measuring the spectrum of CBR beyond the obstacles of the Earth's atmosphere.

The launch was announced at a press conference on November 17, 1989. Larry Caroff of NASA stated that COBE and other planned space observatories were going "to provide us with an unprecedented, unbelievable view of the universe" (Smoot & Davidson, 1993, p. 234). More cautiously, scientist John Maher said that he was "not expecting to overturn big-bang theory with what we see, because it is a good theory and works well. However, we could get a big surprise" (Smoot & Davidson, 1993, p. 235).

The first results of the COBE enterprise were presented before the American Astronomical Society in January 1990. The measurement of CBR matched quite well with the theoretical graph of blackbody radiation. Thanks to this measurement, according to Smoot, "the big bang was still on track" (Smoot & Davidson, 1993, p. 242). Thus, CBR was homogenous, as the picture of an expanding universe required. It was, however, *too*, homogeneous. Contrary to intentions, COBE had in fact failed to detect in the radiation the "ripples," or fluctuations, that are interpreted as the remains of the initial perturbation, the primordial seeds, needed to give rise to the cosmos.

Only some two years later, Smoot announced to his colleagues and the general public that COBE had finally detected fluctuations in CBR on the order of a few parts per million. At 8:00 a.m. on April 23, 1992, he gave his previously mentioned report at the American Astronomical Society. About four hours later, the news was passed to the media, and Smoot and his colleagues spent the rest of the day satisfying the curiosity of journalists from all over the world.

More Big than Bang

Until the second half of the 1980s, criticism of the big-bang model came almost exclusively from scientists such as Fred Hoyle. By the end of that decade, however, the situation was slowly changing. In 1988 Stephen Hawking, who had contributed to consolidating the big-bang model in collaboration with Roger Penrose in the early 1970s, wrote in his bestseller, *A brief history of time*, that he had "changed his mind" about the existence of a singularity at the beginning of the universe (1988/1996, p. 67). Approximately one year later, John Maddox (1989), the influential editor of *Nature*, wrote:

> The Big Bang is an over-simple view of how the universe began, and is unlikely to survive the decade ahead. In all respects, save that of conven-

ience, it is thoroughly unsatisfactory. It is an effect whose cause cannot be identified or even discussed. (p. 425)

Since that time, several contributions have both questioned the big-bang model and responded to such critics. Halton Arp, together with Hoyle and other "well-known dissenters from the Big Bang" (Miller, 1994, p. 446) wrote an article for *Nature* (Arp, Hoyle, Burbidge, Narlikar, & Wickramasinghe, 1990) in which criticism of the model was based on experimental results with which it was judged to be inconsistent, namely, exceedingly high values for the Hubble constant (figures that would imply a much younger age of the universe than that used in the big-bang model) and COBE's initial failure to detect fluctuations in CBR. In the same year, physicist Eric J. Lerner published his book under the eloquent title of *The big bang never happened*. Even the second COBE announcement, in which the detection of ripples in CBR were reported, did not seem to set the matter straight. Debate continued on the basis of new data gathered by the Hubble Space Telescope, from which three different teams estimated the age of the universe to be 8 million years—much younger than the known ages of some stars.

Whereas some researchers now claim that the COBE results and related ones further support a model that was already solid, others seem more critical of such explanations. As Maddox (1995) contended:

> The result, the third of its kind in under a year, makes a nonsense of the standard Big Bang view of how the universe began. But even those who were persuaded on other grounds that the Big Bang is a fairy story, perhaps for no better reason than that it is "too good to be true" had better conduct themselves with circumspection in the months ahead. . . . In short, the new measurements of H_0 are not the death of the Big Bang, but merely a further sign of its fragility. . . . The minute it is suggested that these structures are so exceptional that they must be survivors from an earlier stage in the evolution, the Big Bang will have given way to continuous creation. That will be a turn-up for the book. (p. 99)

Some individuals responded to these attacks, as did Allan Sandage of Carnegie Observatories in Pasadena when he declared, "I think the damn big bang model works" (as quoted in Powell, 1995, p. 19). Others have claimed more cautiously that the big-bang theory need not be patently accepted or discarded but should instead be integrated with additional theories, such as inflationary models (Ostriker & Steinhardt, 1995). Finally, researchers have even questioned whether the occurrence of a big bang should be the primary focus of investigation (Barrow, 1994). One might easily agree with Bradley E. Schaefer, who concluded there was simply "no consensus out there" (as quoted in Powell, 1995, p. 19). In short, the ambiguous status of the big-bang theory is another factor making it difficult to establish whether it is still generally accepted by cosmologists. Should it be regarded as a single, definite model, as a group of theories and models, or as a general framework of thought, a paradigm, that lies behind a number of different theories and data collections? (See for instance Barrow, 1994, p. 42.)

The Big Bang and the Public

In the second half of the 20th century, the big-bang model has been the dominant model used by the general public to understand the origins of the universe. From the 1950s to the early 1960s, when the scientific community was still far from a consensus on the big bang, the public clearly had a role in establishing the model as the leading explanation within cosmology (McConnell, 1998). The idea of a singularity, an instantaneous burst of creation, was far more appealing to common sense than any other model, such as Hoyle's steady-state theory. According to Boslough (1992),

> Hoyle's idea of a static universe never had a grip on the public. His name for it, the steady-state universe, may have been too mundane. In any event, the "big bang" stuck, with Hoyle opposing the very notion from the day he dreamed it up. (p. 39)

The big-bang model was also easier to reconcile with religious beliefs. Upon being informed of the big-bang hypothesis in 1951, Pope Pious XII proclaimed that it was in perfect accordance with the Christian belief in religious creation. "True science is increasingly discovering God, just as though God were waiting behind every door that is opened by science" (Pontificia Academia Scientiarum, 1993, p. 81; my translation). In an address to the Pontifical Academy of Sciences, Pope Pious XII wrote:

> In fact, it seems that present-day science, with one sweeping step back across millions of centuries, has succeeded in bearing witness to that primordial "Fiat Lux" uttered at the moment when, along with matter, there burst forth from nothing a sea of light and radiation, while the particles of the chemical elements split and formed into millions of galaxies. . . . Hence, creation took place in time, therefore, there is a Creator, therefore, God exists! (Pontificia Academia Scientiarum, 1993, p. 93; transl. in Lerner, 1991, p. 385)

For at least a decade, scientist and theologian Stanley L. Jaki has forcefully argued in a number of books and articles against researchers who still question the big-bang model through their alliterative conception of a steady and eternal universe. As Jaki (1989) stated, it is precisely the contingency that the big bang assigns to the universe that makes it possible to believe in the Creation. And Lerner (1991) noted that "while the big bang as a scientific theory is less and less supported by data, its prominence in our culture has increased" (p. 54). It should come as no surprise, then, that the public was heavily involved in the big-bang controversy that ensued in the following years and that the COBE story became was of its hallmarks.

A Public Explosion

"Controversy", however, might not be the appropriate term for describing the treatment of the big-bang COBE story in the daily press. Despite the claims of some researchers involved in the issue,[2] early COBE findings (those indicating homogeneous radiation with no ripples) and criticism directed at the big bang theory had not really generated public distrust or suspicion of the model. According to the daily press, for example: "To many people, the finding [of ripples in CBR] may seem less than earth shattering" (Hawkes, 1992a, p. 1), and "The Big Bang theory was so well developed in other ways that it would have been more exciting, in one respect, if the small signals had not been present" (Wolfendale, 1992, p. 13). Similar opinions appeared elsewhere as well: "In a way, it would have been more exciting not to have found it" (Jasper Wall, as quoted in Darbyshire & Lockwood, 1992, p. 4), with COBE data described as "A triumph for conservatism" (Wolfendale, as quoted in Darbyshire & Lockwood, 1992, p. 4).

In January 1992, more than three months before the COBE announcement, a review of D. Overbye's 1991 book, *Lonely hearts of the cosmos*, provided an opportunity to state the acceptance of the big-bang theory confidently: "Today . . . the idea that everything began in a Big Bang has become part of our culture" (Wilson, 1992, p. 27). Four months later, a harsh review of Lerner's 1991 book went even further: "Even before the discovery in April by the COBE satellite, the consensus of cosmologists was that the universe had a beginning in the form of a gigantic explosion, the big bang, which created space, time and matter" (Berry, 1992c, xxvi).

I do not mean to suggest that the discovery was generally held to be unimportant. Quite the contrary, its elucidation of the formation of the universe

---

[2] See, for instance, Smoot & Davidson (1993, pp. 246–247), where media hostility was clearly part of the authors' discursive strategy of presenting the COBE discovery as a major breakthrough and a much needed rescue of both the big-bang theory and the reputation of cosmology in general. In fact, most of Smoot and Davidson's cited articles and headlines critical of the big-bang model were drawn from specialist and interspecialist literature (e.g., *Astronomy, Science*). In terms of Brannigan's (1981) folk theories of discovery, Smoot and his colleagues and supporters constantly strove to balance the discourse of genius (i.e., discovery as an act of individual creativity) with the discourse of cultural maturation (i.e., discovery as a natural product of a cumulative research tradition).

after the big bang was of particular interest. The big-bang model itself, however, was rarely discussed. Instead, it was taken as an established fact upon which other reflections and debates could be based. In addition, the big-bang was set as a reference point in time, a particular concern being to inquire into what came before the big bang and what would happen to the universe at its end (a big crunch?).

> Most astronomers cringe if you ask them what came before the big bang. (Watts, 1992, p. 2)

> Emboldened by this evidence that theories were right all along, astronomers are already speculating on what happened before the big bang. . . . With the big bang theory itself on a secure foundation, it is questions like those that now provide scope for cosmological speculation, and which will make headline news in the years to come. (Gribbin, 1992, p. 5)

The fact that the big-bang concept was ready to be decontextualized and black-boxed for different purposes can be seen from the number of occasions on which it was used to elucidate other, nonscientific issues. "Prof. Hawking's big bang came in 1988," read an article on the huge success of the English mathematician's bestseller, *A brief history of time* (Thynne, 1992, p. 3). The Labour party's defeat in the election of April 9 was considered a political big bang, and the City's deregulation of the securities market was referred to as a financial big bang (Riley, 1992). There was a big bang in the prison service ("Big bang," 1992) and a big bang in former Yugoslavia (Mortimer, 1992).

This widespread tendency to capitalize on the appeal of the big bang was also lampooned:

> "Clearly, the whole basis of the universe is devolution," the Reverend Murdo McGurr, the hard-left minister of Letterbeg and Girnmore, told reporters last night. "What we see in the universe is a metaphor for an assembly with tax-raising powers. . . . What the scientists call big bang is the original multi-option referendum, even though it only lasted a fraction of a second." . . . Provost Willie McFouter, leader of the ruling Labour group of Dunfreeble Dunstable College, took a similar view. In a 564-page fax from the Seychelles . . . [he] claimed that the ever-expanding universe was a symbol of man's destiny to travel even farther afield. "We are all atoms in transit," he said. . . . Sadie McClarty, the crusading Easterhouse grandmother, was more openly dismissive: "Oarigins of the universe? A load o'bluidy havers! Whit'a that god tae dae wi' the price o' mince? See thae Scientists—they're a bunch of bluidy scanners." (Cockburn, 1992, p. 8)

Therefore, a first-order interpretation of the COBE story might lead to the conclusion that communication at the public level provided an opportunity to counteract the problems and criticisms that the big-bang paradigm encountered at the specialist and interspecialist levels. The public sphere offered a discursive forum with fewer constraints and controls than did the specialist arenas. Among members of the public, the big bang was still supported because it strongly resonated with broad cultural and religious explanations and, even more important, it was still an issue, an irreplaceable source of momentous questions. (For example, what came before the big bang? Who made the big bang happen? Will another big bang mark the end of the universe?) Quite acutely, David Whitehouse from the BBC argued that "the COBE story made the splash it did because the scientific community needed the satellite's results to preserve a cherished theory" (as quoted in Miller, 1994, p. 446).

One might elaborate on this argument by taking into account the necessity of coupling this need with the exigencies and constraints of public discourse. In short, the general paradigm that coordinated an entire constellation of measurements, models, and theories (e.g., the Hubble constant, dark matter, and inflation in its original formulation) mobilized its most immediate and appealing image and linguistic label, the big bang, in order to activate public support against the unbelievers.[3]

The COBE results, referred to as the big bang (which, in turn, is often a synonym for the Creation and, ultimately, God), represented an opportunity to strengthen and defend the boundaries of the general paradigm. The big bang, then, became a sort of metonymy and played a pivotal role in anchoring the general paradigm in public culture and, therefore, in defending as a whole all those elements that are usually subject to separate and detailed criticism within specialist realms.

It is thus possible to identify initial similarities with and differences from the other previously cited case of deviation, cold fusion. The similarities mainly lie in the informational role that the public media played for scientists, especially in the first phase of the reporting. In the COBE saga, the press conference compelled scientists to attend to the issue and, often, to discuss the results of the undertaking without any official publication at hand. As explained by British astronomer Michael Rowan-Robinson (1992): "Almost uniquely in the history of big discoveries, the first news reached the scientists via the Associated Press wire services. . . . Scientists had to

---

[3] A number of metaphors that both insiders and outsiders employ to describe the debate on the big bang are taken from military jargon. Smoot and Davidson (1993) described their most distinguished critics (e.g., Jayant Narlikar and Geoffrey Burbidge) as "veteran foes of the Big Bang" (p. 246) and the sharp contrast between supporters and critics as a fight with no quarter (pp. 247–248). Swedish physicist Hannes Alfvén often referred to the leading big-bang theorists as the big-bang Mafia (S. Lindquist, personal communication, October 15, 1996).

respond to journalists off the cuff, without knowing the details of the COBE announcement" (p. 27).

As mentioned earlier, the cold-fusion case differed from the circumstances of the big-bang story primarily because cold fusion involved controversy in public, with one group (the physicists opposing Pons and Fleischmann's findings) clearly devoting intense efforts to getting the issue back into the specialist arena. The COBE story, by contrast, involved a case of deviation in which the public became a source of consensus to help settle, or at least back up, a debate at the specialist level. This difference helps explain why the COBE example of deviation was not condemned and ultimately sanctioned as in the case of cold fusion. The distinction between "proper" public communication of science and inappropriate spectacularization is in itself a political resource available to scientists. Smoot, for instance, took great care to present his own turn to the public as authorized, appropriate communication with the public. He emphasized (a) the high quality of the media coverage, something that is rarely recognized by scientists, and (b) his largely passive role in the process, with media pressure almost forcing him to speak. As one journalist reported:

> He had expected the scientific community would be excited. He had not expected the public reaction. "I walked round the corner and came into this exhibition hall. There was this row of TV cameras, and there were as many press people as there were scientists to hear my talk. Then I realised it was a little out of hand." ... He was plunged into a welter of questions, hyperbole, telephone calls and TV shows. Oddly enough, he was understood. "I was quite impressed by how well the press did in getting the point across. That may be because cosmologists are used to talking in grandiose terms and the press like it. It's amazing how well the press did cover it, particularly the British press." (Radford, 1992a, p. 3)

However, the internal struggle around a paradigm was not the only discursive boundary dealt within the public arena. In the daily press, COBE and the big bang were usually equated with cosmology as a whole, and the reported findings enhanced the perception of the discipline's prestige and achievements. "Cosmological theory, like the universe itself, develops slowly" (Wilkie, 1992, p. 52).

Cosmology is a relatively young field that still lacks certain features of institutional recognition. There are no chairs in cosmology yet. Nor is there a possibility for graduate students to earn a doctorate in cosmology. Such a degree exists only in, say, astrophysics, a subdiscipline of astronomy. Nevertheless, these researchers define themselves as cosmologists in the media, and they are defined as such by journalists. In their comments and articles, cosmologists lucidly trace the rather short history of their field, from Hubble's discovery to the COBE findings:

It is worth recalling the other great landmarks in 20th century cosmology: the discovery of the expansion of the universe by Edwin Hubble in 1929; the invention of the hot big-bang model by George Gamow in 1945; the discovery of the cosmic microwave background radiation by Arno Penzias and Bob Wilson in 1965; the explanation of the abundance of the primordial light elements, helium, deuterium and lithium by Bob Wagoner, Willy Fowler, [and] Fred Hoyle; and the discovery of the motion of our galaxy through microwave background radiation in 1979 by George Smoot, David Wilkinson and colleagues at Berkeley and Princeton. (Rowan-Robinson, 1992, p. 27)

In this early phase of the process of defining cosmology, communication between the scientific community and the public acquires particular significance in at least three senses. First, it helps orient and coordinate the converging approaches and interests of researchers from different fields. This contribution would not be possible through sectorial and specialist communication alone. (It would, however, have to be made in other disciplinary contexts, for cosmology still lacks its own institutional channels of communication). In the case of cosmology, for instance, it is important to address physicists, astronomers, and mathematicians simultaneously. Moreover, in this phase of institutional fluidity, the public forum is also used by different specialists to negotiate their authority over a common topic, such as the origin of the universe. This topic, originally the terrain of astronomers, has increasingly been invaded by physicists and mathematicians since the end of the 19th century. It is, therefore, not difficult to understand astronomers' celebration of the COBE results as one of the most important discoveries ever, one in which Smoot (an astronomer) could finally join Hawking (a mathematician) and Davies (a physicist) at a public roundtable discussion on the universe.

Second, communication between the scientific community and the public can foster the public recognition and support that is crucial for establishing a discipline and acquiring funding, doctoral programs, university chairs, and other resources. It is, therefore, only an apparent paradox that the identity of cosmology as a discipline appears to be defined and established more in media accounts than in specialist discourses within the scientific community alone. This public character of cosmology's image has other implications as well. By its nature cosmology, unlike nuclear physics or physiology, cannot be legitimated on the grounds of technical or economic importance. There is no promise that cosmology will benefit society in terms of cheaper welfare or enhanced physical well-being. Now that the Cold War is over, cosmology cannot even offer opportunity to demonstrate a country's political and military strength. What cosmology *can* bring to the public arena is more subtle and immaterial than claims of material relevance and realpolitik. It provides an answer to humanity's biggest speculative questions: Where do we come from? What will become of us in the end? By claiming the ability to advance answers to such questions, cosmologists not only need to have their field appear in public as science but must also distinguish their disciplinary pursuits from other sources of interpretations of the meaning of existence—religion and art.

This last point leads to the third, most general, and perhaps most interesting boundary being negotiated in the COBE issue: the boundary between science and other practices, particularly religion. The expressions used by researchers to describe the COBE findings—"sign of creation" (Berry, 1992a, p. 1), "evidence of creation" (Berry, 1992a, p. 1), and, as previously cited, "the Holy Grail of cosmology" and "if you are religious, it's like seeing God"—clearly suggest that the significance of the findings, and hence of cosmology, reach far beyond their value in confirming or disproving scientific theory:

> The Big Bang that launched creation (Radford, 1992b, p. E6)

> It is difficult to know what an appropriate reaction to such mind-expanding discoveries should be except to get down on one's knees in total humility and give thanks to God or Big Bang, or both. ("World beyond," 1992, p. 18)

> The first definitive evidence of the Creation has been discovered by astronomy. (Berry, 1992b, p. 4)

> The whole universe is sacred, from the Big Bang that began it to every particle of what he calls energy/matter. (W. Schwartz, 1992, p. 26)

The most recent headlines about the Big Bang have brought physicists even closer to what Buddhists and other mystics have known all along; that all forms of life are connected and interdependent (Thompson, 1992, p. 37).

Religious readers and representatives of the church were prepared to incorporate scientific findings into religious beliefs and adjust the findings accordingly, as shown by the Vatican's acceptance of the big-bang model in 1951. Several letters written by newspaper readers also reflected this selective appropriation of COBE results and big-bang theory within a religious framework. According to one writer, "the second law of thermodynamics and big-bang cosmology point more toward a universe that evolved from purpose than to one originated by chance out of chaos without the intervention of some 'outside creative force'" (Bryce-Smith, 1992, p. 18). Despite such accommodation, however, leaders, followers, and representatives of the church were careful to contain scientists' aspirations to predominate Creationist discourse, designs thought to be detectable in other articles appearing in the British daily press: "The second verse of the Bible, that [*sic*] speaks of space at the beginning as being 'void'[,] is therefore inaccurate. The great explosion some 15,000 years did not emerge into empty space. It emerged into 'nothingness.'" (Berry, 1992b, p. 4). The church community was not willing to accept the assertions of Adrian Berry and others when they quoted scientists who claimed that wondering about the cause of the big bang or about what came before it was just a pointless, unscientific exercise. As one reader showed, the boundary between science and religion did not remain undefended:

> Sir, would you please point out to Adrian Berry that the Bible is accurate in its second verse. Mr. Berry is inaccurate. . . . The Genesis account of creation does not deal with the creation of the matter, but the later-in-time action of taking a formless earth, and, in several stages, making it into a home for man and animal life. Not all scientists believe it wrong to talk about the 'cause' of the Big Bang and where it came from. . . . Why does the universe reveal such order and harmony? Why are there transcendent laws ruling it? . . . The conclusion that there must be a supernatural lawgiver is confirmed by our improving understanding of the origin of the universe. (Gaidon, 1992, letter to the editor)

Taking a different tack, Rev. Bill Westwood, bishop of Peterbourough, responded to comments such as Berry's by declaring: "This doesn't make any difference to God. If anything, it makes Him even more amazing" (McGourty, 1992, p. 4).

Rev. Dr. John Polkinghorne, a former professor of physics who is now an ordained priest, the president of Queen's College at Cambridge University, and a key actor in this debate, clearly restated the respective roles of science and religion:

> Scientifically, the [COBE] result is very very interesting. It helps to explain why the universe is so lumpy when it started out so smooth. Theologically, it's not very significant. We're concerned with *why* the universe began, not *how*. Our view is not greatly affected by this discovery. What's more important is the future of the universe. It's been known for a long time that it's facing a very dismal future. But Christians have never believed that our purpose is fulfilled only in this life. Our universe will have a destiny beyond its death. (McGourty, 1992; p. 4; emphasis added)

This division of labor—assigning science the task of inquiring into the how and religion the task of inquiring into the why—has been a common argumentative strategy of religious figures throughout the debate.

> At bottom there is no argument between science and religion. Science is about what can or may be known or demonstrated: scientists have to leave God out of the equation. ("World beyond," 1992, p. 18)

> To my mind there is no conflict between science and religion. Rather, scientific discovery enhances theology. Science asks how things happen, theology asks why. Faith is not a matter of choosing between science and religion, of believing either in God the creator or the Big Bang. . . . It takes a two-eyed vision of religion and science to see the world as a whole. . . . Therefore

religion has nothing to fear from science. Science liberates rather than constricts or conflicts with the Bible. . . . Our new knowledge of the beginnings of the universe fills a larger piece of the scientific jigsaw, but theology is concerned with the whole picture. (Polkinghorne, 1992, pp. 2, 5)

The comforting point is that the cosmologists still have no answers. Each scientific step towards explaining the ultimate creation always leaves the ultimate questions wide open. (Herbert, 1992, p. 16)

If you're there, God, come back. New big bang theories about creation are seeing scientists again trying to deal with a Creator. But he is still moving in mysterious ways. And thanks to them, further away from us. (Weldon, 1992, p. 19)

If a creator God existed for the sake of big bang and then had no further role, . . . then the search for a relationships with God is a waste of time. . . . In my foolishness, I shall continue to prefer a faith I cannot entirely understand. (Austin, 1992, p. 12)

Was the touch-paper for the Big Bang lit by God? That is the sort of question the faithful are facing, but the scientists, without knowing it, are having even a harder time. On the one hand, they are proclaiming the existence of a 'dark cold matter' while on the other refuting the existence of a light warm God. Science appears to be increasingly steeped in the devout acceptance of its own faith. (Flintoff, 1992, p. 28)

Scientists in general, and some of the researchers specifically involved in this controversy, were made out to be high priests of scientism: "Physicists may be the high priests of reductionism, but nevertheless they conjure up in the name of God with surprising frequency" (Hawkes, 1992b, p. 1). With discussion no longer limited to the relation between science and religion, much of it turned into a general questioning of science's role in society. A debate that received extensive coverage in the British daily press was held at the London Institute of Education, where novelist Fay Weldon "accused scientists of the responsibility for the death of religion, asserting that belief in cosmological theories such as the Big Bang would leave mankind diminished and wretched" ("Institute of Education Science," 1992, p. 9).

Lewis Wolpert, a biologist who has become famous in recent years as an advocate of the scientific enterprise who is ever ready to battle its critics (including many historians and sociologists of science), strongly reacted to such attacks in a speech published in its entirety by *The Times*. His defense was largely based on the aesthetic value of scientific experience:

No less remarkable and beautiful has been the progress in physics, in areas such as cosmology, the Big Bang and the search for fundamental theories based on superstrings. I use the word beautiful intentionally . . . [U]nderstanding the principles in no way makes it less wonderful. (Wolpert, 1992, p. 9)

Thus, science is not just useful, it inspires deep wonder and reveals great beauty just as religion and art do. As formulated by Smoot and Davidson (1993):

> Cosmologists and artists have much in common: both seek beauty, one in the sky and the other on canvas or in stone. When a cosmologist perceives how the laws and principles of the cosmos begin to fit together, how they are intertwined, how they display a symmetry that ancient mythologies reserved for their gods—indeed, how they imply that the universe must be expanding, must be flat, must be all that it is, then he or she perceives pure, unadulterated beauty. (p. 297)

In Rowan-Robinson's words, "the Cold Dark Matter theory is a very beautiful one" (Watts & Wilkie, 1992, p. 1). In the same spirit, Sir Fred Hoyle stated: "I have an aesthetic bias against the big bang" (Berry, 1992d, p. 12).

In cosmology, therefore, there is an attempt to ground the field's legitimation on its cultural relevance at large. Physicist J. Schwartz (1992) expounded on this viewpoint as follows:

> It is now very difficult for us to understand and experience science in the same way we see other aspects of our culture—music, writing, poetry, architecture; cultural endeavours we engage with and to which we bring strong aesthetic and emotional responses. (p. 4)

Smoot's comment in an interview expressed the thought at a personal level: "I have sacrificed a lot to science, just like people do to their concepts of art or religion. . . . I do think that cosmology is about philosophy and religion as well as science. Our work changes culture, like the first Moon landing did: people see themselves in a different perspective" (Wheeler, 1992, p. 14). In the same interview, he spelled out what he saw as the need for cosmology (which he equated with science in general) to adapt to the new demands of public opinion, announcing his objective of "improving the public image of science. I want to make it more popular, to get away from the idea that technology equals weapons and to help school children get involved in astrophysics" (Wheeler, 1992, p. 14).

The Big Bang as a Boundary Object

The COBE story enables one to contemplate at once several features of the deviation of scientific discourse to the public. As I have shown, the COBE deviation was, above all, an opportunity to avoid ever greater criticism from specialists by targeting an audience for whom big-bang orthodoxy was still strong and appealing. It was a level of communication almost impervious to the subtleties of critics and alternative models (e.g., chaotic inflation). However, deviation not only served the interests of a group of astronomers but also enabled different actors to engage in multilayered boundary work. Being at the core of a metonymy chain that begins with COBE measurements, the big bang turned out in this respect to be an ideal "boundary object." I borrow this concept from an insightful contribution by Star and Griesemer (1989), who described boundary objects as

> both plastic enough to adapt to local needs and the constraints of the several parties employing them, yet robust enough to maintain a common identity across sites. They are weakly structured in common use, and become strongly structured in individual-site use. . . . They have different meanings in different social worlds but their structure is common enough to make them recognisable, a means of translation. The creation and management of boundary objects is a key process in developing and maintaining coherence across intersecting social worlds. (p. 393)

In situations involving local interaction, boundary objects may coincide with concrete objects that are used and exchanged by different categories of actors. For example, in a scientific institution such as a museum there may be files, specimens, forms, or entire libraries that facilitate cooperation between different practitioners. In the communicative situations that I address in this chapter, boundary objects may be thought of as the pivotal discursive elements that lie at the heart of boundary negotiation in public. They make communication possible without necessarily requiring consensus, for an object may be interpreted and used in quite different ways at different levels of communication and by different groups of actors at a single level. For instance, specialists use the term big bang quite differently that nonspecialists. The same is true of proponents and opponents of the big-bang model.[4] "Gene" and "DANN" are familiar examples of boundary objects. They are labels employed at different levels of scientific communication and thereby provide a common language, though they are understood differently in a laboratory conversation than in a car advertisement (Nelkin & Lindee, 1995). The role of boundary objects may be played by nonverbal elements such as visual images, which also may be employed in different ways at different levels of communication. A typical example is a fractal image or a photograph taken from a satellite. Either object may serve as a precious source of information for the specialist or simply as an aesthetically pleasing picture for laypeople.

The concept of boundary object makes it possible to salvage contributions that come from the old canonical perspective on the public communication of science (e.g., the social representations approach; see Farr, 1993;

---

[4] Linguistic elements that are common to both specialist and popular expositions are defined by Jacobi (1987) as *termes-pivot*.

Moscovici, 1961, 1984) by putting them within a new framework: "flexible continuity" between core scientific practice and public discourse about science. For instance, a boundary object such as "gene" may function as a paradigm at the specialist level and as what Moscovici (1961) called a "zero-degree" symbol (i.e., the center of a theory's social representation) at the public level.

In addition, the very term boundary object has important advantages because it emphasizes two points. First, such objects bridge communication at the public level with communication at other levels. That is, they cross the boundaries separating one level from another or separating one audience from another within each level. Second, boundary objects are employed by different actors in order to publicly negotiate around the internal and external boundaries of science. As a boundary object, the term big bang was used to draw a line between proponents and opponents of a certain view of the universe's origin, between physics and astronomy, and between science and religion. In this function, the big bang has allowed different categories of actors to interact at the public level by providing a common set of images, a reservoir of rhetorical practices to be used at several levels of discussion and boundary negotiation.

One level of this work has been concerned with the boundaries of the big-bang paradigm as such, that is, its status and its ability to incorporate new models. Another layer of boundary work has been concerned with ownership rights and competencies pertaining to the discipline of cosmology. Because the discipline is still young, boundary work must be performed in public in order to enable the participation of researchers with different backgrounds (e.g., astronomers, physicists, and mathematicians) and to achieve the public recognition and support necessary for the field to become established.

Lastly, at the most general level, the big bang metonymically stands for science at large, for it is taken as a "touchstone of scientific outlook" with a power that only evolution theory possesses. For instance, in a 1990 survey the scientific literacy of the American and British publics, one of the two items that were used to gauge acceptance of the scientific view of the universe and human life was "The universe began with a huge explosion." The other was "Human beings developed from earlier species of animal" (Lerner, 1991, p. 54). Another example is the news article in which Hoyle's reluctance to accept the big-bang theory was criticized by a writer who claimed that Hoyle also "disbelieves Darwin's theory of evolution by natural selection" (Berry, 1992d, p. 12).

The big bang is also used to question, define, and defend the boundaries between science and other forms of knowledge such as religion. As a kind of watershed, it distinguishes between what should be studied by science and what should be left to religion. As one science correspondent wondered, "Is this theology or physics?" (Hawkes, 1992b, p. 1). And when a conference on cosmology was organized by Jesuits in Vatican City in 1981, the Pope received the participating scientists by urging that they inquire into what had happened after the big bang but not recommending that they investigate the big bang itself, for it was the moment of creation and therefore the work of God (Hawking, 1988/1996, p. 137). As the Pope explained:

> Every scientific hypothesis on the origin of the world as a primeval atom from which the physical universe would come leaves open the problem concerning the beginning of the universe. Science alone cannot solve such a question: It requires the knowledge of the person elevated above physics and astrophysics, namely, metaphysics; it requires in particular the knowledge that comes from God's revelation. (Pontificia Academia Scientiarum, 1993, p. 178; my translation)

In this sense, public debate triggered by the COBE findings fit into a long-standing tradition of boundary negotiations between science and religion centered on the big bang model. Authors writing for the general public either have aimed at shaping the Bible according to scientific theories, as Asimov (1981) did when he "rewrote" Genesis according to the scientific view,[5] or have aimed at placing scientific research within the general framework of religious beliefs, as Schroeder (1990) did when he wrote that "astrophysics is part of the supersearch for the Creator" (p. 71). The essence of this tradition was succinctly pointed out by Italian physicist Tullio Regge (1981), who found in the biblical term *Berescit* the equivalent of *big bang*.

In order to delegitimate the big-bang model, critics had to try to make it publicly trespass on the realm of faith and mythology. Hannes Alfvén, a Nobel prize-winning physicist, claimed, for instance, that "big-bang supporters, being aware of the problems that the model has, are cautious. A Christian does not go to Mecca and become a convert overnight" (*The New York Times*, February 28, 1989, as quoted in Jaki, 1989, p. 82). "What theory is this," Hoyle wondered once, referring to Lemaître's initial formulation of the big bang and to the favorable attitude of Pious XII, "which has been 'proposed by a priest and endorsed by the Pope?'" (Boslough, 1992, p. 88) One reader writing to *The Independent* summed up this view by stating that "all cultures have creation myths and the Big Bang is ours" (Goldberg, 1992, p. 20). Conversely, supporters of the big-bang model had to try to make alternative models seem to trespass on the realm of science cosmology:

---

[5] The first three verses of Genesis could actually be paraphrased in this way, to conform them to the scientific opinion on the original of the universe: "In the beginning, fifteen billion years ago, the universe: was a cosmic egg with no structure that exploded with an immense release of energy." (Asimov, 1981, p. 34)

> At this level the personal *theologies* prevail of those who want to organise the
> cosmos according to a great design. We have thus been provided with cycli-
> cal universes that contract or expand periodically or stationary cosmologies
> variously revised such as Hoyle's and Narlikar's. (Regge, 1981, p. 75)

As a boundary object, the big bang gives continuity to the different stages of
scientific discourse. At the disciplinary, interdisciplinary, and public levels
alike (not to mention the pedagogical level, where the big bang is considered
the uncontested source of the universe), the big bang is a valid currency and,
to use actor-network terminology, an "obligatory point of passage" (Law,
1987) for all the participants in the debates, whatever their views and
approaches. From the perspective adopted in this chapter, however, it is not
as clear as the actor-network model suggests whether a scientific actor, in
order to support his or her own views, can fully control the functioning of
boundary objects by forcing other actors to pass through them. Boundary
objects are translated with different intensities and meanings by the different
audiences involved. To cosmologists, the term big bang may refer to a number of models addressing the problem of the origin of the universe. To scientists who are not directly involved in this research area, it may stand for a
single explanation. And to lay readers, it may mean a benevolent act of creation. Possessing such semantic range, boundary objects have far less to do
with supporting a theory or enlisting allies than with facilitating communication between levels and arenas without necessarily requiring consensus.
As stated by Sir Michael Atiyah, the president of the Royal Society, shortly
before the COBE announcement: "The standard of exposition and illustration of science in the press is now so high that even biologists can understand all about the Big Bang" (Atiyah, 1992, p. 12) Hoyle's previously mentioned experience with the term big bang demonstrates the flexibility of
boundary objects and the fact that they can sometimes escape their progenitors. Indeed, the adaptation of boundary objects to a group's specific need
may invert them or even turn them against the actors originally responsible
for them. The story of the expression big bang also shows that such labels
often originate at the public level or are shaped there. The acronym AIDS, as
another example, was preceded by GRID (Gay Related Immunodeficiency
Disease), a name that was quickly discarded after gay organizations pro-
tested its use (see, for instance, Epstein, 1996).

The universal value of the big bang as a boundary object is made even
clearer by the fact that the metonymical chain illustrated in this chapter
(from COBE to "Science") is not the only one extending from it. True, the
initial emphasis on the COBE results was taken by NASA as an opportunity
to deviate in an attempt to improve its public reputation in the aftermath of
the Shuttle tragedy in 1986 and of the Hubble telescope's initial flaws.

Granted, the COBE results were exploited by astronomers to reassert their authority within cosmology, by cosmologists to increase their prestige within science, and by scientists in general to enhance their image in society. But the term big bang was also translated in a number of other ways to provide a handy interpretative framework for political or economic events. The COBE discovery itself could indeed be seen as the spark that animated latent networks of discursive strategies and negotiations within the British public arena. Concepts such as "sensitization of public opinion," however, would do not do justice to the active and multifold manipulation of such scientific findings at the public level.

References

Arp, H. C., Hoyle, F., Burbidge. G., Narlikar J. V., & Wickramasinghe, N. C. (1990). The extragalactic universe: An alternative view. *Nature, 366,* 807–812.

Asimov, I. (1981). *In the beginning . . .* New York: Crown.

Atiyah, M. (1992, February 3). The unlikeliest bash of the season. *The Daily Telegraph* (London), p. 12.

Austin, B. (1992, August 3). Why belief in scriptures must lie at the heart of Christianity. *The Times* (London), p. 12.

Barrow, J. D. (1994). *The origins of the universe.* London: Widenfeld & Nicolson.

Berry, A. (1992a, April 24). Astronomers find Holy Grail of the cosmos in first sign of creation. *The Daily Telegraph* (London), p. 1.

Berry, A. (1992b, April 25). A question that has no answer. *The Daily Telegraph* (London), p. 4.

Berry, A. (1992c, May 9). No time for aliens to evolve. *The Daily Telegraph* (London), p. xxvi.

Berry, A. (1992d, May 11). Whatever happened to the steady-state universe? *The Daily Telegraph* (London), p. 12.

Biezunski, M. (1985). Popularization and scientific controversy. In R. Whitley (Managing Ed.) & T. Shinn & R. Whitley (Vol. Eds.), *Sociology of the sciences: Vol. 9. Expository science: Forms and functions of popularization* (pp. 183–193). Dordrecht, Reidel.

Big bang in the prison service. (1992, August 6). *The Independent* (London), p. 20.

Boslough, J. (1992). *Masters of time.* Reading: W. Patrick.

Brannigan, A. G. (1981). *The social basis of scientific discoveries.* Cambridge, England: Cambridge University Press.

Bryce-Smith, D. (1992, March 30). There is meaning in our universe [Letter to the editor]. *The Daily Telegraph* (London), p. 18.

Bucchi, M. (1996a). La scienza e I mass media: la 'fusione fredda' nei quotidiani italiani [Science and the Mass Media: "Cold Fusion" in the Italian Daily Press]. *Nuncius, 2,* 38–59.

Bucchi, M. (1996b). When scientists turn to the public: Alternative routes in science communication. *Public Understanding of Science, 5,* 375–394.

Bucchi, M. (1998). *Science and the media: Alternative routes in science communication.* London: Routledge.

Chown, M. (1993). *Afterglow of creation: From the fireball to the discovery of cosmic ripples.* London: Arrow.

Clemens, E. (1986). Of asteroids and dinosaurs: The role of the press in shaping the scientific debate. *Social Studies of Science, 16,* 421–456.

Cloître, M., & Shinn, T. (1985). Expository practice: Social, cognitive and epistemological linkages. In R. Whitley (Managing Ed.) & T. Shinn & R. Whitley (Vol. Eds.), *Sociology of the sciences: Vol. 9. Expository science: Forms and functions of popularization* (pp. 31–60). Dordrecht: Reidel.

Cockburn, H. (1992, April 26). Big bang enjoys universal appeal. *The Times* (London), p. 8.

Collins, H. M. (1987). Certainty and the public understanding of science: Science on television. *Social Studies of Science, 17,* 689–713.

Cooter, R., & Pumfrey, S. (1994). Science in popular culture. *History of Science, 32,* 237–267.

Darbyshire, N., & Lockwood, C. (1992, April 25). Earth dweller's guide to the universe. *The Daily Telegraph* (London), p. 4.

Eddington, A. S. (1931). Presidential address to the Mathematical Association. *Nature, 127,* 447–453.

Epstein, S. (1996). *Impure science: AIDS, activism, and the politics of knowledge.* Berkeley: University of California Press.

Farr, R. M. (1993). Common sense, science and social representations. *Public Understanding of Science, 2,* 189–204.

Fleck, L. (1979). *Genesis and development of a scientific fact* (F. Bradley & T. J. Trenn, Trans.). Chicago: University of Chicago Press. (Original work published in 1935)

Flintoff, I. (1992, May 9). One man's endless search for God. *The Guardian* (London), p. 28.

Gaidon, K. (1992, April 30). Void that is filled by the Bible [Letter to the editor]. *The Daily Telegraph.*

Gieryn, T. F. (1983). Boundary work in professional ideology of scientists. *American Sociological Review, 48,* 781–795.

Gieryn, T. F. (1995). Boundaries of science. In S. Jasanoff, G. E. Markle, J. C. Petersen, & T. Pinch (Eds.), *The handbook of science and technology studies* (pp. 393–443). Thousand Oaks: Sage Publications.

Goldberg, A. (1992, April 29). Letter to the editor. *The Independent* (London), p. 20.

Gribbin, J. (1992, April 26). Proof positive for the big bang believers. *The Times* (London), p. 5.

Hawkes, N. (1992a, April 25). Hunt on for dark secret of universe. *The Times* (London), p. 1.

Hawkes, N. (1992b, May 7). Scientists, the obscure objects of desire. *The Times* (London), "Life and Times" Supplement, p. 1.

Hawking, S. (1992, April 24). The discovery of all time. *The Daily Mail,* p. 6.

Hawking, S. (1996). *A brief history of time* (updated and expanded ed.). New York: Bantam Books. (Original work published 1988)

Herbert, P. (1992, April 27). What caused us? [Letter to the editor]. *The Daily Telegraph* (London), p. 16.

Hilgartner, S. (1990). The dominant view of popularization: Conceptual problems, political issues. *Social Studies of Science, 20,* 519–539.

Holmquest, A. (1990). The rhetorical strategy of boundary-work. *Argumentation, 4,* 235–258.

Institute of Education Science. (1992, May 12). *The Times* (London), p. 9.

Jacobi, D. (1987). *Textes et images de la vulgarisation scientifique* [Texts and images of the popularization of science]. Bern: Peter Lang.

Jaki, S. L. (1989). *God and the cosmologists.* Edinburgh: Scottish Academic Press.

Law, J. (1987). Technology and heterogeneous engineering: The case of the Portuguese expansion. In W. Bijker, T. Hughes, & T. Pinch (Eds.), *The social construction of technological systems: New directions in the sociology and history of technology* (pp. 111–134). Cambridge, MA: Massachusetts Institute of Technology (MIT) Press.

Lerner, E. J. (1991). *The big bang never happened.* New York: Simon & Schuster.

Lewenstein, B. V. (1992). Cold fusion and hot history. *Osiris, 7* (2nd series), 135–163.

Lewenstein, B. V. (1995). From fax to facts: Communication in the cold fusion saga. *Social Studies of Science, 25,* 403–436.

Maddox, J. (1989). Down with the big bang. *Nature, 340,* 425.

Maddox, J. (1995). Big bang not yet dead but in decline. *Nature, 377,* 99.

McConnell, C. (1998). *The Big Bang–Steady State controversy: Cosmology public and scientific forums.* Unpublished doctoral dissertation, University of Wisconsin, Madison.

McGourty, C. (1992, April 25). 'Discovery of the century if not of all time.' *The Daily Telegraph* (London), p. 4.

Miller, S. (1994). Wrinkles, ripples and fireballs: Cosmology on the front page. *Public Understanding of Science, 3,* 445–453.

Mortimer, E. (1992, November 25). Big bang but deaf ears. *Financial Times* (London), p. 19.

Moscovici, S. (1961). *La psychanalyse—son image, son public* [Psychoanalysis—Its image, its public]. Paris: Presses Universitaires de France (PUF).

Moscovici, S. (1984). The phenomenon of social representations. In R. M. Farr & S. Moscovici, *Social representations* (pp. 58–76). Cambridge, England: Cambridge University Press.

Nelkin, D., & Lindee, S. M. (1995). *The DNA mystique: The gene as a cultural icon.* New York: Freeman.

Ostriker, J. P., & Steinhardt, P. (1995). The observational case for a low-density universe with a non-zero cosmological constant. *Nature, 377,* 600–602.

Overbye, D. (1991). *The star gazers who wait and measure: Lonely hearts of the cosmos.* London: MacMillan.

Polkinghorne, J. (1992, April 26). Scientists no threat to God the creator. *The Sunday Times* (London), pp. 2, 5.

Pontificia Academia Scientiarum. (1993). *Discorsi dei papi alla pontificia accademic delle scienze.* Vatican City.

Powell, C. S. (1995). Crisis, what crisis? *Scientific American (June),* 19–22.

Radford, T. (1992a, May 12). Coping with the fall-out from big-bang theory. *The Guardian* (London), p. 3.

Radford, T. (1992b, June 16). Universe: Cosmic ripple. *The Guardian* (London), p. E6.

Regge, T. (1981). Ultime tendenze in cosmologia [Recent trends in cosmology]. In P. Rossi (Ed.), *La Nuova Ragione. Scienza e cultura nella società contemporanea* (pp. 69–76). Bologna: Il Mulino.

Riley, B. (1992, September 16). Securities institute is shaped by the Big Bang. *Financial Times* (London), p. 12.

Rowan-Robinson, R. (1992, May 1). Yesterday's ripples, today's galaxies. *The Guardian* (London), p. 27.

Rowan-Robinson, R. (1993). *Ripples in the cosmos: A view behind the scenes of the new cosmology.* Oxford: Freeman.

Schroeder, G. L. (1990). *Genesis and the big bang.* New York: Bantam Books.

Schwartz, J. (1992, June 13). Relative values: defiance of science. *The Guardian* (London), p. 4.

Schwartz, W. (1992, April 25). Face to faith. *The Guardian* (London), p. 26.

Shapin, S. (1990). Science and the public. In R. C. Olby, G. N. Cantor, J. R. R. Christie, & M. J. S. Hodge (Eds.), *Companion to the history of modern science* (pp. 991–1007). London: Routledge.

Shinn, T., & Whitley, R. (Vol. Eds.). (1985). Expository science: Forms and functions of popularization. In R. Whitley (Managing Ed.), *Sociology of the sciences (Vol. 9).* Dordrecht: Reidel.

Smoot, G., & Davidson, K. (1993). *Wrinkles in time.* New York: William Morrow.

Star, S. L., & Griesemer, J. R. (1989). Institutional ecology, 'Translations' and boundary objects: Amateurs and professionals in Berkeley's Museum of Vertebrate Zoology, 1907–1939. *Social Studies of Science, 19,* 387–420.

Sullivan, D. L. (1994). Exclusionary epideictic: NOVA's narrative excommunication of Fleischmann and Pons. *Science, Technology, and Human Values, 19,* 283–306.

Thompson, E. (1992, May 20). On a mission to save the planet. *The Guardian* (London), p. 37.

Thynne, J. (1992, February 27). A brief history of Hawking's success. *The Daily Telegraph* (London), p. 3.

Watts, S. (1992, April 25). Scientists set to work on ripples in space. *The Independent* (London), p. 2.

Watts, S., and Wilkie, T. (1992, April 24). How the universe began. *The Independent* (London), p. 1.

Weldon, F. (1992, May 18). If you're there, God, come back. *The Guardian* (London), p. 19.

Wheeler, S. (1992, May 13). All this and the universe as well. *The Independent* (London), p. 14.

Whitley, R. (1985). Knowledge producers and knowledge acquirers. In R. Whitley (Managing Ed.) & T. Shinn & R. Whitley (Vol. Eds.), *Sociology of the sciences: Vol. 9. Expository Science: Forms and functions of popularization* (pp. 3–28). Dordrecht, Reidel.

Wilkie, T. (1992, May 3). Hunting for holes in Hawking's universe. *The Independent on Sunday* (London), p. 52.

Wilson, G. (1992, January 11). Review of Overbye's Lonely hearts of the cosmos. *The Independent* (London), p. 27.

Wolfendale, A. (1992, April 25). Fitting the cosmic jigsaw. *The Times* (London), p. 13.

Wolpert, L. (1992, May 12). Blind to the beauty in a ball of gas. *The Times* (London), p. 9.

World beyond the limits of the mind, A. (1992, April 24). *The Guardian* (London), p. 18.

CONCLUSION

CHAPTER 14

PUBLIC UNDERSTANDING OF SCIENCE AND TECHNOLOGY: STATE OF THE ART AND CONSEQUENCES FOR FUTURE RESEARCH

Claudia von Grote and Meinolf Dierkes

The presentation of concepts, results, and methods in this investigation into the public understanding of science and technology has sketched a relatively recent field of research that is only vaguely delimited and both conceptually and analytically rather indistinct. Despite this field's amorphous contours, however, Jon D. Miller already noted in 1992 that the study of the public understanding of science and technology had become a "visible and recognized area of scholarship" in which relevant theoretical constructs had been brought together from a variety of disciplines and subdisciplines. Among them are research on technology assessment, innovation and technology development, risk perception and acceptance, media, attitudes, and especially science itself. The fact that the public understanding of science and technology has achieved such visibility as a field of research is due, however, to a point raised by Wynne (1995), an outspoken critic of the present state of the research on this topic. He asserted that the work subsumed under this label lacks an overarching scientific paradigm and is shaped by political interests dominating the context of research questions, especially since the 1980s.

This dominance of political intentions and the analytical fuzziness of some of the concepts used in this field of inquiry are directly related. As a research topic, public understanding of science and technology emerged partly from the need politicians have to legitimate policy on science and technology and partly from the scientific community's interest in knowing the degree of acceptance it can expect from the population. These interests are in no way new. In Germany, for example, they go back to the late 19th century, when scientific and technological developments generated significant resistance and ultimately triggered critical questions about the controllability of technical change (Dierkes, Knie, & Wagner, 1988; Sieferle, 1984). In the late

1920s, to use another example, the impacts that technical change had brought to the world of work prompted initiatives "for the state to take the lead in formulating the scientific definition and treatment of the problem and to fund scientific research on the fields of technology and work" (Dierkes, 1986a, p. 123). The objective was to generate well-founded knowledge about the human side of production processes profoundly changed by the development of technology. The increasing dynamics of scientific and technological change over recent decades meant that such interaction between the political and scientific communities became ever more important in the advanced democratic societies.

Applied research of this sort is obviously not unaffected by political mindsets, especially when it comes to ascertaining "impressions of popular moods" toward science and technology for purposes of legitimation (Dierkes, 1986b, p. 20). When political intentions are such a strong driving force, "theory" is mostly perceived to have no value of its own. Conceptual vagaries are usually not felt to be a problem, either (on inherent limits of scientific subtleties relating to public opinion research, see Noelle-Neumann & Hansen, 1991, p. 30).

These decidedly politicoscientific interests in the emergence of the field influenced the methodological framework of the research as well. Demand for knowledge about the broadest possible section of society led predominantly to the use of standardized methods for conducting large-scale surveys. The origin of the public understanding of science as a research topic is said by Miller (1992, p. 23) to date from 1957, when the first standardized survey of the adult population was conducted to inquire into the use of scientific reports in the media and the attitude toward science and technology policy. Survey research, therefore, has dominated the field since its inception and still represents a significant part of the work undertaken today.

Taking this status into consideration, we begin this cursory historical outline by presenting the canonized variant of research on public understanding of science and technology—the approach that made this expression familiar to the general public. It will then be possible to elaborate on the details, extensions, and change that research in other fields, primarily in the theory of science, in risk theory, in technology assessment, as well as on the media and their influence, has brought to the key issues in the public understanding of science and technology as a field of inquiry. It will then be possible to examine the research agendas that could follow from such enrichments.

From the study of science, research on the dynamics of science and on the relation between the public's and the academic community's understandings of science are specifically relevant in this context (see Jasanoff, Markle, Petersen, & Pinch, 1995). Also pertinent are investigations into the cognitive, emotional, and cultural aspects of the perception and acceptance of risks (see Bechmann, 1993) and into the public's attitudes and behavior toward science and technology in everyday contexts (see Wynne, 1992b).

In the sociology of science and the research on risk perception, risk communication, and the assessment and acceptance of technology, developments in theoretical concepts and empirical inquiries have emancipated researchers from risk-centered and knowledge-based objectivism (Halfmann, 1995). Research design has thereby been liberated from a view according to which risks and technology impacts are defined solely as characteristics of specific technologies and their technical dimensions. As a result of this change in perception, more and more researchers began to discover social determinants of people's attitudes toward science and technology and their ways of dealing with scientific and technological developments. They increasingly included sociocultural and psychological variables that led to the question of how individuals, groups, or entire social systems "construct" risks of technologies and, even more generally, technological facts.

It is the latter issue that figures so prominently in the field of research known as public understanding of science and technology, in which the public's cognitive, perceptual, and evaluative responses to society's scientific developments are placed at center stage. Furthermore, work on the acceptance and people's assessment of technologies led to distinctions that became relevant points of departure for the field. For example, the difference between acceptance and acceptability, a distinction important to the impact assessment of innovative technologies (Dierkes, 1982), indicates that acceptance is not only a matter of "rationally" evaluating a particular technology. Acceptance is affected by general values and goals that underlie specific attitudes of people and enter into their judgments even before a technology comes into use (Dierkes & Marz, 1991). The range, volume, and diversity of these further theoretical and empirical developments have led to corrections and enrichments of the field's original conceptualization.

The Development of the Research Approach

A brief comparison of research traditions in Germany, the United States, and the European Union shows that the United States has the longest tradition of research on the population's scientific education. In the 1970s this line of inquiry was melded with research on the population's attitudes toward science and technology. In Germany, academic tradition in this field has been fairly limited; public opinion surveys on science and technology have been conducted mainly by institutes for applied social sciences, among which the

Institute for Public Opinion Research in Allensbach has been a leader. According to Noelle-Neumann and Hansen (1991, p. 32), the time span of that organization's data series on public attitudes toward technology is more extensive than that of any other organization in Germany. For the period from 1966 through 1991, the institute's 1993 yearbook contains data on the population's attitude toward technical progress. The two best-known questions were (a) "Do you believe that technical progress is making life easier and easier for people or harder and harder?" and (b) "Do you believe that technology, all in all, is more a blessing or more a curse for people?" (Noelle-Neumann & Köcher, 1993).

In contrast to this valuable but quite limited and technology-centered research, the study of people's attitudes toward science was not initially a topic in its own right. The Allensbach survey on attitudes toward technology did not include science until 1990, when a question was added: "Do you believe that scientific progress will help or hurt people in the long run." This change may be an indication that public attention is also turning to the role of science in society, with people increasingly perceiving a link between technology and science.

Different from these developments in Germany, the original thrust of research conducted as part of Eurobarometer surveys focused, like the inquiries in the United States, on science. In 1977 a comparative Eurobarometer survey on science and European public opinion was carried out in Britain, the Federal Republic of Germany, Italy, and the Netherlands. A subsequent comparative study on the European public's attitudes to scientific and technical development was conducted in France and the Federal Republic of Germany. More than 10 years passed before this work was followed up by the two surveys (1989 and 1992) frequently referred to in this volume. They represent a European variant of U.S. survey research, upon which Miller has had a great impact, and thus link the U.S. and European traditions of research on the public, science, and technology. The preeminence of science over technology in the surveys marks the special nature of this approach, particularly when compared to the technology-focused research in Germany.

In developing the questionnaire used by the National Science Foundation (NSF) (see National Science Board [NSB], 1980), Miller had a decisive role in establishing the field of research on public understanding of science and technology. He drew on two research traditions: investigations into the population's scientific literacy, and research on popular attitudes toward science and technology. The latter line of investigation can be traced back to at least the 1930s, when Dewey claimed to explore the extent to which intellectual openness, the willingness to challenge opinions, and similar norms of the science system had been adopted as behavioral standards by the population at large and whether they could be communicated to the general public (Dewey, 1934). In the 1960s the National Assessment of Educational Progress began collecting such data on the understanding of scientific standards, processes, and the cognitive subject matter of scientific disciplines in order to ascertain the level of scientific literacy among students. Conducted for the first time in 1957, survey research on attitudes toward science and technology was resumed and systematized in 1972 by the National Science Foundation, which decided to collect data on science and the public as part of its science indicators. It has thereby been possible to conduct time-series analyses of attitudes and knowledge pertaining to the understanding of science and technology in the general public and specific subgroups.

This research rested on an emphatic understanding of the relation between democracy and science. Basic knowledge of science and a positive attitude toward science and technology were taken to be essential to each citizen's ability to participate in a democracy. Scientific literacy was expected to enable citizens to understand scientific and technological issues, probe opinions, and take part in making policy decisions. The popularization of science, therefore, was considered to be the central means of acquiring that competence.

Miller carried this early work further by modifying the traditional concept of scientific literacy and, since 1979, by structuring the attitude research behind the science indicators by means of a distinction between the broad, general public and a specific segment consisting of an interested and knowledgeable public, which he called the attentive public (Miller, 1983, p. 45). The key modification of the traditional concept of scientific literacy (see the chapter by Miller & Pardo in this volume; see also Pardo, 1998) was the inclusion of a new dimension. The two traditional measures of scientific literacy have long been (a) the understanding of basic scientific constructs and a vocabulary of scientific terms and concepts (e.g., *atom, radiation*, and *physical laws),* and (b) the understanding of the scientific approach, that is, the processes of science by which one distinguishes what is science from what is not. Miller added a third dimension: the social influence exerted not only by science but by technology as well: "If scientific literacy is to become truly relevant to our contemporary situation, one additional dimension must be added: awareness of the impact of science and technology on society and the policy choices that must inevitably emerge" (Miller, 1983, p. 31). The corresponding items in the 1979 NSF survey focused on the possible damages and possible benefits of science and technology in such applications as nuclear power, space exploration, and food additives (NSB, 1980).

Miller's concept of scientific literacy that encompasses the meaning of technology and science also underlies the Eurobarometer questionnaire, which presents similar items, though they are formulated as statements on which respondents are asked to take a position. An example: "Science and technology are making our lives healthier, easier and more comfortable" (Institut National de Recherche Agricole [INRA] & Report International, 1993, p. 183).

This expansion of scientific literacy as a concept largely reflects the changes that occurred in the context of the political discussion in the late 1970s. The more policy-makers turned to scientific rationale for their decisions, the more science figured as a topic of public discussion. Simultaneously, controversies over technology, and the attendant wane of unquestioning acceptance, became an ever more familiar part of daily reality (Dierkes, 1993). A decisive result of pulling these two strands of research together, however, was that attitudes toward science and technology—Miller (1983) spoke of "understanding science policy issues" (p. 41)—became part of "understanding" or "awareness." Questions about assessing the impact of science and technology came to be posed, but they were expected to be answered from a highly cognitive angle. This accentuation was ultimately manifested in the fact that the expression "public understanding" became the very name of the research field.

These developments clearly gave the field a strong cognitive bias. The idealistic impetus of the early studies—the objective of enlightening people and paving the way for their democratic interaction by disseminating scientific norms and knowledge—was still reflected in Miller's expansion of the concept of scientific literacy, but it became much more limited than it once was. He, too, linked the level of the citizen's scientific education with a capacity to shape democracy, but the perspective was confined to the cognitive ability to follow and understand policy decisions in the field of scientific and technological development. The concept did not incorporate the non-cognitive (i.e., emotional and normative) dimensions characterizing real-world scientific discourses. Presumably, this shift resulted from the overt emphasis on cognitive components in the conceptualization of science at that time and the way in which scientific practice, norms, and institutions were taken for granted. The question is whether this cognitive bias also tends to make technology seem like a derivative of science, whether one is led to see technology more as applied science than as products within specific contexts of use. If the second view is taken, technology should warrant study in its own right.

This prevailing line of thought, according to which increased knowledge enhances understanding and improves opportunities for democratic participation in society, has dominated the research agenda of the field down to the present day. It has come to be known as the *deficit model*. As the name suggests, any critical or disapproving position that the public may take on scientific matters or technological innovations is implicitly interpreted in this approach as a lack of scientific knowledge. Knowledge about science is uncritically attributed positive value and is considered a basis for positive attitudes toward science and technology. Understanding and attitudes therefore are latently treated as two sides of the same coin.

Miller's approach has come to serve as a blueprint for national and international surveys. As mentioned earlier, the 1989 and 1992 Eurobarometer surveys entitled "Europeans, Science and Technology: Public Understanding and Attitude" have been mostly a European adaptation of the American model. The main motive has been to promote the comparability of data from the United States, Japan, and western European countries (see Durant, Miller, Tschernia, & van Deelen, 1991), even though the dimensions of the survey were expanded to include the understanding of the institutional structures of science (see Miller & Pardo in this volume).

However, as the study of science sharpened attention to technology (see Jasanoff et al., 1995, p. 225) and as sociologists developed a keen interest in concepts relating to risks and technology (Dierkes, 1986b), the established type of research on the public understanding of science and technology came under increasing criticism for its heavy reliance on survey-based analyses. The appearance of theoretical and empirical work on technology discourses (Jungermann, Rohrmann, & Wiedemann, 1991) and consensus-building strategies (Dierkes & Marz, 1991) also played a role in this process. The key weakness of the dominant survey approach was seen by the critics as grounded in the deficit model.

The Objections to the Deficit Model: Toward an Interactive Approach to the Public Understanding of Science and Technology

The widely voiced criticism of the deficit model (see most of the authors in this volume) stems from challenges to the basic assumptions of this kind of research. The demonstrably low level of knowledge about scientific facts is not disputed. Nor is there any question about the legitimacy of efforts to enlighten the public or of attempts to popularize science. Instead, the concern is that the educational approach in general and the items in the surveys in specific treat so narrow a segment in the relationship between science and the public that the resulting statements about deficiencies in the population's understanding of science are questionable, as is the significance attached to those shortcomings.

It is equally problematic when opinions about and attitudes toward the impacts that scientific and technical innovations have on society are taken as an expression of a general relationship to science and technology. Generalizations about "science" and "technology" overlook significant dimensions of the public's experience, in particular the contexts within which technologies emerge and are used. Similarly, a broad concept like "people" obscures the various social roles in which one and the same person can experience and judge a given scientific or technological development very differently (Dierkes, 1986b).

From the Public as Recipient to the Public as User

In the traditional approach the public is seen as a passive recipient of scientific information. Each survey, through its items, defines which aspects of knowledge are deemed relevant in a cognitive or political sense. It then collects data on how much of this knowledge is present in the population under study. In this way in the NSF survey and the Eurobarometer determine what scientific knowledge the public has acquired. It is only a small step from that point to the political conclusion that the public should be offered more information conveyed in a better way so as to increase its level of knowledge and thereby, implicitly at least, raise the level of support for scientific and technological innovation. The educational model inherent in this approach entailed two closely related basic assumptions.

1. The linear model of a sender and a receiver. This model is clearly the source of the question posed in communications research about how the sender—whether a scientist, a science journalist, or some other agent—can transmit cognitively demanding knowledge appropriately and understandably to a passive receiver. This assumption of a linear relationship also pervades studies presenting direct conclusions about the effects that positive or negative reports by the media about science and technology have on the public's knowledge and attitude (see Noelle-Neumann & Hansen, 1991).

2. The polarity between the scientist and the layperson, between the specialist and the nonspecialist. This social pairing has attracted a great deal of research attention not only in the theory of science, in which a gap between formal and naive knowledge is assumed to exist, but also in research on the impacts and acceptance of technology, a field in which the

significance of experts in the technology-assessment debate derives from the assertion that different levels of knowledge are responsible for results showing systematic differences between experts' assessments of new technologies and those of laypeople.

This conception of the public as a group of "receivers" and "laypeople" has been criticized repeatedly from different quarters. In the sociology of knowledge, the challenge arose from analysis of the public's processes of negotiation when dealing with scientific information issues in specific relevant contexts (see Wynne, 1995). In the sociology of technology, the critique grew out of the analysis of the social dynamics in the interactions between experts and laypeople (see Dierkes, 1980, pp. 26–29). When, for example, risk assessment is no longer treatable primarily as a technical problem but has instead become a political issue in its own right because grass-roots groups are demanding participation in assessment processes, then factors other than "objective" risks and benefits have to be studied. It becomes necessary to explore the interactions between experts on the one hand and laypeople or grass-roots groups on the other as well as their diverging assessments of risks (Nelkin & Pollak, 1980). Scientific or technical information and knowledge have been found to play less a role in those assessments by various segments of the public than do personal experience and the confirmation of information through the five senses (Otway, 1980).

Critics of the model of the public as a passive receiver are united in their diagnosis that the public plays an active role in the discourse on science and technology and that a deficiency in scientific and technological knowledge does not permit the conclusion that one group is less capable than another of meeting the demands of the scientific and technical living conditions in modern societies.

If such profound conclusions are drawn about what those deficits mean, critical questions have to be posed. For example, is the public really guided by the evaluative criteria of scientists, or do people develop criteria—similar or different—on their own? Do members of the public actively seek out the kind of information that they consider important? What are the contexts in which formal knowledge plays a role, and how does it do so? It is possible that the conditions for accepting scientific and technical innovation are quite different from formally surveyed knowledge. They may tend to be found in factors such as trust in social institutions or personal experience with different forms of social agencies.

Analyses of technology in everyday life such as those by Sørensen et al. (in this volume) have focused on the active, adaptive, and constructive role of the public as users of scientific information and technical innovations.

This public-as-user perspective allows researchers to concentrate on studying the ideas that the public has about the objects of scientific analyses or the outcomes of technological innovations.

Another way of exploring the active role of the public is to look at the interfaces between technical innovations and the everyday working lives, routine consumption, and daily political spheres of individuals. At these interfaces it is possible to study what parts of their knowledge individuals mobilize, what information they seek, and what actions they take in shaping their daily lives as the conditions of technology and science change. For instance, it is possible to look at how large technical systems figure in people's daily lives (see Grote, 1994). Questions in such a research program would include how knowledge is acquired, which possibilities of social access to information exist, and how those possibilities are shaped.

In addition, it becomes evident in projects such as that by Sørensen et al. (in this volume) that daily interaction with science and technology involves important practical and symbolic aspects that usually go unconsidered when the focus is on cognitive aspects. The possibilities that scientists and engineers envision as uses of a technology do not necessarily coincide with the ways it is actually used by people on a day-to-day basis. This phenomenon has received as little recognition in research on the public understanding of science and technology as has the fact that behavior is sometimes the opposite of what is measured in assessments of science and technology. The thesis of technophobia (*Technikfeindlichkeit*), which has been laid to rest many times in scientific studies (see Jaufmann & Kistler, 1988; Hennen, 1994), is still alive in the political sphere and belies, for example, the intense use of new technical appliances by consumers (Dierkes, 1986b).

A second extension of the concept of the public is its heterogeneity. The public cannot be treated as a homogeneous entity. To a certain extent this point has been recognized in the survey-based literature, one example being the differentiation between an "attentive public" and an "uninterested public" (see Miller & Pardo and Einsiedel in this volume), or in the division of social categories into adolescents and adults or into men and women. Nevertheless, these distinctions do not really change anything in the general concept of a public. By being equated with laypeople rather than with a societal being that has numerous social roles to perform, the public as a general concept is imbued with conceptual homogeneity.

The criticisms of this image of a homogeneous and scientifically inept public have come from two different angles. Lévy-Leblond (1992) in particular criticized the conception behind the social pair consisting of the scientist (or expert) and the layperson. Given the increasing specialization of scientific fields, the roles of the expert and the layperson can easily reverse.

Experts in one area can be laypeople in another, just as laypeople in a given scientific or technical field can have acquired amazing scientific expertise or practical knowledge. "It is crucial to stress that if scientists are definitively not universal experts, non-scientists are not universal non-experts" (Lévy-Leblond, 1992, p. 17). By the same token, an individual can develop very different positions on technical innovations, depending on the various roles he or she plays (Dierkes, 1986b). One must also bear in mind that expertise is not limited to formal education and training but rather to the knowledge and experience related to the domains of knowledge in question.

The second thrust of criticism is mainly represented by Wynne (1992b), who argued against the polarity between experts and laypeople on the grounds that it makes cognitive knowledge the sole measure for arriving at different assessments by the two groups. This reduction to a single dimension tends to mask other elements of the assessment process—such as trust in information or ideas about ways of controlling certain scientific and technological developments. These elements can be brought to light only when the knowledge issue in question is treated by the public within its scope for acting on, experiencing, and assessing the matter.

The Multiple Meanings of "Understanding"

For historical reasons the concept of "understanding" in the public understanding of science and technology is particularly vague. The traditional focus on attitudes in the science indicators used in the NSF survey (see Miller, 1983) was subsumed as an element in the overall concept of science literacy. Attitudes toward science and technology therefore tended to become questions of comprehension. Conversely, understanding can thereby be equated with acceptance in discussions about science policy. Depending on the political interests in a given situation, the word understanding can be taken to mean comprehension or acceptance.

However, even when understanding is used to mean comprehension, it is necessary to distinguish analytically between knowing a scientific fact or topic, as measured by the scientific knowledge items in a questionnaire, or comprehending how such knowledge can actually be used and interpreted. All these points suggest, first, that one has to define the meaning of the word understanding precisely; second, that this definition depends directly on the nature of the topic (knowledge of scientific facts and theories versus the technological application of scientific knowledge and technological artifacts); and third, that one must consider the way in which the matter is seen by the actors who are (or are not) attempting to understand.

The specification of the concept of public led to differentiation in how the process of understanding was viewed. When the issue of understanding and assessing scientific and technological innovations is looked at from the perspective of specific users, then it is possible to explore the mental models that people have about science and technology. It becomes important to study how people assimilate scientific and technological innovations into their existing knowledge and which transformations occur in the process. It emerges that certain users operate with a working knowledge (Lévy-Leblond, 1992) that is part of their general culture. The way in which the understanding of a research area such as genetic engineering is embedded in such working knowledge changes dramatically once the possibilities for the scientific application of that working knowledge become reality, as is the case with genetically modified foods and Dolly, the cloned sheep. At that point the processes of understanding can relate to the behavioral level. The need for greater differentiation has also been recognized in research on attitudes. For example, the overarching concept of understanding has been criticized because a general perception or assessment of technology does not necessarily coincide with the perception or assessment of specific technologies. Furthermore, the perception and assessment of a technology can change, depending on the specific area of application in question (see Hamstra in this volume; Brüggemann & Jungermann, 1998).

A limitation of the approaches outlined thus far is that they are focused exclusively on the cognitive strategies laypeople use to acquire new scientific or technological knowledge. Motivational factors need to be included in the analysis of the specific use of science. A general finding in this regard is that "people show an uncanny ability to learn what they need and not more" (Lévy-Leblond, 1992, p. 19). This realization means that researchers must ascertain the varying needs of different subgroups of the public more systematically than in the past and must relate those needs to the relevant images of science. Only then will it be possible to form an impression of the "demand side" of science.

Extension of the Concept of Science and the Turn to Technology

The basic question about scientific literacy relates to its relevance. Such well-known Eurobarometer survey items as "Does the earth revolve around the sun or does the sun revolve around the earth?" and "Do you believe that astrology is a science?" stand for many others designed to test school knowledge about the physical world. Questions about the meaning of concepts like the greenhouse effect or the ozone hole check the respondent's level of knowledge about current controversies. Questions like these may be useful ways of measuring knowledge about the natural sciences taught in schools, but they have yet to be proven relevant as indicators of understanding science in the sense of successful participation in the scientific and technological world.

Lévy-Leblond (1992) called for a more precise definition of what it means to be equipped for a life in a scientifically and technologically complex world. He also went a step further, denying that questions such as those used in surveys constitute a relevant test of scientific literacy. He asserted that the questions in the surveys could not provide a reliable picture of the public understanding of science as long as the relative character of scientific knowledge was not taken into account. In his view the questions had to reflect the fact that science does not provide a universal truth but rather specifies the "conditions of validity" for its statements, including the statement that the earth revolves around the sun. According to Lévy-Leblond, it is not the knowledge about the underlying scientific or technical principles that is most important but rather the active use of a relevant scientific or technical field.

> In more general terms, genuine expertise (scientific, in particular) does not consist in the knowledge of a large body of abstract results (theorems, laws, etc.) but in the ability to master operating statements. It consists less in knowing, than in knowing how to know, where to look, what to read, whom to ask—and why to bother. (Lévy-Leblond, 1992, p. 17)

Lévy-Leblond's criticism leads to the view that

> scientific expertise relies on a very selective and highly organized cognitive scheme. Its criterion is not so much that of truth, or validity, but rather that of interest, or relevance. As such, scientific knowledge (like any form of knowledge) is contextual and the meaning of a question cannot be appreciated in the abstract. (Lévy-Leblond, 1992, p. 18)

If even the knowledge of scientific experts takes on meaning only through specific problems and search strategies, then this criticism of the traditional representation of science as an abstract canon of knowledge applies even more to the public's knowledge. Wynne (1995) underscored this viewpoint in his thesis that people do not experience science in the abstract but always in a social form. As soon as knowledge is required in an information process or a context of use (which is usually how it tends to enter the public sphere), it is not a cognitive process of appropriately understanding scientific statements that is involved but rather social processes in which the relevant knowledge is negotiated or adapted to a specific situation.

The way science is conceived of in these studies must therefore be changed significantly or must at least be expanded to include this important aspect of contextual or situational knowledge. It is the contexts of use that enable a public to take an active part in constructing the relevance of science and technology and in using them. Accordingly, it is less important in such research to collect data on factual knowledge than to concentrate on ascertaining the following additional dimensions: (a) the relevance of this knowledge, (b) the conditions governing this knowledge (or what Dierkes, Edwards, & Coppock, 1980, referred to as frames of reference), and (c) strategies for the active use of knowledge to deal with specific problems.

A second important expansion of the understanding of science was achieved by the sociology of scientific knowledge. "Because science is rooted in scientists' activity, the character of scientific knowledge depends on the nature of scientific work" (Yearley, 1994, p. 245). From this perspective science is a system of action. Two main assumptions are that the public has to deal with scientific knowledge and that science as an institution stands behind this knowledge. On this basis the sociology of scientific knowledge has focused on the dimension of doing, that is, of producing scientific facts in scientific institutions (Shapin, 1992; Yearley, 1994). The object of analysis has been the emergence of scientific facts in the laboratories and in the controversies between researchers. When seen from this angle, scientific facts become (partially at least) the result of processes of negotiation between scientists and scientific institutions, a view that can make a scientific statement seem arbitrary and random to someone who is an outsider to the specific knowledge community. This aspect of the image of science is discussed by Yearley (in this volume) as an outcome of such interactive processes. The credibility of science can be at stake when the differences in assessment that are genuinely part of the scientific process are attributed to parties in public institutions.

Building on this view of understanding science, Wynne (1992a) added another perspective: "[science's] forms of institutional embedding, patronage, organization and control . . . [I]t is this social dimension which permeates all experiences of and responses to science" (Wynne 1992a, p. 42). There is also a degree of external influence on science, so science cannot be understood unless all these sources of influence are considered. Ideas about external factors affect the way science is received. Wynne's research on sheep farmers is a good example of how scientific experts contribute to their own biased perceptions. They did not see a problem in the fact that post-Chernobyl ministerial interventions in farmers' methods of animal husbandry had been communicated as scientifically founded without any attempt to integrate these external interventions into the work, knowledge, and practices of the farmers. The approach made the scientists and the ministry inseparable in the eyes of the farmers. Scientific recommendations lose credibility under such circumstances, for farmers' models of social control for assessing scientific data and recommendations differ from those of scientific experts. Wynne pointed to the experience farmers have in dealing with uncertainty and adapting to uncontrollable factors. "This deep cultural outlook—reflected in their intellectual frameworks as well as in their whole way of life—was simply incompatible with the scientific-bureaucratic cultural idiom of standardization, formal and flexible methods and procedures, and prediction and control" (Wynne, 1992b, p. 296).

Such assessment differences between farmer experts and scientific experts are not a matter of knowledge gaps but of trust and credibility, which must be developed in specific social relations and networks. Considered in this way, the issue of how much the public knows becomes altogether secondary to the question of public trust in science and its institutions. How is trust built? Who is trusted, individuals or institutions? Why do consumer or environmental organizations, for example, enjoy more trust than institutions of science? Why do such organizations appear to be the final authority? What leads to the loss of trust?

The subject of waning popular trust is not new in the discussion of the public's attitudes toward technology. However, research on the impacts of translating scientific and technical knowledge into useful, innovative, acceptable technologies has prompted close analysis of the rather diffuse diagnosis that the population has lost confidence in technical and scientific advancement (see Dierkes & Marz, 1991). This deeper insight has to do with the recognition that the risks associated with new technologies bear on vital popular interests that are less evident in an individual's way of life than at the macrosocial level. One result of this work has been to show that, behind the question about the acceptance of innovative and useful technologies, one finds discussion about whether the objectives associated with the new technologies are socially desirable.

Inevitably, the decision-making organs that deal with matters of funding science and technology are drawn into the issue by these considerations. Analysis of the process shows that the crisis over objectives has been superseded by a crisis within institutions. It is manifested primarily as a shift of decision-making from legislative, executive, and administrative levels to the courts (Dierkes & von Thienen, 1977). Attributing the issue of trust to such underlying mechanisms and their inherent dynamics has created an analytical perspective that greatly sharpens the question about the role of trust. Yearley's chapter in this volume exemplifies how that kind of focus can facilitate efforts to explain the erosion of trust. Because popular skepticism of the decision-making community's ability to function is now often leading to litigation over specific applications of technology, the increasingly overburdened courts are bringing in specialists in the necessary areas. The ever greater involvement and strategic use of scientists as experts by the parties to a suit has set in motion a mechanism that is partly responsible for drawing science into disputes that have cost it public trust.

Although science as an institution and its working methods have been made a part of the surveys on the public understanding of science and technology, circumstances like those described above have yet to be addressed. The aspect of use, which is directly relevant in matters of technology, still has no significance in its own right. Instead, the understanding of technology will tend to be handled as an appendix to the scientific concept of understanding. The trend toward reorientation to technology has done little to change this focus thus far. Science and technology have been closely paired in the title of the research area, but the special nature of technology has not been accounted for and the equating of science and technology has not been legitimated. Undeniably, however, the extensive research literature on the perception and acceptance of technology and on risk communication about new technologies has helped highlight questions pertaining to the appropriation of and debates about scientific and technological innovations. Research has shown that it is often neither technology as such nor technological and scientific progress that the public is questioning, but rather the way in which they are applied and used (see Dierkes & Marz, 1991). Issues relating to the processes by which the public acquires scientific knowledge and uses new technologies should be given more prominence than they presently have in research on the public understanding of science and technology.

Future Agenda for Research on the Public Understanding of Science and Technology

Summarizing the extensions and enrichments of the four dimensions invoked in the name of this research field, one finds a common thread, namely, the conviction that strong emphasis on the cognitive dimension of the topic should be considerably broadened in a wider conception of how the public understands, accepts, and uses science and technology. Interactive elements in the processes of negotiating and acquiring knowledge that is used should become a major focus of study, as should the social contexts in which knowledge is produced. These recommendations also have important implications for the research methodologies chosen.

Spectrum of Methodological Approaches

Many of the substantive and conceptual questions posed above require serious grounding in qualitative research. There is a pent-up need for research approaches that can compensate for the reductionist mode in which public understanding has mainly been studied thus far. The complexity of cognitive, emotional, and social processes involved in understanding, assessing and dealing with technological innovations require appropriate research designs. Standardized surveys like the Eurobarometer remain important and necessary instruments in this context, though they should be modified (see the suggestions offered in this volume by Bauer and by Durant et al.) and thematically enlarged.

For that reason "methodological pluralism" is the key word for the future research agenda. There is an emerging consensus that quantitative surveys and qualitative studies should be linked much more closely than they have been. Qualitative studies are essential for preparing the ground for projects in which researchers seek to capture as much as possible of the complexity of a case. It is then possible, as a next step in the research process, to design surveys capable of differentiating between the topics, pressures, and demands for knowledge that arise in the widely varied daily situations confronting members of a society. In this context, qualitative research also has the important function of providing the background information needed to interpret survey data. For this reason, too, studies on the public understanding of science and technology should bring together and flexibly use the entire spectrum of methods available to the social sciences.

An open-minded view of methods is particularly important if one agrees that the analysis of credibility and trust is the most significant task of future research on the public understanding of science and technology. Although this insight was expressed in a few past studies focused on technology (e.g., Dierkes et al., 1980), research programs thus far have largely lacked the theoretical-methodological basis provided by an interactive model. This foundation is now available in the expanded approaches for studying the public understanding of science and technology. The conceptual and empirical work presented in this volume discredits "the predominant analytical tendency to treat these [trust and credibility] as unambiguous, quasi-cognitive categories of belief or attitude that people supposedly choose to espouse or reject" (Wynne, 1992b, p. 299). Trust and credibility cannot be measured as a unidimensional construct.

Trust and Credibility

Coming from various disciplinary backgrounds, most of the authors in this book have arrived independently of one another at the conclusion that the question about understanding or accepting science and technology centers on the matter of trust in the value and utility of scientific research and new technologies. This question has come to be linked ever more closely with the problem of trust in scientific as well as political institutions. If the risks to which the public feels exposed are seen to be the outcomes of scientific or political decisions, then public trust becomes the key, but precarious, resource of society (Hennen, 1995). It is precarious because experts themselves produce uncertainty and because science ultimately has few answers to questions about the purposefulness of socially desirable goals. All these elements lead to the important question of the role that an expertocracy leaves for the public in the communication of "science" and "technology." Wynne considers it necessary to take two steps:

> The first step is the recognition that the trustworthiness and credibility of the social institutions concerned are basic to people's definition of risks or uptake of knowledge. However, the second step is to recognize that trust and credibility are themselves analytically derivative of social relations and identity negotiations. (Wynne 1992b, p. 300)

Questions about credibility and trust do in fact emerge when scientific knowledge is reconstructed as the result of concrete interactions between researchers and when risk perceptions and assessments of technologies are seen as the product of processes of communication and negotiation. Scientific knowledge derives its authority from moorings in science as an institution. If the public debate in democratic societies permits contradictions between scientific statements to emerge, allows lack of agreement between researchers to become apparent, or brings scientists' seeming identification with the interests of their sponsors to the surface, then the credibility of science as an institution is challenged. Communication about "science in the making" may, however, compensate for loss of confidence if people recognize that controversial assessments and external influences, for example, are only interludes in a long process in which controversies are eventually overcome. One could expect that research will become understandable and acceptable precisely because the public learns to understand "the process by which science is made" as a process in which phases of experimentation and errors are inherently necessary. For this lesson to succeed, however, the selected periods for observing controversies should include phases in which information has consolidated enough to permit well-founded distinction between what is true and what is not.

The role that mass media play, or could play, in this whole process must also be considered. They present science to the public. Why do certain scientific and technological topics make the headlines whereas others do not (see Dierkes et al., 1980)? To what extent does the scientific community use the media for its own internal discussions (see Bucchi in this volume)? These questions and similar ones should be the subject of future research.

A second way to approach the problem of trust has emerged from the analysis of controversies about technologies. If an acceptable consensus is not achieved by the "right" scientific or technical information alone but rather by the manner in which that information is accessed and by the kind of communication that occurs, then the democratic organization of technological discussions takes on enormous significance for generating trust. Transparent and open means of communication and broadly based decision-making processes could be central to consensus-building (see Jelsøe in this volume). Another way to study trust is to focus on consensus-building institutions and processes as mediators between the scientific community, policy-makers, and society (Dierkes & von Thienen, 1977; Daele et al., 1996). How do attempts to achieve a sense of procedural justice create or undermine trust in the negotiation process? How can trust that is won or lost in such processes be communicated to a larger public? In these processes it is not only a narrow concept of "technology impacts" that is treated but usually also questions about socially desirable goals, fair distribution of the societal costs and benefits that result from risky decisions, and the purposefulness of certain scientific and technological innovations. If research on the public understanding of science and technology were to tie into such efforts, it could also monitor broad democratic discourses.

References

Bechmann, G. (Ed.). (1993). *Risiko und Gesellschaft. Grundlagen und Ergebenisse interdisziplinärer Risikoforschung* [Risk and society: Foundations and results of interdisciplinary risk research]. Opladen: Westdeutscher Verlag.

Brüggeman, A., & Jungermann, H. (1998). *The whole and the parts of genetic engineering*. WZB discussion paper FSII 98/111. Wissenschaftszentrum Berlin für Sozialforschung.

Daele, W. van den, Pühler, A., Sukopp, H., Bora, A., Döbert, R., Neubert, S., & Siewert, V. (1996). *Grüne Gentechnik im Widerstreit. Modell einer partizipativen Technikfolgenabschätzung zum Einsatz transgener herbizidresistenter Pflanzen* [Green genetic engineering in conflict: Model of participatory technology impact assessment relating to the introduction of transgenic herbicide-resistant plants]. Weinheim: Verlag Chemie.

Dewey, J. (1934). The supreme intellectual obligation. *Science Education, 18*, 1–4.

Dierkes, M. (1980). Assessing technological risks and benefits. In M. Dierkes, S. Edwards, & R. Coppock, *Technological risk: Its perception and handling in the European Community* (pp. 21–30). Cambridge, MA: Oelgeschlager, Gunn & Hain.

Dierkes (1982). Akzeptanz und Akzeptabilität der Informationstechnologien. *Wissenschaftsmagazin der Technischen Universität Berlin, 1*, 12–15.

Dierkes, M. (1986a). Technikfolgenabschätzung als Interaktion von Sozialwissenschaften und Politik. Die Institutionalisierungsdiskussion im historischen Kontext [Technology impact assessment as interaction between the social sciences and the policy-making community: The discussion about institutionalization in its historical contexts]. In M. Dierkes, T. Petermann, & V. von Thienen (Eds.), *Technik und Parlament* (pp. 115–145). Berlin: sigma.

Dierkes, M. (1986b). Die Einstellung des Menschen zur Technik [People's attitude toward technology]. In International Hightech-Forum (Ed.), *Neue Dimensionen der Information. Perspektiven für die Praxis* (pp. 19–43). Stuttgart: Poler.

Dierkes, M. (1993). Mensch, Gesellschaft, Technik: Auf dem Wege zu einem neuen gesellschaftlichen Umgang mit der Technik. In M. Dierkes, *Die Technisierung und ihre Folgen. Zur Biographie eines Forschungsfelds* (pp. 273–260). Berlin: sigma.

Dierkes, M., Edwards, S., & Coppock, R. (1980). *Technological risk: Its perception and handling in the European Community.* Cambridge, MA: Oelgeschlager, Gunn & Hain.

Dierkes, M., Knie, A., & Wagner, P. (1988). Die Diskussion über das Verhältnis von Technik und Politik in der Weimarer Republik [The discussion about the relation between technology and policy in the Weimar Republic]. *Leviathan, 1*, 1–22.

Dierkes, M., & Marz, L. (1991). Technikakzeptanz, Technikfolgen und Technikgenese. Zur Weiterentwicklung konzeptioneller Grundlagen der sozialwissenschaftlichen Technikforschung [Technology acceptance, technology impacts, and technology development: On the further development of conceptual foundations of technology studies]. In D. Jaufmann & E. Kistler (Eds.), *Einstellungen zum technischen Fortschritt. Technikakzeptanz im nationalen und internationalen Vergleich* (pp. 157–187). Frankfurt on the Main: Campus.

Dierkes, M., & von Thienen, V. (1977). Science court: Ein Ausweg aus der Krise? Mittler zwischen Wissenschaft, Politik und Gesellschaft [Science court as a mediary between science, policy, and society: A way out of the crisis?]. *Wirtschaft und Wissenschaft, 25*(4), 2–14.

Durant, J. R., Miller, J., Tschernia, J.-F., & van Deelen, W. (1991, February). *Europeans, science, and technology.* Paper presented to the Annual Meeting of the American Association for the Advancement of Science, Washington, DC.

Grote, C. von (1994). Anschlüsse an den Alltag. Versuche zu einer Hermeneutik technischer Infrastrukturen [Connecting with everyday life: Toward a hermeneutics of technical infrastructures]. In I. Braun & B. Joerges (Eds.), *Technik ohne Grenzen* (pp. 251–304). Frankfurt on the Main: Suhrkamp.

Halfmann, J. (1995). Risiko, Verantwortung, Vetrauen [Risk, responsibility, trust]. *Soziologische Revue, 18*, 33–38.

Hennen, L. (1994, March). *Ist die (deutsche) Öffentlichkeit 'technikfeindlich'? Ergebnisse der Meinungs- und Medienforschung* [Is the German public "technophobic"? Results of survey and media research] (Report No. 24). Bonn: Büro für Technikfolgenabschätzung beim Deutschen Bundestag.

Hennen, L. (1995, February). *Wissenschaft und Technik in der Diskussion: Technik-folgenabschätzung und öffentliche Technikkontroversen* [Science and technology under discussion: Technology impact assessment and public controversies over technology]. Address delivered at the Technical University of Aachen.

Institut National de Recherche Agricole (INRA), & Report International. (1993, June). *Europeans, science and technology: Public understanding and attitudes* (EUR 15461). Brussels: Commission of the European Communities.

Jasanoff, S., Markle, G. E., & Petersen, J. C., & T. Pinch (Eds.). (1995). *The handbook of science and technology studies.* Thousand Oaks, CA: Sage Publications.

Jaufmann, D., & Kistler, E. (Eds.). (1988). *Sind die Deutschen Technikfeindlich? Erkenntnis oder Urteil* [Are the Germans technophobic? Knowledge or prejudice]. Opladen: Leske & Budrich.

Jungermann, H., Rohrmann, B., & Wiedemann, P. M. (Eds.). (1991). *Risikokontroversen. Konzepte, Konflikte, Kommunikation* [Controversies over risk: Concepts, conflicts, communication]. Berlin: Springer.

Lévy-Leblond, J.-M. (1992). About misunderstandings about misunderstandings. *Public Understanding of Science, 1,* 17–21.

Miller, J. D. (1983). Scientific literacy: A conceptual and empirical review. *Daedalus, 112*(2), 29–48.

Miller, J. D. (1992). Toward a scientific understanding of the public understanding of science and technology. *Public Understanding of Science, 1,* 23–26.

National Science Board (NSB). (1980). *Science Indicators 1980.* Washington, DC: U.S. Government Printing Office.

Nelkin, S., & Pollak, M. (1980). Consensus and conflict resolution: The politics of assessing risk. In M. Dierkes, S. Edwards, & R. Coppock (Eds.), *Technological risk: Its perception and handling in the European Community* (pp. 65–75). Cambridge, MA: Oelgeschlager, Gunn & Hain.

Noelle-Neumann, E., & Hansen, J. (1991). Technikakzeptanz und Medienwirkung [Acceptance of technology and media effects]. In D. Jaufmann & E. Kistler (Eds.), *Einstellungswellungen zum technischen Fortschritt. Technikakzeptanz im nationalen und internationlen Vergleich* (pp. 27–52). Frankfurt on the Main: Campus.

Noelle-Neumann, E., & Köcher, R. (Eds.). (1993). *Allensbacher Jahrbuch der Demoskopie 1984–1992* (Vol. 9). Munich: K. G. Saur.

Otway, H. (1980). The perception of technological risks: A psychological perspective. In M. Dierkes, S. Edwards, & R. Coppock (Eds.), *Technological risk: Its perception and handling in the European Community* (pp. 35–44). Cambridge, MA: Oelgeschlager, Gunn & Hain.

Pardo, R. (1998). *Scientific-technical knowledge and the legitimation of science and technology in Spain.* Manuscript submitted for publication.

Shapin, S. (1992). Why the public ought to understand science-in-the-making. *Public Understanding of Science, 1,* 27–30.

Sieferle, R. P. (1984). *Fortschrittsfeinde. Opposition gegen Technik und Industrie von der Romantik bis zur Gegenwart* [Enemies of progress: Opposition to technology and industry since the early 19th century]. Munich: Beck.

Wynne, B. (1992a). Public understanding of science research: New horizons or hall of mirrors? *Public Understanding of Science, 1,* 37–43.

Wynne, B. (1992b). Misunderstood misunderstanding: Social identities and public uptake of science. *Public Understanding of Science, 1*, 281–304.

Wynne, B. (1995). Public understanding of science. In S. Jasanoff, G. E. Markle, J. C. Petersen, & T. Pinch (Eds.), *The handbook of science and technology* (pp. 361–388). Thousand Oaks: Sage Publications.

Yearley, S. (1994), Understanding science from the perspective of the sociology of scientific knowledge: An overview. *Public Understanding of Science, 3*, 245–258.

List of Authors

Margrethe Aune (Ph.D. in sociology) works as a researcher at the Center for Technology and Societyat the Norwegian University of Science and Technology, Trondheim. Her research includes work on the social studies of energy consumption and everyday life and the social studies of information and communication technologies in education.

Martin Bauer (Ph.D.) is a member of the staff at the London School of Economics, Department of Social Psychology, and lectures on science and technology and research methodology. He is a research fellow at the Science Museum, London. Current projects include biotechnology and the European public; "la longue duree" of popular science, from 1930 to the present; science in the media as cultural indicators; and the relationship between public perception and media coverage of science.

Massimiano Bucchi (Ph.D. in social and political science at the European University, Florence) is a research fellow in the Department of Sociology at the University of Trento. He was awarded the 1997 Nicholas Mullins Prize by the Society for Social Studies of Science and the 1996 Italian Public Television (RAI) Prize for research in mass communications.

Meinolf Dierkes is Director of the Research Unit on Organization and Technology at the Wissenschaftszentrum Berlin für Sozialforschung (WZB). He is also Professor of the Sociology of Science and Technology at the Technical University of Berlin and a visiting professor at the University of California at Berkeley. He is a member of the board of several companies, foundations, and management institutes in Europe. His focus of interest is on international management development, organizational culture and learning, technology assessment, and innovation.

John Durant is Director of Science Communication at the Science Museum, London, and Professor of Public Understanding of Science at Imperial College, London. His main research interests lie in the field of public perception of science and technology, particularly biotechnology and public participation in the policy-making process. Among other affiliations, he is Chairman of the Task Group on Public Perceptions on Biotechnology and a member of the committee for the journal *Public Understanding of Science*.

Edna F. Einsiedel is Professor of Communications Studies in the graduate program in communications at the University of Calgary. Her research interests lie in the social construction of technology, the public understanding of science, and public participation mechanisms in technology issues. She is working with European colleagues on a project on biotechnology and the public.

Ulrike Felt (Ph.D. in theoretical physics) is Associate Professor of Science Studies at the University of Vienna. Since 1991 she has been the coordinator of a research program and a student exchange program of the Network of European Centers for Science and Technology Studies (NECSTS). She also serves on the board of the European Association for the Study of Science and Technology (EASST). Topics of interest to her are public science policy as well as science and women.

George Gaskell is Director of the Methodology Institute and Reader of Social Psychology at the London School of Economics.

Maria E. Gonçalves is Associate Professor at the Instituto Superior de Ciencias do Trabalho e da Empresa in Lisbon. Her fields of interest, research, and teaching are science, technology, and public policy, with a special emphasis on law, technology and the law; and European Union law and institutions. She is a member of the Council of the European Association for the Study of Science and Technology (EASST), a member of the executive council of the International Council for Science Policy Studies, and a member of the board of the Portuguese Federation of Scientific Associations and Societies.

Claudia von Grote (Ph.D. in sociology) is a senior fellow at the WZB. Her fields of interest are medical technology, large technical systems, and the public understanding of science and technology.

Anneke Hamstra has a degree in consumer studies from the Agricultural University in Wageningen, the Netherlands, and has worked as a researcher at the SWOKA Institute for Strategic Consumer Research in Leiden. She is now the head of the consumer test center of Phillips Domestic Appliances and Personal Care in Drachten, the Netherlands, where consumer testing and research is focused on tailoring personal-care products to consumer needs and wishes.

Morten Hatling is a senior fellow in the Department of Industrial Management, SINTEF, in Trondheim, Norway. His main research interests include the social studies of information and communication technologies, knowledge management, technology assessment, and technology policy.

Sheila Jasanoff is Professor of Science and Public Policy at the Kennedy School of Government and the School of Public Health at Harvard University. Her long-standing research interest centers on the interaction of law, science, and politics. She received the Distinguished Achievement Award of the Society for Risk Analysis in 1992 and has served on the Council of the Society for Social Studies of Science as well as on advisory committees of the National Science Foundation. She was also a consultant for numerous science policy organizations at the Office of Technology Assessment and is a fellow of the American Association for the Advancement of Science.

Erling Jelsøe (M.S. in chemical engineering) is Associate Professor in the Department of Environment, Technology, and Social Studies at Roskilde University.

Miltos Liakopoulos is a research assistant in the Department of Social Psychology at the London School of Economics. He is currently working on a thesis about biotechnology in the public realm.

Cees J. H. Midden is Professor of Psychology in the Department of Technology Management at Eindhoven University of Technology, the Netherlands. His research focus is on human–technology interactions in the development of new products and systems, in the introduction of new products and systems into society and the market, and in the consumption and use of products.

Jon Miller is Vice President of the Chicago Academy of Science and is the director of its International Center for the Advancement of Scientific Literacy. He is also a professor at Northwestern University Medical School and the Medill School of Journalism. He is a member of the editorial board of *Public Understanding of Science* and a member of different associations concerning public opinion research. He has studied the development of scientific literacy and attitudes toward science and technology for more than 20 years.

Elisabeth Noelle-Neumann is the founder (1947) and director of the Allensbach Institute for Public Opinion Research, Germany's first such institute, and has worked with Dr. Renate Köcher since 1988. Dr. Noelle-Neumann was Professor of Journalism at the University of Mainz and served as the director of the university's Institute of Journalism until 1983. From 1978 to 1991 she was often a visiting professor of political science at the University of Chicago. In 1997 she was named honorary professor at the Moscow External University of Humanities. She was on the board of the World Association for Public Opinion Research (WAPOR) for 16 years, including 2 years as its president. Since 1989 she has coedited the *International Journal of Public Opinion Research.*

Rafael Pardo is Professor of Sociology at the Institute of Economics (CSIC) in Madrid. Since 1991 he has been the director of the Science, Technology, and Society Center, a Banco Bilbao Vizcaya Foundation. He has been a member of the advisory committee of COTEC (a Spanish business association for the promotion of technological innovation) since 1994, and until 1996 he served as president of the National Commission for the Evaluation of Research Projects in the Social Sciences in the Spanish Ministry of Science and Education. His main research interests are the sociology of scientific and technological culture in advanced societies and the sociology of organizations as new systems of work organization and collective action of business organizations.

Hans Peter Peters (Ph.D. in the social sciences) is a research fellow in the "Humans, Environment, Technology" program group at the Research Center, Jülich. He also lectures on risk communication and empirical communication research at the Institute for Mass Communication at the University of Münster, Germany. His research deals with the relationship between science and the public; interactions between scientific experts and journalists; mass media coverage on science, technology, and the environment; and the reception and effects of media coverage on biotechnology.

Liesbeth Scholten graduated as an engineer from the Technology and Society program at Eindhoven University, the Netherlands. She has specialized in the area of human–technology interaction and is working as a researcher for Philips Design in Eindhoven.

Knut H. Sørensen is a professor at the Center for Technology and Society at the Norwegian University of Science and Technology, Trondheim. His main research interests include the social studies of information and communication technologies, energy and environmental issues, and technology policy.

Steven Yearley is Professor for Sociology at the University of York, England. He is an Economic and Social Research Council (ESRC) fellow for the public understanding of science. His main fields of interest are qualitative studies in the public understanding of science, constructionist sociology of science, and environmental sociology.

Index of Persons and Institutions

Subject Index